"Working on a great agile team can feel magical. Problems, personalities, and politics all fall away through the emphasis and focus on delivering something of value together. But creating such magical teams is hard. It just got easier. *Head First Agile* is the book many of us have been waiting for. A perfect blend of practical advice, supported by the principles of Scrum, Extreme Programming, and Kanban, *Head First Agile* can help anyone create a great, magical, agile team."

— **Mike Cohn, Author of *Succeeding with Agile*, *Agile Estimating and Planning*, and *User Stories Applied***

"If you've ever worked in a software team, you will immediately relate to the accurate and insightful case studies in *Head First Agile*. There's plenty of great advice here that I wish someone had told me earlier in my career. But regardless of how long you've been in the software field, you're guaranteed to find things to learn and new ways of looking at old problems. It was a surprise to me to learn how much more there is to XP than just Pair Programming—and I'm definitely going to be looking at how my team can benefit from these practices and ideas."

— **Adam Reeve, Principal Architect at RedOwl Analytics**

"*Head First Agile* is written in such an easy to digest manner that I found it difficult to put down. Since the book is written as an interaction between people, instead of just providing facts, I feel I have a much better understanding of both the principles and practices of Agile."

— **Patrick Cannon, Senior Program Manager, Dell**

"*Head First Agile* thoroughly explains some very challenging Agile concepts to successfully pass the PMI-ACP exam and re-enforces these concepts with meaningful exercises, real-life stories, and brain-teasing visuals for greater learning retention. The unique format of this book caters to all learning styles and anyone from beginner to expert will benefit from studying this book. Thank you, Andrew and Jennifer, for writing another outstanding Head First book!"

— **John Steenis, PMP, CSM, CSPO**

"*Head First Agile* is great! I love how the authors are able to explain the subject matter in a fun, easy to read and understand format. Kudos to Andrew and Jennifer for a job well done!"

— **Mark Andrew Bond, Network Operations Project Manager at a higher education institution**

"Members of any software development team, in any capacity, Agile based or not, should read this *Head First Agile*. Teams that have adopted Agile in "whole" or part will better understand how to improve their processes. Teams that are not using Agile can really learn what it takes to start the journey to Agile. For me, this book was a good reflection on projects that succeeded and failed over the last 20 years of my career."

— **Dan Faltyn, Chief Security Officer, BlueMatrix**

"Agile has been a buzz word in the industry, which many people talk about and use but without fully understanding the underlying principles and values behind the practices. *Head First Agile* helps demystify the topic and provides a superb guide into the Agile journey and mindset."

— **Philip Cheung, Software Developer**

Praise for *Head First Agile*

"*Head First Agile* shows you that Agile is not a single methodology, but a range of approaches and ways of thinking about the development lifecycle. This book will guide you to find the method that best fits the needs of your team, and to understand how to continually improve over time. Approaches like Kanban, where you establish a pull system by visualizing the workflow and limiting work in progress, can be particularly empowering for teams."

— **Nick Lai, Senior Engineering Manager at Uber Technologies**

"What better way to learn about an innovative way to approach work, Agile, than through this ingenious, creative, and ground-breaking approach to learning that engages the whole mind, emotions, and different learning styles. Chuck the boring text books of the past and pick up *Head First Agile* for a contemporary and intriguing learning experience."

— **Tess Thompson, MS, PMP, PMI-ACP, CSP Agile Transformation Coach and Professor at Saint Mary's University**

"Agile can be a very challenging topic for any PM who is schooled in traditional ways of working. The authors of this book have taken great care to make the topics accessible and they've presented the information in a straightforward, concise way that offers not just valuable information you need to be successful at the PMI-ACP exam, but more importantly information you need to be successful in applying Agile in a practical setting."

— **Dave Prior, Certified Scrum Trainer, PMP, PMI-ACP at LeadingAgile**

"*Head First Agile* thoroughly explores practical problems and solutions using a wide array of techniques from the Agile playbook. The authors adeptly explain different types of Agile approaches, providing a great repository of Agile resources to the reader."

— **Keith Conant, Senior Software Engineer at a payments company**

"*Head First Agile* is a great resource because it teaches people not only the practices and methods of Agile, but more importantly the mindset change. This book is based upon the values and principles outlined in the Agile Manifesto. I highly recommend this book for anyone who wants to learn about Agile delivery."

— **Mike MacIsaac, Scrum Master, MBA, PMP, CSM**

"I am a certified Agile coach at IBM, where I teach and coach people from my organization helping them become Agile. I found *Head First Agile* to be a great addition to the Agile bookshelves, as well as a wonderful resource to complement the studies of those professionals preparing for the PMI-ACP exam"

— **Renato Barbieri, PMP, PMI-ACP, Manager, Agile coach, Kepner-Tregoe Program Leader at IBM**

"I am a Project Manager who used *Head First PMP* to obtain my PMP last year. *Head First Agile* follows the same concept as *Head First PMP*... visual thinking and learning to understand, comprehend, and retain the concepts. In a market where quality books for preparation are limited, this was a great resource to have and use for PMI-ACP exam prep. The practice exam questions and Chapter 7 (including Domain Reviews) were really good preparation. The whole thing went great, and I passed on my first try!"

— **Kelly D. Marce, Project Manager at a financial services firm**

Praise for other *Head First* books

"With *Head First C#*, Andrew and Jenny have presented an excellent tutorial on learning C#. It is very approachable while covering a great amount of detail in a unique style. If you've been turned off by more conventional books on C#, you'll love this one."

— **Jay Hilyard, software developer, coauthor of *C# 3.0 Cookbook***

"Going through this *Head First C#* book was a great experience. I have not come across a book series which actually teaches you so well…This is a book I would definitely recommend to people wanting to learn C#"

— **Krishna Pala, MCP**

"*Head First Web Design* really demystifies the web design process and makes it possible for any web programmer to give it a try. For a web developer who has not taken web design classes, *Head First Web Design* confirmed and clarified a lot of theory and best practices that seem to be just assumed in this industry."

— **Ashley Doughty, senior web developer**

"Building websites has definitely become more than just writing code. *Head First Web Design* shows you what you need to know to give your users an appealing and satisfying experience. Another great Head First book!"

— **Sarah Collings, user experience software engineer**

"*Head First Networking* takes network concepts that are sometimes too esoteric and abstract even for highly technical people to understand without difficulty and makes them very concrete and approachable. Well done."

— **Jonathan Moore, owner, Forerunner Design**

"The big picture is what is often lost in information technology how-to books. *Head First Networking* keeps the focus on the real world, distilling knowledge from experience and presenting it in byte-size packets for the IT novitiate. The combination of explanations with real-world problems to solve makes this an excellent learning tool."

— **Rohn Wood, senior research systems analyst, University of Montana**

Other related books from O'Reilly

Learning Agile

Beautiful Teams

Head First PMP®

Applied Software Project Management

Making Things Happen

Practical Development Environments

Process Improvement Essentials

Other books in O'Reilly's *Head First* series

Head First PMP

Head First C#

Head First Java

Head First Object-Oriented Analysis and Design (OOA&D)

Head First HTML with CSS and XHTML

Head First Design Patterns

Head First Servlets and JSP

Head First EJB

Head First SQL

Head First Software Development

Head First JavaScript

Head First Physics

Head First Statistics

Head First Ajax

Head First Rails

Head First Algebra

Head First PHP & MySQL

Head First Web Design

Head First Networking

Head First Agile

Wouldn't it be dreamy if there were a book to help me learn about Agile that was MORE FUN THAN GOING TO THE DENTIST? It's probably nothing but a fantasy...

Andrew Stellman
Jennifer Greene

Beijing · Boston · Farnham · Sebastopol · Tokyo

O'REILLY®

Head First Agile

by Andrew Stellman and Jennifer Greene

Copyright © 2017 Andrew Stellman and Jennifer Greene. All rights reserved.

Published by O'Reilly Media, Inc., 1005 Gravenstein Highway North, Sebastopol, CA 95472.

O'Reilly Media books may be purchased for educational, business, or sales promotional use. Online editions are also available for most titles (*http://oreilly.com/safari*). For more information, contact our corporate/institutional sales department: (800) 998-9938 or *corporate@oreilly.com*.

Series Creators:	Kathy Sierra, Bert Bates
Editor:	Nan Barber
Design Editor:	Louise Barr
Cover Designers:	Karen Montgomery, Louise Barr
Production Editors:	Melanie Yarbrough
Indexer:	Lucie Haskins
Proofreader:	Jasmine Kwityn
Brain Image on Spine:	Eric Freeman
In Our Hearts Forever:	Quentin the whippet and Tequila the pomeranian

Printing History:

September 2017: First Edition.

The O'Reilly logo is a registered trademark of O'Reilly Media, Inc. The *Head First* series designations, *Head First Agile*, and related trade dress are trademarks of O'Reilly Media, Inc.

PMI-ACP, PMP, and PMBOK are registered marks of Project Management Institute, Inc.

Many of the designations used by manufacturers and sellers to distinguish their products are claimed as trademarks. Where those designations appear in this book, and O'Reilly Media, Inc., was aware of a trademark claim, the designations have been printed in caps or initial caps.

While every precaution has been taken in the preparation of this book, the publisher and the authors assume no responsibility for errors or omissions, or for damages resulting from the use of the information contained herein.

ISBN: 978-1-449-31433-0
[LSI]   [2018-10-19]

For Nisha and Lisa

the authors

> THANKS FOR BUYING OUR BOOK! WE REALLY LOVE WRITING ABOUT THIS STUFF, AND WE HOPE YOU **GET A KICK OUT OF READING IT...**

> ...BECAUSE WE KNOW YOU'RE GOING TO **DO GREAT WORK** WITH AGILE!

Andrew

Photo by Nisha Sondhe

Jenny

Andrew Stellman is a developer, architect, speaker, trainer, agile coach, project manager, and expert in building better software. Andrew is an author and international speaker, with top-selling books in software development and project management, and world-recognized expert in transforming and improving software organizations, teams, and code. He has architected and built large-scale software systems, managed large international software teams, and consulted for companies, schools, and corporations, including Microsoft, the National Bureau of Economic Research, Bank of America, Notre Dame, and MIT. Andrew's had the privilege of working with some pretty amazing programmers during that time, and likes to think that he's learned a few things from them.

Jennifer Greene is an enterprise agile transformation leader, an agile coach, development manager, project manager, speaker, and authority on software engineering practices and principles. She's been building software for over 20 years in many different domains, including media, finance, and IT consulting. She's led large-scale agile adoption efforts supporting development teams around the world and helped individual team members get the most out of agile practices. She looks forward to continuing to work with talented teams solving interesting and difficult problems.

Jenny and Andrew have been building software and writing about software engineering together since they first met in 1998. Their first book, **Applied Software Project Management**, was published by O'Reilly in 2005. They published their first book in the Head First series, **Head First PMP**, and their second one, **Head First C#**, both in 2007. Both books have gone on to third and, soon, fourth editions. Their fourth book, **Beautiful Teams**, was released in 2009, followed by their fifth, **Learning Agile**, in 2014.

They founded Stellman & Greene Consulting in 2003—their first project as a consulting company was a really fascinating software project for scientists studying herbicide exposure in Vietnam veterans. And when they're not building software or writing books, they do a lot of speaking at conferences and meetings of software engineers, architects, and project managers.

Check out their website, **Building Better Software**, at http://www.stellman-greene.com.

Table of Contents (Summary)

Intro ... xix

PART 1: UNDERSTANDING AGILE
1. What is agile? *Principles and practices* ... 1
2. Agile values and principles: *Mindset meets method* ... 23
3. Managing projects with Scrum: *The rules of Scrum* ... 71
4. Agile planning and estimation: *Generally accepted Scrum practices* ... 117
5. XP (Extreme Programming): *Embracing change* ... 177
6. Lean/Kanban: *Eliminating waste and managing flow* ... 245

PART 2: PMI-ACP® CERTIFICATION GUIDE

7. Preparing for the PMI-ACP® exam: *Check your knowledge* ... 307
8. Professional responsibility: *Making good choices* ... 377
9. Practice makes perfect: *Practice PMI-ACP exam* ... 391

Table of Contents (the real thing)

Intro

Your brain on agile. Here *you* are trying to *learn* something, while here your *brain* is doing you a favor by making sure the learning doesn't *stick*. Your brain's thinking, "Better leave room for more important things, like which wild animals to avoid and whether naked snowboarding is a bad idea." So how *do* you trick your brain into thinking that your life depends on knowing enough to really "get" agile—and maybe even get through the PMI-ACP certification exam?

Who is this book for?	xx
We know what you're thinking	xxi
Metacognition: thinking about thinking	xxiii
Here's what YOU can do to bend your brain into submission	xxii
Read me	xxvi
The technical review team	xxvii
Acknowledgments	xxviii

table of contents

> ### PART 1: UNDERSTANDING AGILE
> In part 1, you'll learn all about agile and how it helps your teams work better together and build better software. First we'll take you through the basics of agile principles and practices. Then we'll do a deeper dive on the most common forms of agile found on teams today: Scrum, XP, Lean, and Kanban. By the end of Part 1, you'll have a solid foundation in the ideas that drive agile teams... and a toolbox that you can use to put agile in place on your own team today.

1

What is agile?
Principles and practices

It's an exciting time to be agile! For the first time, our industry has found a real, sustainable way to solve problems that generations of software development teams have been struggling with. Agile teams use simple, straightforward **practices** that have been proven to work on real-life projects. But wait a minute... if agile is so great, why isn't everyone doing it? It turns out that in the real world, a practice that works really well for one team causes serious problems for another team, and the difference is the team **mindset**. So get ready to change the way you think about your projects!

The new features sound great...	2
...but things don't always go as expected	3
Agile to the rescue!	4
Kate tries to hold a daily standup	5
Different team members have different attitudes	6
A better mindset makes the practice work better	8
So what is agile, anyway?	10
Scrum is the most common approach to agile	12
The PMI-ACP certification can help you be more agile	18

In a daily standup meeting, everyone on the team stands for the duration of the meeting. That keeps it short, sweet, and to the point.

But is this guy really paying attention to what his teammates are saying?

x

2

Agile values and principles
Mindset meets method
There's no "perfect" recipe for building great software.

Some teams have had a lot of success and seen big improvements after adopting agile practices, methods, and methodologies, while others have struggled. We've learned that the difference is the mindset that the people on the team have. So what do you do if you want to get those great agile results for your own team? How do you make sure your team has the right mindset? That's where the **Agile Manifesto** comes in. When you and your team get your head around its **values and principles**, you start to think differently about the agile practices and how they work, and they start to become *a lot more effective*.

Something big happened in Snowbird	24
The Agile Manifesto	25
Adding practices in the real world can be a challenge	26
Individuals and interactions over processes and tools	27
Working software over comprehensive documentation	28
Customer collaboration over contract negotiation	31
Responding to change over following a plan	32
Question Clinic: The "which-is-BEST" question	36
They think they've got a hit...	38
...but it's a flop!	39
The principles behind the Agile Manifesto	40
The agile principles help you deliver your product	42
The agile principles help your team communicate and work together	52
The new product is a hit!	56
Exam Questions	58

table of contents

3 Managing projects with Scrum
The rules of Scrum

The rules of Scrum are simple. Using it effectively is not so simple.

Scrum is the most common agile methodology, and for good reason: the **rules of Scrum** are straightforward and easy to learn. Most teams don't need a lot of time to pick up **the events, roles, and artifacts** that make up the rules of Scrum. But for Scrum to be most effective, they need to really understand **the values of Scrum** and the Agile Manifesto principles, which help them get into an effective mindset. Because while Scrum seems simple, the way a Scrum team constantly **inspects and adapts** is a whole new way of thinking about projects.

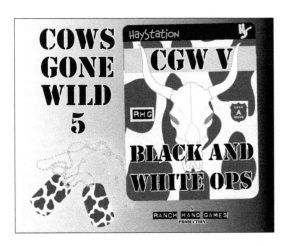

Meet the Ranch Hand Games team	73
The Scrum events help you get your projects done	74
The Scrum roles help you understand who does what	75
The Scrum artifacts keep the team informed	76
The Scrum values make the team more effective	82
Question Clinic: The "which-comes-next" question	90
A task isn't done until it's "Done" done	92
Scrum teams adapt to changes throughout the Sprint	93
The Agile Manifesto helps you really "get" Scrum	96
Things are looking good for the team	102
Exam Questions	104

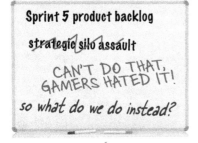

With a new Product Owner, the team should be able to figure out the most valuable features to include in the next sprint.

table of contents

4 Agile planning and estimation
Generally Accepted Scrum Practices

Agile teams use straightforward planning tools to get a handle on their projects. Scrum teams plan their projects together so that everybody on the team commits to each sprint's goal. To maintain the team's **collective commitment**, planning, estimating and tracking need to be simple and easy for the whole team to do as a group. From **user stories** and **planning poker** to **velocity** and **burndown charts**, Scrum teams always know what they've done and what's left to do. Get ready to learn the tools that keep Scrum teams informed and in control of what they build!

Meanwhile, back at the ranch…	118
So… what's next?	121
Introducing GASPs!	122
No more 300-page specs… please?	124
User stories help teams understand what users need	125
Story points let the team focus on the relative size of each story	126
The whole team estimates together	132
No more detailed project plans	134
Taskboards keep the team informed	136
Question Clinic: The red herring	140
Burndown charts help the team see how much work is left	143
Velocity tells you how much your team can do in a sprint	144
Burn-ups keep your progress and your scope separate from each other	147
How do we know what to build?	148
Story maps help you prioritize your backlog	149
Personas help you get to know your users	150
The news could be better…	152
Retrospectives help your team improve the way they work	154
Some tools to help you get more out of your retrospectives	156
Exam Questions	164

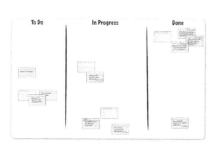

table of contents

5 XP (extreme programming)
Embracing change

Software teams are successful when they build great code.

Even really good software teams with very talented developers run into problems with their code. When small code changes "bloom" into a series of **cascading hacks**, or everyday code commits lead to hours of fixing merge conflicts, work that *used to be satisfying* becomes **annoying, tedious, and frustrating**. And that's where **XP** comes in. XP is an agile methodology that's focused on building cohesive teams that **communicate** well, and creating a **relaxed, energized environment**. When teams build code that's **simple**, not complex, they can *embrace change* rather than fear it.

Meet the team behind CircuitTrak	178
Late nights and weekends lead to code problems	180
XP brings a mindset that helps the team and the code	181
Iterative development helps teams stay on top of changes	182
Courage and respect keep fear out of the project	184
Teams build better code when they work together	190
Teams work best when they sit together	192
XP teams value communication	194
Teams work best with relaxed, rested minds	196
Question Clinic: The "which-is-NOT" question	200
XP teams embrace change	204
Frequent feedback keeps changes small	205
Bad experiences cause a rational fear of change	206
XP practices give you feedback about the code	208
XP teams use automated builds that run quickly	210
Continuous integration prevents nasty surprises	211
The weekly cycle starts with writing tests	212
Agile teams get feedback from design and testing	214
Pair programming	216
Complex code is really hard to maintain	223
When teams value simplicity, they build better code	224
Simplicity is a fundamental agile principle	225
Every team accumulates technical debt	226
XP teams "pay down" technical debt in each weekly cycle	227
Incremental design starts (and ends) with simple code	228
Exam Questions	234

table *of* contents

6 Lean / Kanban
Eliminating waste and managing flow

Agile teams know that they can always improve they way they work. They use the **Lean mindset** to find out where they are spending time on things that aren't helping them **deliver value**. Then they get rid of the **waste** that's slowing them down. Many teams with a lean mindset use **Kanban** to set **work in progress limits** and create **pull systems** to make sure that people are not getting sidetracked by work that doesn't amount to much. Get ready to learn how to seeing your software development process as a **whole system** can help you build better software!

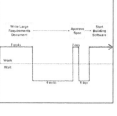

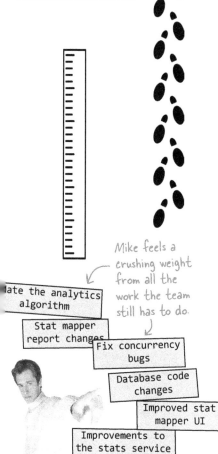

Mike feels a crushing weight from all the work the team still has to do.

Trouble with Audience Analyzer 2.5	246
Lean is a mindset (not a methodology)	248
Lean principles help you see things differently	249
More Lean principles	250
Some thinking tools you haven't seen before	254
More Lean thinking tools	256
Categorizing waste can help you see it better	260
Value stream maps help you see waste	264
Trying to do too many things at once	267
Anatomy of an Option	270
Systems thinking helps Lean teams see the whole	272
Some "improvements" didn't work out	273
Lean teams use pull systems to make sure they're always working on the most valuable tasks	274
Question Clinic: Least worst option	278
Kanban uses a pull system to make your process better	280
Use Kanban boards to visualize the workflow	281
How to use Kanban to improve your process	282
The team creates a workflow	284
The team is delivering faster	291
Cumulative flow diagrams help you manage flow	292
Kanban teams talk about their policies	293
Feedback loops show you how it's working	294
Now the whole team is collaborating on finding better ways to work!	295
Exam Questions	300

xv

table *of* contents

PART 2: PMI-ACP® CERTIFICATION GUIDE

The first part of this book was all about agile. It turns out that learning about the principles and practices of agile and drilling down into Scrum, XP, Lean, and Kanban gets you 90% of the way to preparing for the PMI-ACP® certification! The second part of this book gets you that last 10%, so you're 100% prepared to take the PMI-ACP® exam... and the next step in your career as an agile practitioner.

7

Preparing for the pmi-acp® exam
Check your knowledge
Wow, you sure covered a lot of ground in the last six chapters!

There's **more to the PMI-ACP® exam** than just understanding agile tools, techniques, and concepts really ace the test, you'll need to explore how teams **use them in real-world situations**. So let's g brain a *fresh look at agile concepts* with a **complete** set of exercises, puzzles, and practice que (along with some new material) specficially constructed to help prepare you for the PMI-ACP® exa

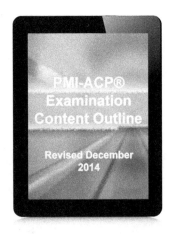

The PMI-ACP® certification is valuable...	308
The PMI-ACP® exam is based on the content outline	309
"You are an agile practitioner..."	310
A long-term relationship for your brain	313
Domain 1: Agile Principles and Mindset	314
Domain 1: Exam Questions	316
Domain 2: Value-Driven Delivery	322
Agile teams use customer value to prioritize requirements	325
Value calculations help you figure out which projects to do	326
Domain 2: Exam Questions	330
Domain 3: Stakeholder Engagement	336
Domain 4: Team Performance	337
Domain 3: Exam Questions	338
Domain 4: Exam Questions	339
Domain 5: Adaptive Planning	348
Adapt your leadership style as the team evolves	349
A few last tools and techniques	351
Domain 6: Problem Detection and Resolution	360
Domain 7: Continuous Improvement	361
Domain 5: Exam Questions	372
Domain 6: Exam Questions	373
Domain 7: Exam Questions	374
Are you ready for the final exam?	376

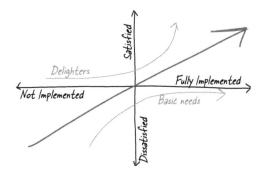

8 Professional responsibility
Making good choices

It's not enough to just know your stuff. You need to make good choices to be good at your job. Everyone who has the PMI-ACP credential agrees to follow the **Project Management Institute Code of Ethics and Professional Conduct**, too, because the Code helps you with **ethical decisions**. There may be a few questions based on this material scattered through the PMI-ACP® exam. Luckily, most of what you need to know is **really straightforward**, and with a little review, you'll do great with this material.

Doing the right thing	378
Keep the cash?	380
Fly business class?	381
New software	382
Shortcuts	383
A good price or a clean river?	384
We're not all angels	385
Exam Questions	386

AWESOME. I'VE BEEN WANTING TO GO SHOPPING FOR A WHILE. AND WHAT ABOUT THAT VACATION? ACAPULCO, HERE WE COME!

I WOULD NEVER ACCEPT A GIFT LIKE THAT. DOING A GOOD JOB IS ITS OWN REWARD!

I'M SORRY, I CAN'T ACCEPT THE GIFT. I REALLY APPRECIATE THE GESTURE, THOUGH.

Practice makes perfect

9 Practice PMI-ACP® Exam

Bet you never thought you'd make it this far! It's been a long journey, but here you are, ready to review your knowledge and get ready for exam day. You've put a lot of new information about agile into your brain, and now it's time to see just how much of it stuck. That's why we put together this full-length, 120-question PMI-ACP® practice exam for you. **We followed the exact same PMI-ACP® Examination Content Outline** that the experts at PMI use, so it looks *just like the one you're going to see* when you take the real thing. Now's your time to flex your mental muscle. So take a deep breath, get ready, and let's get started.

Complete PMI-ACP® Practice Exam	391
Before you look at the answers...	423

how to use this book

Intro

In this section, we answer the burning question: "So why <u>DID</u> they put that in a book about agile?"

how to use this book

Who is this book for?

If you can answer "yes" to any of these questions:

1. Are you a **developer, project manager, business analyst, designer**, or other member of a team, and you're looking to improve your projects?

2. Is **your team going agile**, but you're not really sure what that means or how you fit in?

3. Are you thinking about a job search, and want to understand **why employers are asking for agile experience**?

4. Do you prefer **stimulating dinner-party conversation** to **dry, dull, academic lectures**?

this book is for you.

> This book includes a complete guide to preparing for the PMI-ACP® certification exam, and has 100% coverage of the exam material. But the most effective way to prepare for it is to **learn agile principles, practices, and ideas**. So the entire first part of this book is focused on **learning agile**, not on preparing for the exam.

The PMI-ACP® (Agile Certified Practitioner) is one of the fastest-growing certifications in the world that employers are increasingly demanding.

Who should probably back away from this book?

If you can answer "yes" to any of these:

1. Are you **completely new** to working on any kind of team or working with other people to achieve something?

2. Are you a "go-it-alone" loner who feels that working with other people on a team is always a **waste of time**?

3. Are you **afraid to try something different**? Would you rather have a root canal than mix stripes with plaid? Do you believe that a technical book can't be serious if agile concepts, tools, and ideas are anthropomorphized?

this book is not for you.

[Note from marketing: this book is for anyone with a pulse.]

> ### Are you preparing for the PMI-ACP® exam?
>
> *Then this book is **definitely** for you! We built this book to get the ideas, concepts, and practices of agile into your brain—and we made sure to include 100% coverage of every topic that appears on the exam. We included a lot of exam preparation material, including a full-length simulated exam that's **as close to the real thing** as you can get!*

If you've never been on any kind of team before, then many of the ideas in agile will feel foreign. Just to be clear, we're not necessarily talking about a software team—experience on any kind of team will be just fine!

We know what you're thinking.

"How can *this* be a serious book on agile?"

"What's with all the graphics?"

"Can I actually *learn* it this way?"

And we know what your *brain* is thinking.

Your brain craves novelty. It's always searching, scanning, *waiting* for something unusual. It was built that way, and it helps you stay alive.

So what does your brain do with all the routine, ordinary, normal things you encounter? Everything it *can* to stop them from interfering with the brain's *real* job—recording things that *matter*. It doesn't bother saving the boring things; they never make it past the "this is obviously not important" filter.

How does your brain *know* what's important? Suppose you're out for a day hike and a tiger jumps in front of you, what happens inside your head and body?

Neurons fire. Emotions crank up. *Chemicals surge.*

And that's how your brain knows…

This must be important! Don't forget it!

But imagine you're at home, or in a library. It's a safe, warm, tiger-free zone. You're studying. Getting ready for an exam. Or trying to learn some tough technical topic your boss thinks will take a week, 10 days at the most.

Just one problem. Your brain's trying to do you a big favor. It's trying to make sure that this *obviously* unimportant content doesn't clutter up scarce resources. Resources that are better spent storing the really *big* things. Like tigers. Like the danger of fire. Like how you should never again snowboard in shorts.

And there's no simple way to tell your brain, "Hey brain, thank you very much, but no matter how dull this book is, and how little I'm registering on the emotional Richter scale right now, I really *do* want you to keep this stuff around."

Your brain thinks THIS is important.

GREAT. ONLY 458 MORE DULL, DRY, BORING PAGES.

Your brain thinks THIS isn't worth saving.

how to use this book

We think of a "Head First" reader as a learner.

So what does it take to *learn* something? First, you have to *get* it, then make sure you don't *forget* it. It's not about pushing facts into your head. Based on the latest research in cognitive science, neurobiology, and educational psychology, *learning* takes a lot more than text on a page. We know what turns your brain on.

Some of the Head First learning principles:

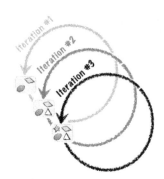

Make it visual. Images are far more memorable than words alone, and make learning much more effective (up to 89% improvement in recall and transfer studies). It also makes things more understandable.
Put the words within or near the graphics they relate to, rather than on the bottom or on another page, and learners will be up to *twice* as likely to solve problems related to the content.

Use a conversational and personalized style. In recent studies, students performed up to 40% better on post-learning tests if the content spoke directly to the reader, using a first-person, conversational style rather than taking a formal tone. Tell stories instead of lecturing. Use casual language. Don't take yourself too seriously. Which would *you* pay more attention to: a stimulating dinner-party companion, or a lecture?

I ASKED AMY A SIMPLE 10-SECOND QUESTION, AND I'VE BEEN STUCK WAITING FOR TWO HOURS FOR HER TO GET BACK TO ME.

Get the learner to think more deeply. In other words, unless you actively flex your neurons, nothing much happens in your head. A reader has to be motivated, engaged, curious, and inspired to solve problems, draw conclusions, and generate new knowledge. And for that, you need challenges, exercises, and thought-provoking questions, and activities that involve both sides of the brain and multiple senses.

Get—and keep—the reader's attention. We've all had the "I really want to learn this, but I can't stay awake past page one" experience. Your brain pays attention to things that are out of the ordinary, interesting, strange, eye-catching, unexpected. Learning a new, tough, technical topic doesn't have to be boring. Your brain will learn much more quickly if it's not.

Touch their emotions. We now know that your ability to remember something is largely dependent on its emotional content. You remember what you care about. You remember when you *feel* something. No, we're not talking heart-wrenching stories about a boy and his dog. We're talking emotions like surprise, curiosity, fun, "what the...?", and the feeling of "I rule!" that comes when you solve a puzzle, learn something everybody else thinks is hard, or realize you know something that "I'm more technical than thou" Bob from engineering *doesn't*.

Metacognition: thinking about thinking

If you really want to learn, and you want to learn more quickly and more deeply, pay attention to how you pay attention. Think about how you think. Learn how you learn.

Most of us did not take courses on metacognition or learning theory when we were growing up. We were *expected* to learn, but rarely *taught* to learn.

But we assume that if you're holding this book, you really want to learn about project management. And you probably don't want to spend a lot of time. And since you need to use this on a real project (and <u>especially</u> if you're going to take an exam on it!) you need to *remember* what you read. And for that, you've got to *understand* it. To get the most from this book, or *any* book or learning experience, take responsibility for your brain. Your brain on *this* content.

The trick is to get your brain to see the new material you're learning as Really Important. Crucial to your well-being. As important as a tiger. Otherwise, you're in for a constant battle, with your brain doing its best to keep the new content from sticking.

So just how *DO* you get your brain to think that the material about agile is a hungry tiger?

There's the slow, tedious way, or the faster, more effective way. The slow way is about sheer repetition. You obviously know that you *are* able to learn and remember even the dullest of topics if you keep pounding the same thing into your brain. With enough repetition, your brain says, "This doesn't *feel* important to him, but he keeps looking at the same thing *over* and *over* and *over*, so I suppose it must be."

The faster way is to do **anything that increases brain activity,** especially different *types* of brain activity. The things on the previous page are a big part of the solution, and they're all things that have been proven to help your brain work in your favor. For example, studies show that putting words *within* the pictures they describe (as opposed to somewhere else in the page, like a caption or in the body text) causes your brain to try to makes sense of how the words and picture relate, and this causes more neurons to fire. More neurons firing = more chances for your brain to *get* that this is something worth paying attention to, and possibly recording.

A conversational style helps because people tend to pay more attention when they perceive that they're in a conversation, since they're expected to follow along and hold up their end. The amazing thing is, your brain doesn't necessarily *care* that the "conversation" is between you and a book! On the other hand, if the writing style is formal and dry, your brain perceives it the same way you experience being lectured to while sitting in a roomful of passive attendees. No need to stay awake.

But pictures and conversational style are just the beginning.

how to use this book

Here's what WE did:

We used **pictures**, because your brain is tuned for visuals, not text. As far as your brain's concerned, a picture really *is* worth a thousand words. And when text and pictures work together, we embedded the text *in* the pictures because your brain works more effectively when the text is *within* the thing the text refers to, as opposed to in a caption or buried in the text somewhere.

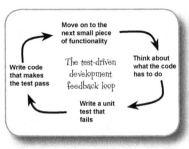

We used **redundancy**, saying the same thing in *different* ways and with different media types, and *multiple senses*, to increase the chance that the content gets coded into more than one area of your brain.

We used concepts and pictures in **unexpected** ways because your brain is tuned for novelty, and we used pictures and ideas with at least *some* **emotional** *content*, because your brain is tuned to pay attention to the biochemistry of emotions. That which causes you to *feel* something is more likely to be remembered, even if that feeling is nothing more than a little **humor**, **surprise**, or **interest**.

We used a personalized, **conversational style**, because your brain is tuned to pay more attention when it believes you're in a conversation than if it thinks you're passively listening to a presentation. Your brain does this even when you're *reading*.

We included more than 80 **activities**, because your brain is tuned to learn and remember more when you **do** things than when you *read* about things. And we made the exercises challenging-yet-doable, because that's what most people prefer.

We used **multiple learning styles**, because *you* might prefer step-by-step procedures, while someone else wants to understand the big picture first, and someone else just wants to see an example. But regardless of your own learning preference, *everyone* benefits from seeing the same content represented in multiple ways.

We include content for **both sides of your brain**, because the more of your brain you engage, the more likely you are to learn and remember, and the longer you can stay focused. Since working one side of the brain often means giving the other side a chance to rest, you can be more productive at learning for a longer period of time.

And we included **stories** and exercises that present **more than one point of view,** because your brain is tuned to learn more deeply when it's forced to make evaluations and judgments.

We included **challenges**, with exercises, and by asking **questions** that don't always have a straight answer, because your brain is tuned to learn and remember when it has to *work* at something. Think about it—you can't get your *body* in shape just by *watching* people at the gym. But we did our best to make sure that when you're working hard, it's on the *right* things. That **you're not spending one extra dendrite** processing a hard-to-understand example, or parsing difficult, jargon-laden, or overly terse text.

We used **people**. In stories, examples, pictures, and so on, because, well, because *you're* a person. And your brain pays more attention to *people* than it does to *things*.

the intro

Here's what YOU can do to bend your brain into submission

So, we did our part. The rest is up to you. These tips are a starting point; listen to your brain and figure out what works for you and what doesn't. Try new things.

Cut this out and stick it on your refrigerator.

---

① **Slow down. The more you understand, the less you have to memorize.**

Don't just *read*. Stop and think. When the book asks you a question, don't just skip to the answer. Imagine that someone really *is* asking the question. The more deeply you force your brain to think, the better chance you have of learning and remembering.

② **Do the exercises. Write your own notes.**

We put them in, but if we did them for you, that would be like having someone else do your workouts for you. And don't just *look* at the exercises. **Use a pencil.** There's plenty of evidence that physical activity *while* learning can increase the learning.

③ **Read the "There are No Dumb Questions"**

That means all of them. They're not optional sidebars—***they're part of the core content!*** Don't skip them.

④ **Make this the last thing you read before bed. Or at least the last challenging thing.**

Part of the learning (especially the transfer to long-term memory) happens *after* you put the book down. Your brain needs time on its own, to do more processing. If you put in something new during that processing time, some of what you just learned will be lost.

⑤ **Drink water. Lots of it.**

Your brain works best in a nice bath of fluid. Dehydration (which can happen before you ever feel thirsty) decreases cognitive function.

⑥ **Talk about it. Out loud.**

Speaking activates a different part of the brain. If you're trying to understand something, or increase your chance of remembering it later, say it out loud. Better still, try to explain it out loud to someone else. You'll learn more quickly, and you might uncover ideas you hadn't known were there when you were reading about it.

⑦ **Listen to your brain.**

Pay attention to whether your brain is getting overloaded. If you find yourself starting to skim the surface or forget what you just read, it's time for a break. Once you go past a certain point, you won't learn faster by trying to shove more in, and you might even hurt the process.

⑧ **Feel something!**

Your brain needs to know that this *matters*. Get involved with the stories. Make up your own captions for the photos. Groaning over a bad joke is *still* better than feeling nothing at all.

⑨ **Create something!**

Apply this to your daily work; use what you are learning to make decisions on your projects. Just do something to get some experience beyond the exercises and activities in this book. All you need is a pencil and a problem to solve…a problem that might benefit from using the tools and techniques you're learning in this book.

how to use this book

Read me

This is a learning experience, not a reference book. We deliberately stripped out everything that might get in the way of learning whatever it is we're working on at that point in the book. Once you've read this book, you'll definitely want to keep it on your shelf, so you can revisit useful ideas, tools, and techniques. But the first time through, you need to begin at the beginning, because the book makes assumptions about what you've already seen and learned.

The redundancy is intentional and important.

One distinct difference in a Head First book is that we want you to *really* get it. And we want you to finish the book remembering what you've learned. Most reference books don't have retention and recall as a goal, but this book is about *learning*, so you'll see some of the same concepts come up more than once.

The Brain Power exercises don't have answers.

For some of them, there is no right answer, and for others, part of the learning experience of the Brain Power activities is for you to decide if and when your answers are right. In some of the Brain Power exercises, you will find hints to point you in the right direction.

The activities are NOT optional.

The exercises and activities are not add-ons; they're part of the core content of the book. Some of them are to help with memory, some are for understanding, and some will help you apply what you've learned. **Don't skip the exercises.** Even crossword puzzles are important—they'll help get concepts into your brain. But more importantly, they're good for giving your brain a chance to think about the words and terms you've been learning in a different context.

Try the exam questions—even if you're not studying for the exam!

Some of our readers are preparing for the three-hour, 120-question exam PMI-ACP® certification exam. Luckily, the most effective way to prepare for the exam is **to learn agile**. So even if you're not interested in the PMI-ACP® certification, this book is still for you. But you should still try the practice exam questions at the end of each chapter, because answering the exam questions is *a really effective way* to get agile concepts into your brain.

The technical review team

Technical reviewers:

Dave Prior has been managing technology projects for over 20 years and he has been focusing exclusively on agile since 2009. He is a Certified Scrum Trainer and works for LeadingAgile. His spirit animal is Otis Redding and if he could ingest only one food substance, it would be coffee.

Keith Conant has developed software for 20 years as a software engineer, project manager, and group manager. He currently leads a team enhancing a point-of-sale payment application used by universities around the world. Away from the office, Keith can be found composing music, playing drums, guitar, or keyboards in a band, or challenging himself physically kayaking, running, hiking, or cycling.

Philip Cheung has been developing software for 15 years and he has been exclusively using agile since 2013 to manage and deliver projects. He works in the financial industry involved in creating various enterprise-level applications. Philip likes to escape into the rural English countryside and one day hopes to retire in a charming country cottage.

Kelly D. Marce, PMP®, PMI-ACP® has more than nine years in project management. Kelly is an agile trainer, certified project manager, and PMP® mentor at a leading financial services company in Canada. In his spare time, he acts as a producer of live events within his community and tries to keep up with his four-year-old son, Jacob.

And, as always, we were lucky to have **Lisa Kellner** return to our tech review team. Lisa was awesome, as usual. Thanks so much, everyone!

how to use this book

Acknowledgments

Our editor:

We want to thank our editor, **Nan Barber**, for editing this book. Thanks!

Nan Barber

The O'Reilly team:

There are so many people at O'Reilly we want to thank that we hope we don't forget anyone.

We'll definitely be thanking the production team in this space. In the meantime, here are a few people we definitely want to acknowledge.

And as always, we love **Mary Treseler**, and can't wait to work with her again! And a big shout out to our other friends and editors, **Mike Hendrickson**, **Tim O'Reilly**, **Andy Oram**, **Laurel Ruma**, **Lindsay Ventimiglia**, **Melanie Yarbrough**, **Ron Bilodeau**, **Lucie Haskins**, and **Jasmine Kwityn**. And if you're reading this book right now, then you can thank the greatest publicity team in the industry: **Marsee Henon**, **Kathryn Barret**, and the rest of the folks in Sebastopol.

O'Reilly Safari®

Safari (formerly Safari Books Online) is a membership-based training and reference platform for enterprise, government, educators, and individuals.

Members have access to thousands of books, training videos, Learning Paths, interactive tutorials, and curated playlists from over 250 publishers, including O'Reilly Media, Harvard Business Review, Prentice Hall Professional, Addison-Wesley Professional, Microsoft Press, Sams, Que, Peachpit Press, Adobe, Focal Press, Cisco Press, John Wiley & Sons, Syngress, Morgan Kaufmann, IBM Redbooks, Packt, Adobe Press, FT Press, Apress, Manning, New Riders, McGraw-Hill, Jones & Bartlett, and Course Technology, among others.

For more information, please visit *http://oreilly.com/safari*.

More Praise for *Head First Agile*

We asked organizational transformation expert and part-time international rock star Mike Monsoon to review an early release of this book and give us a quote for the "Praise for..." page. Instead, he wrote a song!

Praise Quote

I've read so many agile books before
But they all come off the same,
A bunch of pretentious preachy solemn righteous stuff,
I just find them all lame.

Break it down (you) break it down for me.
(If I could T-shirt size it
It'd be an extra-large)

Meta meta meta cognition
I finally found the right information.
I really appreciate the book that you wrote.
But I can't make a praise quote.

– **Mike Monsoon, International Rock Star**

Listen to the song here: *https://bit.ly/head-first-agile-song*

PART 1: UNDERSTANDING AGILE

In part 1, you'll learn all about agile and how it helps your teams work better together and build better software. First we'll take you through the basics of agile principles and practices. Then we'll do a deeper dive on the most common forms of agile found on teams today: Scrum, XP, Lean, and Kanban. By the end of Part 1, you'll have a solid foundation in the ideas that drive agile teams... and a toolbox that you can use to put agile in place on your own team today.

> Part 1 of this book has just a little bit of material to help you prepare for the PMI-ACP® exam. But we were really careful to design that material so that it's focused less on preparing for the exam, and more on getting agile into your brain. So even if you're not even thinking about taking the PMI-ACP® exam, you should still do those parts because they'll help you learn the core material in this book.

1 what is agile?

Principles and practices

It's an exciting time to be agile! For the first time, our industry has found a real, sustainable way to solve problems that generations of software development teams have been struggling with. Agile teams use simple, straightforward **practices** that have been proven to work on real-life projects. But wait a minute...if agile is so great, why isn't everyone doing it? It turns out that in the real world, a practice that works really well for one team causes serious problems for another team, and the difference is the team **mindset**. So get ready to change the way you think about your projects!

looks like we won't get that **bonus**

The new features sound great...

Meet Kate. She's a project manager at a successful Silicon Valley startup. Her company builds software that's used by video and music streaming services and internet radio stations to analyze audiences in real time and choose programming suggestions that make their viewers or listeners happy. And now Kate's team has an opportunity to deliver something that will really help the company.

Kate's a project manager for a software team.

Ben is the product owner. His job is to talk to clients, figure out what they need, and come up with new features that they'll use.

...but things don't always go as expected

Kate's discussion with the project team didn't go nearly as well as she'd hoped. What's she going to tell Ben?

Mike is the lead programmer and architect.

> IT **SOUNDS** LIKE WE'VE GOT A REAL OPPORTUNITY TO MAKE OUR CUSTOMERS HAPPY.

Kate: So if we can get those new advanced audience analytics features into the next release, we'll all get a big bonus.

Mike: Well, that sounds like it would be great.

Kate: Fantastic! So we can count on you guys?

Mike: Hold on! Not so fast. I said it **sounds** like it **would** be great. But there's no way it's happening.

Kate: Wait, what?! Don't mess with me, Mike.

Mike: Look, if we'd known about this change four months ago when we were designing the audience data analysis service, this would be easy. But now we'd have to rip out a huge chunk of code and replace it with... well, I don't want to get into technical details.

Kate: Good. I don't want to hear them.

Mike: So... are we done here? Because my team's got a lot of work to do.

standing keeps the meeting short

Agile to the rescue!

Kate's been reading about agile, and she thinks it might help her get those features into the next release. Agile's gotten really popular with software teams because the ones that have "gone agile" often talk about the great results they get. The software they build is better, which makes a big difference to them *and* their users. Not only that, but when agile teams are effective, they have a much better time at work! Things are more relaxed, and the working environment is a lot more enjoyable.

So why has agile gotten so popular? Lots of reasons:

- ★ When teams go agile, they find that it's a lot easier to meet their deadlines.
- ★ They also find that they can really cut down on bugs in their software.
- ★ Their code is a lot easier to maintain—adding, extending, or changing their codebase is no longer a headache.
- ★ The users are a lot happier, which always makes everyone's lives easier.
- ★ And best of all, when agile teams are effective, the team members' lives are better, because they can go home at a reasonable hour and rarely have to work weekends (which, for a lot of developers, is a first!).

A daily standup is a good starting point

One of the most common agile practices that teams adopt is called the **daily standup**, a meeting that happens every day, during which team members talk about what they're working on and their challenges. The meeting is kept short by making everyone stand for the duration. Adding a daily standup to their projects has brought success to a lot of teams, and it's often the first step in going agile.

In a daily standup meeting, everyone on the team stands for the duration of the meeting. That keeps it short, sweet, and to the point.

But is this guy really paying attention to what his teammates are saying?

Kate tries to hold a daily standup

To Kate's surprise, not everyone on Mike's team shares her excitement for this new practice. In fact, one of his developers is angry that she's even suggesting that they add a new meeting, and seems to feel insulted by the idea of attending a meeting every day where he's asked prying questions about his day-to-day work.

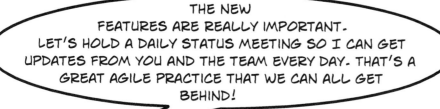

> THE NEW FEATURES ARE REALLY IMPORTANT. LET'S HOLD A DAILY STATUS MEETING SO I CAN GET UPDATES FROM YOU AND THE TEAM EVERY DAY. THAT'S A GREAT AGILE PRACTICE THAT WE CAN ALL GET BEHIND!

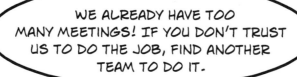

> WE ALREADY HAVE TOO MANY MEETINGS! IF YOU DON'T TRUST US TO DO THE JOB, FIND ANOTHER TEAM TO DO IT.

↑ Kate thinks Mike and his team are being irrational, but maybe they have a point. What do you think?

So what's going on here? Is Mike being irrational? Is Kate being too demanding? Why is this simple, well-accepted practice causing a conflict?

mindset *meets* **methodology**

Different team members have different attitudes

Kate ran into problems adopting agile almost as soon as she started—and she's not alone.

The truth is that a lot of teams simply haven't had as much success with agile as they'd hoped they would. Did you know that well over half of companies that build software have experimented with agile? Despite the success stories—and there are many of them!—a lot of teams try agile, but end up with results that they're not particularly happy with. In fact, they feel a little ripped off! It seemed like agile came with a promise of big changes, but trying to get agile working on their own projects never seemed to work out.

And that's what's happening to Kate. She created a plan all on her own, and now she wants to get status updates from her team. So she's started dragging a reluctant team to her daily standup meeting. She's able to get them in the room. But will it really make a difference? She's worried about how people are deviating from her plan, so she'll concentrate on getting a status update from each person. Mike and his developers, on the other hand, want the meeting to end as quickly as possible so they can get back to "real" work.

In Kate's less-than-effective daily standup, each person tunes out everyone else while they're waiting to give their updates, then says as little as possible when it's time to speak. She still gets some useful information, but at the cost of conflict and boredom—and nobody in the meeting is collaborating.

Two people can see the same practice very differently. If they don't <u>**both**</u> feel like they're getting something out of it, it can make the practice much less effective.

6 Chapter 1

what is *agile?*

> THAT'S JUST HOW SOFTWARE PROJECTS ARE, RIGHT? THINGS THAT WORK IN TEXTBOOKS DON'T REALLY WORK OUT IN REAL LIFE. THERE'S NOTHING WE CAN DO ABOUT IT, RIGHT?

No! The right mindset makes practices more effective.

Let's be clear: the way Kate is running her standups is how many daily standups are run. And while it's not optimal, a daily standup that's run this way *will still produce results*. Kate will find out about problems with her plan, and Mike's team will benefit in the long run because those problems that do affect them can be taken care of sooner rather than later. The whole thing doesn't take much time every day, and that makes it worth doing.

But there's a big difference between an agile team that goes through the motions and one that gets great results. The key to that difference is the **mindset** the team brings to each project. Believe it or not, the attitude that each person takes toward a practice can make it much more effective!

The attitude that each team member brings to a practice like the daily standup makes a huge difference in how effective it is. But even when everyone tunes out, the meeting is still effective enough to make it worth doing, even if it's boring.

you are here ▸ 7

keep mindset in mind

A better mindset makes the practice work better

What would happen if Kate and Mike had a different mindset? What if each person on the team approached the daily standup *with an entirely different attitude*?

For example, what would happen if Kate felt like everyone on the team **worked together** to plan the project? Then she would genuinely listen to every single developer. If Kate changes her attitude toward the standup, she'll stop trying to figure out how they've deviated from *her* plan so that *she* can correct them. The focus of the meeting changes for her: now it's about understanding the plan that everyone on the team worked together to create, and her job is about helping the whole team do their work more effectivly.

That's a very different way of thinking about planning, one that Kate was never taught in any of her project management training courses. She was always taught that it was her job to come up with a project plan and basically dictate it to the team. She had tools that let her measure how well the team followed her plan, and strict processes that she would enforce to make changes to it.

Now things are totally different for her. She realized that the only way she could make the daily standup work is if **she puts effort into working with the team** so that everyone on the team can work together to figure out the best approach to the project. Then the daily standup turns into a way for the whole team to work together to make sure everyone is making solid decisions, and that the project is on track.

I DON'T HAVE ALL THE ANSWERS. WE NEED THIS MEETING SO WE CAN PLAN THE PROJECT TOGETHER!

Kate used to get really frustrated when she discovered changes to her project plan, because it was usually too late for the team to effectively change course.

Now that the daily standup is in place, the whole team works with her every day to find those changes, so they can make them much earlier. That's a lot more effective!

> SO THE DAILY STANDUP MEANS YOU'LL **LISTEN TO ME AND MY TEAM** AND ACTUALLY CHANGE THE WAY THE PROJECT RUNS?

And what if Mike felt like this meeting wasn't just about giving status updates, but about *understanding how the project is going*, and coming together every day to find ways that everyone can work better? Then the daily standup becomes important to him.

A good developer almost always has opinions not just about his own code, but about the whole direction of the project. The daily standup becomes his way to make sure that the project is run in a sensible, efficient way—and Mike knows that in the long run that will make his job of coding more rewarding for his team, because the rest of the project is being run well. And he knows that when he brings up a problem with the plan during the meeting, ***everyone will listen***, and the project will run better because of it.

Things work best when Mike (and the rest of the team) figure out that the daily standup helps them plan the next day's worth of work—and every single person on the team is part of the planning process.

> THAT MAKES SENSE! PLANNING A PROJECT WORKS A LOT BETTER WHEN EVERYONE ON THE TEAM IS ENGAGED. BUT I BET THIS ONLY WORKS IF EVERYONE AT THE MEETING IS TUNED IN THE WHOLE TIME.

How do you change the mindset in a team or an individual? Can you think of examples from your own projects where someone's mindset changed—maybe your own?

ke perfect

...ile, anyway?

...**ods and methodologies** that are optimized to help with specific problems ... run into, and kept simple so they're relatively straightforward to implement.

...nd methodologies address all of the areas of traditional software engineering, ...t management, software design and architecture, and process improvement. Each of those ... ...ds and methodologies consists of **practices** that are streamlined and optimized to make them as easy as possible to adopt.

> I SPENT ALL THIS TIME COMING UP WITH **MY** PLAN, BUT THE TEAM KEEPS DEVIATING FROM IT. I CAN USE THE DAILY STANDUP TO MAKE SURE THEY DO **EVERYTHING** I TELL THEM TO DO.

Mindset versus methodology

Agile is *also* a **mindset**, and that's a new idea for a lot of people who haven't worked with agile before. It turns out that each team member's attitude toward the practices they use *can make a huge difference* in how effective those practices are. The agile mindset is focused on helping people share information with each other, which makes it much easier for them to make important project decisions (rather than just relying on a boss or project manager to make those decisions). It's about opening up planning, design, and process improvement to the *entire* team. To help everyone get into an effective mindset, each agile methodology has its own set of **values** that team members can use as a guide.

> IF WE **ALL WORK TOGETHER** TO PLAN OUR PROJECT, THEN WE CAN USE THE DAILY STANDUP TO MAKE COURSE CORRECTIONS ALONG THE WAY.

What happens if one team member checks out during the daily standup and doesn't really listen to his or her teammates?

Sharpen your pencil

Here are a few problems that Kate, Ben, and Mike brought up during a daily standup. Across from them, we've written down the names of a few different practices that you'll often see used on agile teams. Don't worry if you haven't run across some of them—you'll learn a lot more about them later in the book, so we included a brief description of each practice to help you out. See if you can match each problem with a practice that might help fix it.

"WE JUST WASTED HOURS GRINDING THROUGH SPAGHETTI CODE TO FIND THAT BUG!"

A **retrospective** is a meeting in which everyone talks about how the last part of the project went, and talks about what lessons can be learned.

"OK, WE'VE GONE OVER THE USER STORIES. NOW LET'S FIGURE OUT HOW THEY FIT TOGETHER SO WE CAN PLAN OUT THE NEXT FEW WEEKS OF WORK."

A **user story** is a way to express one very specific need that a user has, usually written out as a few sentences on a sticky note or index card.

"WE ALWAYS SEEM TO RUN INTO THE SAME KIND OF PROBLEMS OVER AND OVER AGAIN WITH EVERY RELEASE."

A **task board** is an agile planning tool in which user stories are attached to a board and categorized into columns based on their status.

"I JUST DEMOED A NEW FEATURE TO ONE OF OUR VIDEO STREAMING USERS. SHE SAID IT WON'T ACTUALLY FIX THE PROBLEM IT'S SUPPOSED TO ADDRESS FOR HER."

A **burndown chart** is a line chart, updated daily, that tracks the amount of work left on the project, "burning" down to zero when the work is done.

"I THOUGHT WE'D BE DONE UPDATING THE SONG DATABASE CODE BY FRIDAY. NOW YOU'RE TELLING ME THAT IT'LL BE THREE MORE WEEKS?"

Developers fix code problems by constantly **refactoring** their code, or improving the code structure without changing its behavior.

It's OK if you haven't run across these practices yet. You'll learn more about them in the next few chapters.

a methodical framework

Scrum is the most common approach to agile

There are many ways that teams can be agile, and there's a long list of methods and methodologies that agile teams use. But there have been many surveys done over the years that have found that the most common approach to agile is **Scrum**, a software development framework focused on project management and product development.

When a team uses Scrum, every project follows the same basic pattern. There are three main roles on a Scrum project: the **Product Owner** (like Ben) works with the team to maintain a **Product Backlog**; the **Scrum Master** helps guide the team past roadblocks; and the **Development Team members** (everyone else on the team). The project is divided into **sprints**, or cycles of equal length (often two weeks or 30 days) that follow the Scrum pattern. At the start of a sprint, the team does **sprint planning** to determine which features from the Product Backlog they'll build during the sprint. This is called the **Sprint Backlog**, and the team works throughout the sprint to build all of the features in it. Every day the team holds a short meeting called the **Daily Scrum**. At the end of the sprint, working software is demonstrated to the product owner and stakeholders in the **sprint review**, and the team holds a **retrospective** to figure out lessons they've learned.

We'll cover Scrum in depth in Chapters 3 and 4, and we'll not only teach you how it helps teams build better software and run more successful projects, but we'll also use it to explore important concepts and ideas shared by all agile teams.

XP and Lean/Kanban

While Scrum is the most popular agile methodology, many teams take other approaches. The next most popular methodology is **XP**, a methodology focused on software development and programming that's often used in combination with Scrum. Other teams approach agile with **Lean** and **Kanban**, a mindset that gives you tools to understand the way you build software today and a method that helps you evolve to a better state tomorrow. You'll learn about XP and Lean/Kanban in Chapters 5 and 6.

Getting a little overwhelmed with new vocabulary?

We'll highlight new vocabulary words in **boldface** when we first introduce them. We did that a lot on this page—and if there are a few that you're not familiar with, that's OK! Seeing new ideas in context now will help them sink into your brain when you learn about them in detail later. That's part of the *Head First* neuroscience that makes this book brain-friendly!

Sharpen your pencil
Solution

Teams can avoid making the same mistakes over and over again by looking back at the project and talking about what went right and what could be improved.

"ONE OF MY PROGRAMMERS JUST WASTED HOURS GRINDING THROUGH SPAGHETTI CODE TO FIND A BUG!"

A **retrospective** is a meeting in which everyone talks about how the last part of the project went, and talks about what lessons can be learned.

A task board is a great way for everyone on the team to see the same big-picture view of the project.

"OK, WE'VE GONE OVER THE USER STORIES. NOW LET'S FIGURE OUT HOW THEY FIT TOGETHER SO WE CAN PLAN OUT THE NEXT FEW WEEKS OF WORK."

A **user story** is a way to express one very specific need that a user has, usually written out as a few sentences on a sticky note or index card.

"WE ALWAYS SEEM TO RUN INTO THE SAME KIND OF PROBLEMS OVER AND OVER AGAIN WITH EVERY RELEASE."

A **task board** is an agile planning tool in which user stories are attached to a board and categorized into columns based on their status.

"I JUST DEMOED A NEW FEATURE TO ONE OF OUR VIDEO STREAMING USERS. SHE SAID IT WON'T ACTUALLY FIX THE PROBLEM IT'S SUPPOSED TO ADDRESS FOR HER."

When everyone on the team understands the users and what they need, they do a better job of building software that users love.

A **burndown chart** is a line chart, updated daily, that tracks the amount of work left on the project, "burning" down to zero when the work is done.

This is an XP practice. Some project managers are surprised the first time they discover agile practices that are focused on code, and not just on planning and executing the project.

"I THOUGHT WE'D BE DONE UPDATING THE SONG DATABASE CODE BY FRIDAY. NOW YOU'RE TELLING ME THAT IT'LL BE THREE MORE WEEKS?"

Developers fix code problems by constantly **refactoring** their code, or improving the code structure without changing its behavior.

agile is not just *new names for old things*

there are no Dumb Questions

Q: It sounds like Scrum, XP, and Lean/Kanban are very different from each other. How can they all be agile?

A: Scrum, XP, and Lean/Kanban focus on very different areas. Scrum is mainly focused on project management: what work is getting done, and making sure that it's in line with what the users and stakeholders need. XP is focused on software development: building high-quality code that's well-designed and easy to maintain. Lean/Kanban is a combination of the Lean mindset and the Kanban method, and teams use it to focus on continually improving the way that they build software.

In other words, Scrum, XP, and Lean/Kanban are focused on three different areas of software engineering: project management, design and architecture, and process improvement. So it makes sense that they would have different practices—that's how they differ.

In the next chapter you'll learn about what they have in common: **shared values and principles** that help teams adopt an agile mindset.

Q: Isn't this all just stuff I know already, only with a new name? Like, Scrum sprints are really just milestones and project phases, right?

A: When you come across an agile methodology like Scrum for the first time, it's really common to look for the parts of it that are similar to things you already know—and that's a good thing! If you've been working on teams for a while, then a lot of agile *should* feel familiar. Your team builds something, and you and your teammates are almost certainly doing a lot of things well that you don't want to change (yet!).

However, it's very easy to fall into the trap of thinking that a familiar-seeming part of agile is exactly the same thing as something you already know. For example, Scrum sprints are *not the same thing as project phases*. There are many differences between phases or milestones in traditional project management and sprints in Scrum.

For example, in a typical project plan, all of the project phases are planned at the beginning of the project; in Scrum, only the next sprint is planned in detail. This difference can feel very strange to a team accustomed to traditional project management.

You'll learn a lot more about how Scrum planning works, and how it may be different from what you're used to. In the meantime, keep an open mind—and try to catch yourself when you have the thought, "This is just like something I already know!"

BULLET POINTS

- Many teams that want to adopt agile start with the **daily standup**, a meeting with the whole team where everyone stands in order to keep it short.

- Agile is a set of **methods and methodologies**, but it's *also* a **mindset**, or an attitude that's shared by everyone on the team.

- The daily standup meeting is much more effective when everyone on the team has the right **mindset**—everyone listens to each other, and they all work together to make sure the project is on track.

- Every agile methodology comes with a set of **values** to help the team get into the most effective mindset.

- When team members follow shared **principles** and share the same set of **values**, it can make the method that they use <u>much more effective.</u>

- **Scrum**, a framework focused on project management and product development, is the most common approach to agile.

- In a Scrum project, the team breaks the work into **sprints**, or cycles of equal length (often 30 days) that follow the Scrum pattern.

- Every sprint starts with a **sprint planning** session to determine what they'll build.

- During the sprint the team works on the project, and every day they hold a short meeting called a **daily scrum.**

- At the end of the sprint, the team holds a **sprint review** with the stakeholders to demo the working software that they built.

- To finish the sprint, the team holds a **retrospective** to look back at how the sprint went and discuss ways that they can improve together as a team.

Watch it! Don't just dismiss the idea that mindset matters!

A lot of people—especially hardcore developers—tune out as soon as they start hearing words like mindset, values, and principles. That's especially true of coders who have a habit of locking themselves in a room and never talking to anyone. If you're starting to think this way, really try to give these ideas the benefit of the doubt. After all, a lot of great software has been built this way, so there has to be something to it... right?

Sharpen your pencil

Which of these scenarios are examples of applying practices, and which are examples of applying principles? Don't worry if you haven't seen some of these practices yet, just use the context to try to figure out the right answer. (That's a good skill to work on for taking a certification exam!)

1. Kate knows that the most effective way to communicate important information about the project to her team is with a **face-to-face conversation**.

    ☐ Principle   ☐ Practice

2. Mike and his team know that the users will probably change their minds later, and those changes can wreak havoc with their code, so they use **incremental design** to make sure the code that they build is easy to change later.

    ☐ Principle   ☐ Practice

3. Ben uses a **persona** to model a typical user because he knows that the more the team understands the users, the better job they'll do of building software.

    ☐ Principle   ☐ Practice

4. Mike always makes sure that his team is working on something that he can demonstrate to Kate and Ben, because he knows that **working software** is the best way to show the team's progress.

    ☐ Principle   ☐ Practice

5. Kate wants to improve the way that the team builds software, so she gets them all to **improve collaboratively and evolve experimentally** by coming up with changes that they can make to their process together, and using data to figure out if those changes made things better.

    ☐ Principle   ☐ Practice

6. Mike and his team **embrace change** by building code that's easy to change down the road.

    ☐ Principle   ☐ Practice

➝ Answers on page 20

more visibility is better **(right?)**

> WOW! WE'VE NEVER WORKED TOGETHER THIS WELL BEFORE. THAT DAILY STANDUP MEETING CHANGED EVERYTHING!

Kate: This project's gone so much better than ones in the past. And all because of one little meeting every day!

Mike: Well, I wouldn't say that.

Kate: Come on, Mike! Don't be such a pessimist.

Mike: No, seriously. Look, you didn't really think you're the first person to try to solve our project problems by adding meetings, did you?

Kate: Well, I... um...

Mike: We got some really good results, so I'll be honest with you here. When you started holding those daily standups, almost everyone on the team was unhappy.

Kate: Really?

Mike: Yeah. Don't you remember how for the first week and a half, most of us just stared at our phones the whole time?

Kate: Well, sure. I guess that wasn't particularly useful. If I'm honest with myself, I was actually thinking about calling the whole thing off.

Mike: But then one of my coders brought up that serious architecture issue. Everyone listened because she's really good, and they all respect her opinion.

Kate: Right. We had to make a major change, and I cut two of the features out of the release to make room for it.

Mike: Yes! That was really important. Normally when we run into a problem like that, we have to work late nights to deal with the aftermath. Like when we found out that the listener feedback analysis algorithm had a serious flaw in it.

Kate: Ugh, that was awful. I usually find out about problems like that after we've all promised things we couldn't deliver. This time we caught the problem early, and I could work with Ben to manage our users' expectations and get you guys the time you needed to come up with a new approach.

Mike: We'll definitely bring up problems like that every time they come up.

Kate: Wait—what? That kind of problem happens a lot?!

Mike: Are you kidding? I've never had a project that didn't run into at least one nasty surprise like that once we started coding. That's how software projects work in the real world, Kate.

Looks like Kate discovered that software projects are a lot less clean and simple in real life than they are on paper. Before, she could just build her plan and then force the team to work it... and when things went wrong, it was their fault, not hers.

On the other hand, this project went a lot better than her last ones. She had to work a lot harder to deal with problems, but she got better results!

Agilecross

what is agile?

Solve this crossword and get these agile ideas to stick in your brain!
How many words you can get without looking back at the chapter?

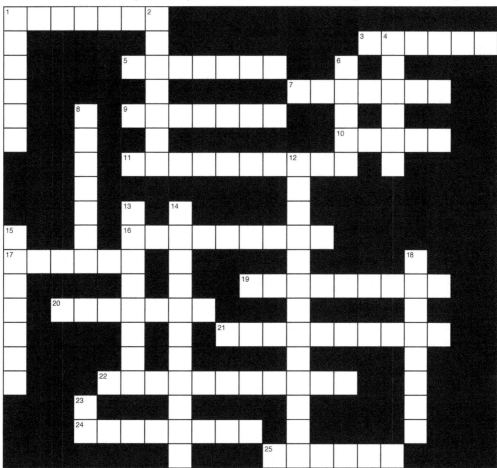

Across
1. A daily _____ can be valuable, but it really works best if everyone on the team has the right mindset
3. In the Kanban method, teams improve collaboratively and _____ experimentally
5. Kanban is an agile method focused on _____ improvement
7. Holds the features that haven't been built yet
9. The Scrum team always demos _____ software
10. Who the Scrum team does the demo for
11. What Scrum teams do every day
16. Helps the team understand their users' needs
17. The Scrum _____ guides the team past roadblocks and helps them implement Scrum
19. Agile planning tool
20. The _____ Owner on a Scrum team maintains a backlog
21. The most effective way to communicate
22. _____ design helps XP teams make code easy to change
24. Scrum teams get together to do this at the start of the project
25. Scrum team's demo at the end of the project

Down
1. How Scrum teams divide up their projects chronologically
2. Helps the team understand who their users are
4. These help teams understand the mindset of a methodology
6. Framework focused on project management and product development
8. Makes a big difference when adopting practices
12. When the team gets together to figure out what lessons they've learned
13. Chart that tracks the amount of work left on a project
14. What XP teams do constantly to improve their code structure
15. What XP teams do with change
18. A tool or technique used by a team
23. Methodology focused on code and software design

you are here ▶ 17

The PMI-ACP certification can help you be more agile

The Agile Certified Practitioner (PMI-ACP)® certification was created by the Project Management Institute to meet the needs of project managers who have increasingly found themselves working with agile methods, methodologies, practices, and techniques. And just like with the PMP certification, PMI has constructed an exam based on real-world tasks, tools, and practices used by agile teams every day.

The PMI-ACP certification is for anyone who works on agile teams, or in an organization that's moving toward adopting agile.

The exam doctor is here to help you get in the best shape possible for taking the exam.

> IF YOU'RE PLANNING ON TAKING THE EXAM, WE'LL HELP YOU PREPARE FOR IT BY INCLUDING PRACTICE EXAM QUESTIONS TO REVIEW THE MATERIAL THAT YOU LEARNED IN EACH CHAPTER.

The exam is focused on real-world agile knowledge.

The PMI-ACP exam is designed to reflect the way teams work in the real world. It covers the most common methods and methodologies, including Scrum, XP, and Lean/Kanban. The exam questions are based on knowledge and real-world tasks that teams use every day.

Part I of this book has just a little bit of material specific to the PMI-ACP® because focusing on learning about agile is the most effective way to prepare for the exam.

That's why this book is, first and foremost, **built to teach you agile**: because understanding agile methods, methodologies, practices, values, and ideas is the most effective way to prepare for the PMI-ACP certification.

In addition to teaching you all about agile, we will also spend some time focusing specifically on exam material. This book has **100% coverage of the PMI-ACP exam content**, and includes many practice questions, test-taking tips, and exam preparation exercises, including a complete, full-length practice exam that mimics the real thing.

> AND EVEN IF YOU **AREN'T USING THIS BOOK** TO PREPARE FOR THE PMI-ACP® CERTIFICATION, THE PRACTICE QUESTIONS WILL GIVE YOU A DIFFERENT WAY OF APPROACHING THE MATERIAL. THAT'S A GREAT WAY TO **SEAL IT INTO YOUR BRAIN!**

Chapters 2 through 7 each end with a set of practice questions. They also include "Question Clinic" sections that break down different types of questions that you'll run across on the exam.

Learning to recognize different kinds of questions that you'll see on the exam is really useful because when you see something familiar it helps your brain relax, which can help the answer come more quickly. We call this one the **"Just the facts, ma'am" question**. It looks like it's just asking for some basic information, but *read all of the answers carefully*! There's often a misleading answer that looks like it might be right, but isn't.

We'll use practice questions to give you tips and exam strategy...

> 39. Which of the following is used by teams to understand the progress of their project?
>
> A. Refactoring ← Some answers will clearly be wrong. Refactoring is about improving code, not understanding project progress.
>
> B. Retrospective ←
> Some answers can be misleading! The retrospective helps the team understand their project better, but it doesn't keep track of progress because it mainly looks back at work that's been done.
>
> C. Burndown chart
> ↑ Here's the right answer! The burndown chart is a tool that shows how the project has been doing, and how much work is left.
>
> D. Continuous integration ←

You haven't seen this term yet—it's a practice that XP teams use. Some exam questions will have answers you don't recognize. And that's OK! Just relax, and concentrate on the other answers. In this case, one of the others is correct. But if none of them are, then you can eliminate them and take an educated guess!

> YOU SHOULD DO THE "QUESTION CLINIC" SECTIONS AND PRACTICE QUESTIONS—EVEN IF YOU'RE NOT PREPARING FOR THE PMI-ACP®! THEY'RE DESIGNED TO GET AGILE CONCEPTS INTO YOUR BRAIN QUICKLY.

exercise solutions

Sharpen your pencil Solution

This is one of the principles behind agile: that a face-to-face conversation is the most effective way to convey information to and within a software team.

1. Kate knows that the most effective way to communicate important information about the project to her team is with a **face-to-face conversation**.

   ☒ Principle　　　　　　　　　　☐ Practice

   Incremental design is an XP practice in which the team grows the codebase incrementally over time.

2. Mike and his team know that the users will probably change their minds later, and those changes can wreak havoc with their code, so they use **incremental design** to make sure the code that they build is easy to change later.

   ☐ Principle　　　　　　　　　　☒ Practice

   A persona is a practice in which the team creates an fictional user with a name (and often a fake photo) to better understand who will be using their software.

3. Ben uses a **persona** to model a typical user because knows that the more the team understands the users, the better job they'll do of building software.

   ☐ Principle　　　　　　　　　　☒ Practice

   An important agile principle is that working software is the primary measure of progress in the project, because it's the most effective way for everyone to gauge exactly what the team has accomplished.

4. Mike always makes sure that his team is working on something that he can demonstrate to Kate and Ben, because he knows that **working software** is the best way to show the team's progress.

   ☒ Principle　　　　　　　　　　☐ Practice

5. Kate wants to improve the way that the team builds software, so she gets them all to **improve collaboratively and evolve experimentally** by coming up with changes that they can make to their process together, and using data to figure out if those changes made things better.

   ☐ Principle　　　　　　　　　　☒ Practice

   This is one of the core practices of Kanban. The team uses the scientific method to see if their improvements actually work in practice.

6. Mike and his team **embrace change** by building code that's easy to change down the road.

   ☒ Principle　　　　　　　　　　☐ Practice

   An important value shared by all effective XP teams is that they embrace change, rather than try to prevent or resist it.

what is agile?

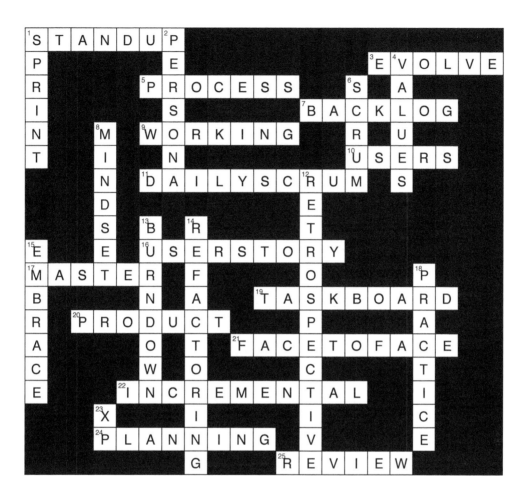

this page *intentionally left blank*

2 agile values and principles

Mindset meets method

*I KNOW THIS SPECIFICATION HAS PROBLEMS. ON THE OTHER HAND, I CAN'T REMEMBER THE LAST TIME ANYONE ON THE TEAM **ACTUALLY READ A SPEC** BEFORE WRITING CODE. SO... I GUESS IT'S A WASH?*

There's no "perfect" recipe for building great software.

Some teams have had a lot of success and seen big improvements after adopting agile practices, methods, and methodologies, while others have struggled. We've learned that the difference is the mindset that the people on the team have. So what do you do if you want to get those great agile results for your own team? How do you make sure your team has the right mindset? That's where the **Agile Manifesto** comes in. When you and your team get your head around its **values and principles**, you start to think differently about the agile practices and how they work, and they start to become *a lot more effective*.

what agile teams *value*

Something big happened in Snowbird

In the 1990s there was a growing movement throughout the software development world. Teams were growing tired of the traditional way of building software using a **waterfall process**, where the team first defines strict requirements, then draws up a complete design, and builds out all of the software architecture on paper before the code is written.

← *People on waterfall teams weren't always 100% clear on exactly why they didn't like their process, but there was a lot of agreement that it was somehow too "heavyweight" and cumbersome.*

By the end of the decade, there was a growing consensus that teams needed a more "lightweight" way to build software, and several methodologies—especially Scrum and XP—were gaining popularity as the way to accomplish that.

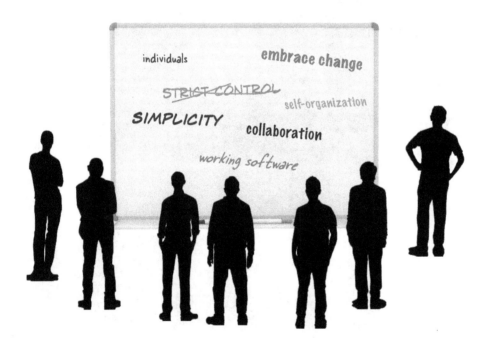

Leaders from across the industry got together to figure out if there was anything in common among the various and increasingly popular <u>lightweight</u> methods for building software.

Meeting of the minds

In 2001, a group of seventeen open-minded people got together at the Snowbird ski resort in the mountains outside of Salt Lake City, Utah. The group included thought leaders from all across the new "lightweight" world, including the creators of Scrum and XP. They weren't sure exactly what would come out of the meeting, but there was a strong sense that these new lightweight methods for building software had something in common. They wanted to figure out if they were right, and maybe find a way to write it down.

agile values and principles

The Agile Manifesto

It didn't take long for the group to converge on four values that they all had in common. They wrote those values down in what would come to be known as the **Agile Manifesto**.

> **Manifesto for Agile Software Development**
>
> We are uncovering better ways of developing software by doing it and helping others do it. Through this work we have come to value:
>
> **Individuals and interactions** over processes and tools
> **Working software** over comprehensive documentation
> **Customer collaboration** over contract negotiation
> **Responding to change** over following a plan
>
> That is, while there is value in the items on the right, we value the items on the left more.

> IF THOSE GUYS WERE SO SMART, WHY DIDN'T THEY JUST COME UP WITH THE **BEST WAY TO BUILD SOFTWARE**? WHY BOTHER WITH THESE "VALUES" AT ALL?

The idea that there's no "silver bullet" method for building software was first introduced in the 1980s by pioneering software engineer Fred Brooks, in an essay called "No Silver Bullet."

They weren't trying to come up with a "unified" methodology.

One of the fundamental ideas in modern software engineering is that **there's no single "best" way to build software**. That's an important idea that's been around in software engineering for decades. The Agile Manifesto is effective because it *lays out values that help teams get into an agile mindset*. When everyone on the team genuinely incorporates these values into the way they think, it actually helps them build better software.

you are here ▸ **25**

tools are good people are better

Adding practices in the real world can be a challenge

Teams are always looking for ways to improve. We've seen how practices can help. That's especially true of the lightweight practices used by agile teams, which are designed to be simple, straightforward, and easy to adopt. But we've also seen that the team's mindset or attitude can make it much more difficult to adopt them successfully—like when Kate found that the attitude that she, Mike, and the rest of the team had made a big difference when she tried to start holding a daily standup meeting.

MEANWHILE, BACK AT THE SILICON VALLEY STARTUP THAT'S BUILDING SOFTWARE USED BY VIDEO AND MUSIC STREAMING SERVICES TO ANALYZE AUDIENCES...

THE DAILY STANDUP IS A BEST PRACTICE THAT A LOT OF TEAMS USE. **EVERY BOOK I'VE READ ON AGILE** SAYS IT'S A GOOD IDEA.

BUT THIS IS THE **REAL WORLD**, AND WE STILL NEED THE TEAM TO "GET" IT. IF THEY DON'T, WE'LL JUST GO THROUGH THE MOTIONS AND IT WON'T MAKE MUCH DIFFERENCE.

Mike and Kate discovered that it's not enough to have a "best" or "correct" practice. If the people don't buy into it, it'll cause conflict and eventually be thrown out.

The four values of the Agile Manifesto <u>guide the team</u> to a <u>better, more effective mindset</u>

The Agile Manifesto contains four X over Y lines that help us understand what agile teams value. Each of those lines tells us something specific about the values that drive an agile mindset. We can use them to help us understand what it means for a team to be agile.

Let's take a closer look each of those four values ⟶

agile values *and principles*

Individuals and interactions over processes and tools

Agile teams recognize that processes and tools are important. You've already learned about a few practices that agile teams use: daily standups, user stories, task boards, burndown charts, refactoring, and retrospectives. These are all valuable tools that can make a real difference to an agile team.

But agile teams value individuals and interactions even more than processes and tools, because teams always work best when you pay attention to the human element.

You've already seen an example of this—when Kate tried to introduce daily standups, and ended up getting into a conflict with Mike and his development team. That's because a tool that works really well for one team can cause serious problems for another team if the people on the team aren't getting anything out of it, and if it's not directly helping them build software.

NEXT TIME I TRY TO INTRODUCE A NEW TOOL OR PROCESS, I'LL TALK TO THE TEAM AND TRY TO UNDERSTAND IT *FROM THEIR PERSPECTIVE.*

Good idea, Kate!

Processes and tools are important to getting the project done, and they can be really valuable. But the **individual people on the team** are even more important, and any tool that you introduce needs to improve their **interactions** with each other and with their users and stakeholders.

you are here ▶ **27**

it's a feature not a bug

> **Working software** over comprehensive documentation

What does "working" software mean? How do you know if your software works? That's actually a harder question to answer than you might think. A traditional waterfall team starts a project by building comprehensive requirements documents to determine what the team will build, reviews that documentation with the users and stakeholders, and then passes it on to the developers to build.

Most professional software developers have had that terrible meeting where the team proudly demonstrates the software they've been working on, only to have a user complain that it's missing an important feature, or that it doesn't work correctly. It often ends in an argument, like the one that Ben had with Mike after he gave a demo of a feature his team had been working on for a few months:

A lot of people try to fix this problem with comprehensive documentation, but that can actually make the situation worse. The problem with documentation is that two people can read the same page and come away with two very different interpretations.

That's why agile teams value **working software** over comprehensive documentation—because it turns out that the most effective way for a user to gauge how well the software works is to actually use it.

agile *values* **and principles**

BRAIN BARBELL

This little puzzle should be familiar to anyone who read *Head First PMP!*

Lisa is testing the firmware component for the Black Box 3000™. The product is only "working" in one of these scenarios. Can you help Lisa figure out which version of the product has "working" firmware?

Here's a hint: there's an important piece of information that we <u>haven't</u> given you. Without it, this puzzle is <u>really hard</u> to solve.

The Black Box 3000™

Some might even say <u>impossible!</u>

Scenario 1
Lisa presses the button, but nothing happens.

SO... HOW DO I KNOW IF THE FIRMWARE IN THE BOX IS WORKING?

Scenario 2
Lisa presses the button and a voice comes out of the box that says, "You pressed the button incorrectly."

Lisa, our tester, is testing the Black Box 3000™ firmware, but she isn't sure what she's supposed to be testing for.

Scenario 3
Lisa presses the button and the box heats up to 628°F. Lisa drops the box and it shatters into hundreds of pieces.

```
Just in case you don't happen to know the term,
firmware is software that's programmed into
the read-only memory of a piece of hardware.
```

BRAIN BARBELL

EXPLANATION

Here's that crucial missing piece of information that we didn't give you on the previous page: the Black Box 3000™ is the heating element from an industrial oven. So Scenario 3 is the one that demonstrates "working" software.

The best—and sometimes only!—way to tell if software is working is to put it in the hands of the people who need to use it. If they can use the software to do what they need it to do, it's working. But it's not always easy to know exactly what "working" means, which is why agile teams *also* value comprehensive documentation—they just value working software more.

In this case, literally in Lisa's hands. Let's hope that she was wearing heat-proof gloves!

Here's an example of comprehensive documentation that the team actually found useful: the specification for the Black Box 3000™.

BLACK BOX 3000™
Specification Manual

The BB3K™ is a heating element for an industrial oven.

BB3K™ must heat up to exactly 628°F in 0.8 seconds.

BB3K™ must have a large, easy-to-press button.

Sometimes documentation <u>can</u> be useful—like when it's not clear exactly what "working" software is supposed to do.

> LOOKS LIKE SCENARIO #3 IS THE ONE THAT SHOWS WORKING SOFTWARE. GOOD THING I HAD SOME DOCUMENTATION TO TELL ME THAT...BUT IT'S EVEN MORE IMPORTANT TO ME THAT I HAVE THE **ACTUAL PRODUCT** IN MY HANDS.

Now that she knows what "working" means for this software, Lisa can actually test it.

agile *values* and principles

Customer collaboration over contract negotiation

No, this *isn't* about consultants or procurement teams who have to deal with contracts!

When people on agile teams talk about contract negotiation, they often mean an attitude that people take toward their users, customers, or people on other teams. When people on a team have a "contract negotiation" mindset, they feel like they have to come to a strict agreement on what the team will build or do before any work can start. Many companies encourage this mindset, asking teams to provide explicit "agreements" (often documented in specifications, and enforced with strict change control procedures) about what it is they will deliver and when.

Agile teams value customer collaboration over contract negotiation. They recognize that projects change, and that people never have perfect information when starting a project. So instead of trying to nail down exactly what's going to be built before they start, they **collaborate with their users** to try to get the best results.

> Contract negotiation is necessary in cases where the customers are unwilling to collaborate. It's very difficult to genuinely collaborate with someone who's being unreasonable—like a customer who routinely changes the scope of the project, but refuses to give the team enough time to make those changes.

> WHEN WE **COLLABORATE** WITH OUR USERS, IT ALWAYS WORKS OUT BETTER THAN WHEN WE TRY TO NEGOTIATE WITH THEM.

Scrum teams are especially good at this because they have a **product owner** like Ben who is a true member of the team. He might not have been developing code, but he worked hard on the project by talking to users, understanding what they needed, and working with the rest of the team to help them understand those needs and build working software.

your plans will change and it's OK

Responding to change over following a plan

Some project managers have a saying: "Plan the work, work the plan." And agile teams recognize that planning is important. But working a plan that has problems will cause the team to build a product with problems.

Traditional waterfall projects have ways of handling changes, but they usually involve strict and time-consuming change control procedures. This reflects a mindset in which changes are the exception, not the rule.

The problem with plans is that they're built at the start of projects, and that's when the team knows the least about the product they're going to build. So agile teams **expect that their plans will change**.

That's why they typically use methodologies that have tools to help them constantly look for changes and respond to them. You've already seen a tool like that: the daily standup.

> In agile projects, your product is developed step by step, each new step drawing knowledge from the previous step. When a plan (or requirement, or anything else you use in a project) is developed this way, it's called *progressive elaboration*.

> BEFORE WE STARTED HAVING THE DAILY STANDUP, I DIDN'T FIND OUT ABOUT PROBLEMS WITH THE PROJECT PLAN UNTIL IT WAS TOO LATE TO DEAL WITH THEM.

Once Kate and Mike sorted out their differences, they both realized that the daily standup was a way for everyone to look at the plan every day, and work together to respond to any changes that were needed. When everyone on the team worked together to respond to change, they were able to update the plan that they all built together without introducing chaos into the project.

It's important to plan your project, but it's even more important to recognize that those plans will change once the team starts working on the code.

32   Chapter 2

agile values and principles

Watch it!

Responding to change is important to agile teams, but they <u>still value</u> following a plan.

Take another look at the last line of the Agile Manifesto:

> That is, while there is value in the items on the right, we value the items on the left more.

Each of the four values in the Agile Manifesto contains two parts: something (on the righthand side) that agile teams value, and then something else (on the lefthand side) that agile teams value more.

*So when agile teams say they value responding to change over following a plan, that **doesn't** mean that they don't value planning—in fact, it means the opposite! They absolutely value following a plan. They just value responding to change <u>more</u>.*

In fact, Scrum teams actually do <u>more</u> planning than traditional teams following a waterfall process! But since they're really good at responding to change, it doesn't feel like it to the people on the team.

BULLET POINTS

- The **Agile Manifesto** was created in 2001 by people who came together to find common ground between different "lightweight" methods, methodologies, and approaches to building software.

- The Agile Manifesto has **four values** that help agile teams get into the right mindset.

- Agile teams **value processes and tools** because they help the team get organized and work effectively.

- But they **value people and interactions** *more* because teams work best when you pay attention to the human element.

- Agile teams **value comprehensive documentation** because it's an effective way to communicate complex requirements and ideas.

- But they **value working software** *more* because it's the most effective way to communicate progress and get feedback from users.

- Agile teams **value contract negotiation** because sometimes it's the only way to work effectively in an office culture where mistakes are punished.

- But they **value customer collaboration more** because it is much more effective for building software than having a legalistic or antagonistic customer relationship.

- Agile teams **value following a plan** because without planning, complex software projects go off the rails.

- But they **value responding to change more** because the team that works the wrong plan ends up building the wrong software.

you are here ▶ **33**

get the values into your brain

Manifesto Magnets

Oops! You had the Agile Manifesto re-created perfectly with refrigerator magnets! But someone slammed the door and they all fell off. Can you put the whole thing back together? See how much of it you can do without flipping back and looking for hints.

Some of the magnets stayed on the fridge. Leave these magnets in place.

```
Manifesto for Agile Software Development

We are uncovering better ways of developing
   software by doing it and helping others do it.
                                                    :
Through this work we have come to

                     over

                     over

                     over

                     over
```

Don't worry about the order of the values.

The four values in the Agile Manifesto are equally important, so there's no particular order for them. Just make sure that each specific value (X over Y) is matched up.

All of the other magnets fell off the fridge. Can you put them back in the right place?

```
That is, while there is   value   in   the items on the        ,   we
      value   the items on the      more      .
```

```
following   individuals   change   left   customer         value      tools    plan
contract              working          and       and    to       comprehensive   negotiation
       processes                                                                  software
  right       collaboration        a         documentation
                    interactions       responding
```

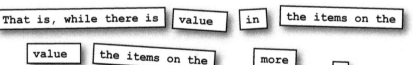

Answers on page 66

agile values and principles

there are no
Dumb Questions

Q: I'm still not clear on what "waterfall" means.

A: "Waterfall" is the name given to a specific way that software companies have traditionally built software. They divide their projects into phases, usually drawn in a diagram that looks like this:

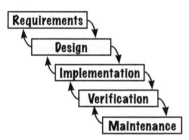

It was given the name "waterfall" by a software engineering researcher in the 1970s who first described it as a less-than-effective way to build software. The team is often expected to come up with a near-perfect requirements document and design before they start building code, because it takes a lot of time and effort to go back and fix the requirements and design when the team finds problems with them.

The problem is that there's often no way to know if the requirements and design are right until the team starts building code. It's very common for a waterfall team to think they got everything right in the documentation, only to discover serious flaws when the developers start implementing the design.

Q: Then why would anyone ever use a waterfall process?

A: Because it works—or, at least, it *can* work. There have been plenty of teams that used a waterfall process and built great software. It's certainly possible to create the requirements and design first so that there are relatively few changes.

More importantly, there are a lot of companies where the culture really lends itself to a waterfall process. For example, if you work for a boss who will severely punish you if he thinks you made a mistake, then having him personally approve fully fleshed-out requirements and design documents before any code is written can help you keep your job. But no matter how you do it, the effort it takes to figure out who's accountable for each decision in the project takes away from effort that could be spent actually building your product.

Q: So is waterfall good or bad? And how is it less "lightweight" than an agile methodology, approach, or framework?

A: Waterfall isn't "good" or "bad," it's just a certain way of doing things. Like any tool, it has its strengths and weaknesses.

However, many teams find a lot more success with agile methodologies like Scrum than they do with a waterfall process. One reason is that they find a waterfall process too "heavyweight" because it imposes a lot of restrictions on how they work: they're required to go through complete requirements and design phases before any code is written. In the next chapter, you'll learn how Scrum teams use their sprints and planning practices to start building the software quickly, which lets them get working software into the hands of their users. This feels a lot more "lightweight" to the team because everything they're doing has an immediate effect on the code that gets built.

Q: Then exactly how should I run my projects? Should my team create documentation or not? Do we work from a complete specification? Should we throw out documentation altogether?

A: Documentation is important to agile teams—but mainly because it can be an effective way to build working software. Documentation is only useful if people read it, and the truth of the matter is that a lot of people just don't read documentation.

When I write a specification requirements document and give it to you and your team to build, the document isn't important. The important thing is that **what's in my head matches what's in your head**, and what's in each of the team members' heads. In some cases, like when there are complex calculations or workflows, *a document can be a really effective* way to the **shared understanding** that leads to great software.

Q: I'm just not sold on this idea that values are important. How do they help me and my team actually write code?

A: The values in the Agile Manifesto help you and your team get into a mindset that helps you build better software. And you've already learned that your mindset—the attitude that you have toward the practices that you use—can make a big difference.

Think about the example in the last chapter, where Kate and Mike were having trouble with the daily standup meeting. Kate only got mediocre results when she used it as a way to dictate her plan to the team and demand status from them. But when everyone had a more collaborative attitude, they got much better results. That's how mindset can have a big effect on real-world results.

you are here ▶

Question Clinic: The "which-is-BEST" question

A great way to prepare for the exam is to learn about the different kinds of questions, and then try writing your own. Each of these Question Clinics will look at a different type of question, and give you practice writing one yourself. Even if you're not using this book to prepare for the PMI-ACP certification exam, give this a shot. It can still be a great way to help get these concepts into your brain!

Take a little time out of the chapter for this Question Clinic. It's here to let your brain have a break and <u>think about something different</u>.

> A LOT OF EXAM QUESTIONS ASK YOU TO CHOOSE WHICH OF THE ANSWERS IS **BEST**. THAT USUALLY MEANS ONE ANSWER IS *PRETTY GOOD*, BUT THERE'S ANOTHER ANSWER THAT'S **BETTER**.

It's not a <u>terrible</u> idea to get senior management involved, but that really <u>ought</u> to be left for resolving serious conflicts. It's better for the team to work with the user and figure things out together, without appealing to authority.

---

82. A user asks the team for a new feature that's very important, but doing the work will make them miss a committed deadline for a different feature that they plan to work on. Which is the BEST thing the team should do first?

   A. Call a meeting with senior management to get an official decision on relative priority

   B. Follow the existing plan so they don't miss the deadline, and prioritize the new feature so they work on it next

   C. Talk to to the user and figure out if the new feature is important enough for them to change direction

   D. Initiate the change control process

Agile teams value following a plan, so this is a good idea... but they value responding to change more. Is there another answer that's more in line with agile values?

This is the BEST answer! Agile teams value responding to change over following a plan. So the best thing to do is respond to the user quickly and get all of the information. Then they can all work together to figure out how the plan is impacted.

If you were studying for the PMP exam, this could be the right answer. And since many agile teams work in companies that have a change control process, they may eventually do this. But the question asked what the team should do <u>first</u>, and this would come later (and maybe not at all!).

> THE "WHICH-IS-BEST" QUESTION HAS MORE THAN ONE GOOD ANSWER, BUT ONLY ONE **BEST** ANSWER.

The BEST answer

HEAD 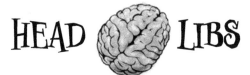 LIBS

Fill in the blanks to come up with your own "which-is-BEST" question about how agile teams value customer collaboration over contract negotiation.

You're a developer on a _____ project. A _____
                                        (an industry)   (a type of user)

wants you to _____, but you need to _____.
              (something the user wants you to do)   (a conflicting thing you want to do)

What is the BEST way to handle this situation?

    A. _____
           (an obviously wrong answer that has nothing to do with this question at all)

    B. _____
           (a good answer that is a good idea, but isn't really relevant to this value)

    C. _____
           (a better answer that's consistent with valuing contract negotiation)

    D. _____
           (the BEST answer that's consistent with valuing customer collaboration)

Ladies and Gentlemen, We Now Return You To Chapter Two

wish we'd *talked about this* **sooner**

They think they've got a hit ...

Mike's been working with the development team for almost a year on the latest killer feature, and he's really excited that they're finally done.

agile values **and principles**

... but it's a flop!

Uh-oh. It looks like Mike and his team wasted a year working on a product that nobody wants. What happened?

THIS WOULD HAVE BEEN GREAT IF WE'D DELIVERED IT A **YEAR AGO**. BUT THERE'S NO WAY WE CAN USE THIS TODAY!

Mike: What?! We've been working on this for a year. Are you telling me we just wasted our time?

Ben: I have no idea what you've been doing with your time. But I'm telling you right now, I don't see *any* of our clients using this.

Mike: But what about that big presentation we gave at that conference last year? Every client we talked to said they would love to figure out exactly who their listeners are and start marketing directly to them.

Ben: Right. And nine months ago, three of those clients were named in a lawsuit for violating privacy laws. Now none of them would touch this.

Mike: But... but this is a huge innovation! You have no idea how many technical problems we had to solve. We even brought in a specialized AI consulting company to help us do advanced customer analysis!

Ben: Look, Mike, I don't know what to tell you. Maybe you can repurpose the code for something else?

Mike: We're going to have to salvage what we can. But I'll tell you right now, **whenever we tear out chunks of code, <u>we always have bugs</u>**.

Ben: Ugh. I wish you'd talked to me about this sooner.

> This makes sense! A major source of bugs is <u>rework</u>, or taking code that was already built and modifying it for another purpose.

If agile teams value responding to change but changes often cause rework, then how do they keep that rework from always causing bugs?

a principled approach

The principles behind the Agile Manifesto

The four values in the Agile Manifesto do a really good job of capturing the core of the agile mindset. But while those four values are great at giving you a high-level understanding of what it means to "think agile," there are a lot of day-to-day decisions that every software team needs to make. So in addition to the four values, there are **twelve principles behind the Agile Manifesto** that are there to help you *really* understand the agile mindset.

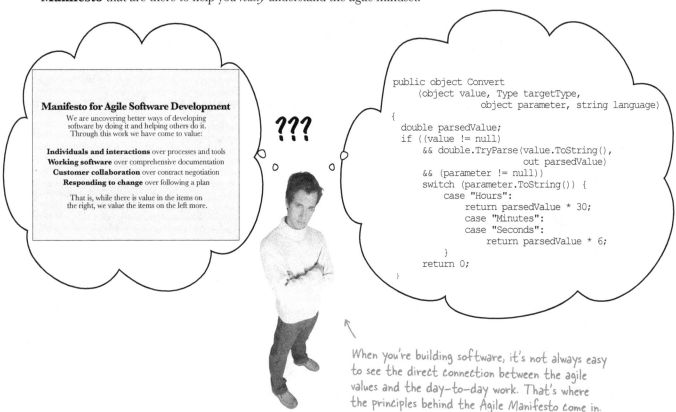

When you're building software, it's not always easy to see the direct connection between the agile values and the day-to-day work. That's where the principles behind the Agile Manifesto come in.

Behind the Scenes

The group at Snowbird came up with the four values pretty quickly, but it took them a few days of deep discussion to agree on the twelve principles behind the Agile Manifesto—and even after they left Utah, they still hadn't finalized the wording. The version they first came up with is a little different. The final version is on the next page (and also at http://www.agilemanifesto.org, the official website of the Agile Manifesto). But even though the wording changed slightly over the first few years, the <u>ideas</u> laid out in the twelve principles haven't.

agile *values* and principles

Principles Behind the Agile Manifesto

We follow these principles:

- Our highest priority is to satisfy the customer through early and continuous delivery of valuable software.

- Welcome changing requirements, even late in development. Agile processes harness change for the customer's competitive advantage.

- Deliver working software frequently, from a couple of weeks to a couple of months, with a preference to the shorter timescale.

- Business people and developers must work together daily throughout the project.

- Build projects around motivated individuals. Give them the environment and support they need, and trust them to get the job done.

- The most efficient and effective method of conveying information to and within a development team is face-to-face conversation.

- Working software is the primary measure of progress.

- Agile processes promote sustainable development. The sponsors, developers, and users should be able to maintain a constant pace indefinitely.

- Continuous attention to technical excellence and good design enhances agility.

- Simplicity—the art of maximizing the amount of work not done—is essential.

- The best architectures, requirements, and designs emerge from self-organizing teams.

- At regular intervals, the team reflects on how to become more effective, then tunes and adjusts its behavior accordingly.

what makes software valuable

The agile principles help you deliver your product

The first three principles are all about delivering software to your users. And the most effective way to deliver the best software possible is to make sure that it's **valuable**. But what does "value" really mean? How do we make sure that we've got our users, stakeholders, and customers' best interests in mind when we're building software? These principles help us understand those things.

> THE SOFTWARE DOES *SOMETHING*, BUT NOT WHAT WE NEED IT TO DO.

A product owner like Ben works with users and customers to really understand what they need in the software. He can usually spot a feature that the users won't actually use... which happens a lot more often than you might think!

- Our highest priority is to satisfy the customer through <u>ear</u>ly and <u>continuous delivery</u> of valuable software.

So what does that mean, exactly? It means that
early delivery **and** continuous delivery
add up to satisfied users:

Early delivery

Getting the first version of the software into users' hands as early as possible so you can get early feedback

+

Continuous delivery

Constantly getting updated versions to the users so they can help the team build software that solves their most important problems

=

Satisfied users

The users help the team stay on track by making sure that the most important features are added first

agile *values* and principles

> BUT WE ALREADY BUILT THE CODE! THIS IS GOING TO BE A **REAL PAIN** TO CHANGE.

This is how a lot of teams act when someone points out a major change to the code. It's understandable, because a change that could have been found earlier now requires a lot of work that can be slow and highly annoying.

When the team delivers software early and often to the users, stakeholders, and customers, that gives everyone lots of chances to <u>find changes</u> early, when they're much easier to make.

- <u>Welcome changing requirements</u>, even late in development. Agile processes harness change for the customer's competitive advantage.

How does the team react when someone points out that they need to make a change that will affect a lot of the code? Every developer has been through this, and it can be a lot of (often difficult) work. So how does the team react? It's natural to resist a big change. But if the team can find a way to not just accept but **welcome** that change, it means that they're *putting the users' long-term needs ahead of their own short-term annoyance*.

"Requirements" just means what the software is supposed to do... but sometimes the users' needs change, or the programmers misunderstand what's needed, and that can lead to changing requirements.

you are here ▸ **43**

deliver frequently and only make small changes

> - Deliver working software frequently, from a couple of weeks to a couple of months, with a preference to the shorter timescale.
> - Working software is the primary measure of progress.

When developers push back against changes, it's not an irrational response: if they've spent many months working on a feature, changing that code can be a slow, painful, and error-prone process. One reason is that when teams do **rework** (that's when they change existing code to do something new) it almost always leads to bugs, often ones that are really nasty and difficult to track down and fix.

So how does the team avoid rework? **Deliver working software to the users frequently**. If the team is building a feature that isn't useful or does the wrong thing, the users will spot it early, and the team can make the change before too much code is written... and preventing rework prevents bugs.

THAT CHANGE WAS A LOT OF WORK, BUT NOW THE SOFTWARE IS A LOT MORE **VALUABLE**.

People *actually* say stuff like this when they welcome changing requirements. What does it mean for software to be "valuable"?

AND NOW THAT WE'RE GETTING WORKING SOFTWARE OUT TO THE USERS EVERY FEW WEEKS, WE CAN AVOID **NASTY SURPRISES** IN THE FUTURE.

agile *values* and principles

*I'M SURE THAT ALL SOUNDS GOOD **ON PAPER**, BUT I DON'T SEE HOW THESE PRINCIPLES CAN MAKE A DIFFERENCE IN THE REAL WORLD.*

Principles make the most sense in practice.

Most of us have been on teams that have struggled at one point or another, and the most common way to handle that situation is to adopt a new practice. But some practices that work really well for some teams only get marginal results with other teams—just like we saw with the daily standup in Chapter 1.

So what makes the difference between the team that only gets so-so results with a practice and the one that gets really great results? More often than not, it has a lot to do with *the mindset of the team*, and the attitude they bring to the practice. And that's what these principles are about: helping teams find the best mindset that makes their practices as effective as possible.

You'll see an example on the next page...

iteration and backlog

Principles in Practice

The first three principles behind the Agile Manifesto talk about early and continuous delivery of software, welcoming changing requirements, and delivering working software on a short timescale. So how do teams do that in the real world? With great practices, like **iteration** and using a **backlog**.

Iteration: repeatedly performing all of the project activities to continuously deliver working software

The team gets together at the start of the iteration to plan out which features they'll build. They try to include only work that they can actually get done during the iteration.

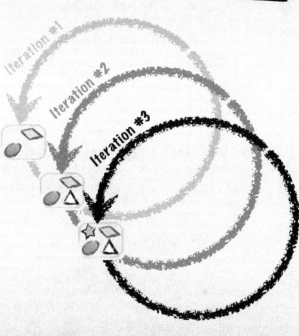

Iterations are timeboxed, so if you and your team discover partway through an iteration that there's too much work planned for it, you'll push features to the next one.

DICTIONARY DEFINITION

timeboxed, adjective

setting a hard deadline for an activity to be completed, and adjusting the scope of that activity to meet the deadline

*The team couldn't fit all of the requested features into the current **timeboxed** iteration, so they concentrated on the most valuable ones.*

---

46 Chapter 2

If you read our book *Learning Agile* then you'll recognize these iteration and backlog illustrations!

agile values and principles

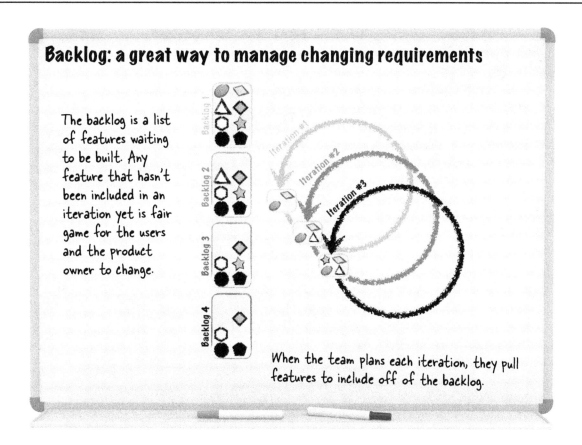

*HEY, WAIT A MINUTE! WE LEARNED ABOUT HOW SCRUM TEAMS USE A BACKLOG EARLIER. DOES THAT MEAN **SCRUM SPRINTS** ARE A FORM OF **ITERATION**?*

Yes! Scrum uses an iterative approach.

The Scrum practice of using sprints is a classic example of how teams use iteration in real life to deliver working software early and frequently. The Scrum team has a Product Owner who works with the users and stakeholders to understand their needs. Everyone learns more with each new version of the working software, and the Product Owner uses that new knowledge to add or remove features from the backlog.

We'll talk a lot more about Scrum in the next chapter.

you are here ▶   47

bet we can do better than "better than nothing"

Fireside Chats

Tonight's talk: **Practice meets principle**

Principle:

I've been looking forward to this debate for a while.

Practice:

I'm not sure there's going to be much of a debate, if you want to know the truth.

This again? There he goes, back on his "nothing gets done without practices" kick.

Well, you have to admit, it's a pretty good point. After all, where would an approach like Scrum be without me? Take away the sprints, backlogs, retrospectives, sprint reviews, Daily Scrum meetings, and sprint planning sessions, and what are you left with? Chaos!

Yeah? OK. Let's talk about those practices for a minute. Like the Daily Scrum, for example—

Hold it right there. I already know what you're going to say next. It's that whole thing about the Daily Scrum getting "mediocre" results if the team doesn't "get" the principles.

Yep! And it's not just the Daily Scrum. Let's talk about iteration.

A fantastic practice, thank you very much.

Indeed. But what happens if the people on the team don't really, genuinely believe in the principle of delivering working software frequently?

There will still be iterations! And you know what? It'll be better than it was before they added the practice.

This is true! Even when the team doesn't really understand the principles, adding iterations is still usually an improvement... not much of one, but enough to justify doing it.

Yes. But will they *really* deliver **working** software? Or will they cut corners just to push something out the door before the iteration ends? Will they *really* delay a feature until the next iteration because it won't fit? Or will adding iteration to the project make everyone on the team feel like they're just "going through the motions"?

Have you been on a team that tried to go agile but ended up with mediocre results? If so, this might remind you of your own experience.

Well, at least they'll have *something* to show for their effort. Even if they only get a marginal improvement, it's better than nothing!

Dictionary drill

Here are a bunch of definitions for words that you've seen in this chapter. Can you fill in the words that each definition belongs to?

_____, noun

work done by the team to change previously written code to make it function differently or serve a different purpose, often considered risky by teams due to its increased likelihood of introducing bugs

_____, adjective

setting a hard deadline for an activity to be completed, and adjusting the scope of that activity to meet the deadline

_____, noun

a practice used by teams in which the team works with users, customers, and/or stakeholders to maintain a list of features that will be built in the future, often prioritized with the most valuable features at the top

_____, noun (two words)

developing a project artifact (like a plan) in steps, using knowledge gained from the previous step to improve it

_____, adjective

a kind of methodology in which teams break the project down into smaller parts, delivering working software at the end of each one, and possibly changing direction based on the feedback from the working software

_____, adjective

a type of model, process, or method for building software in which the entire project is broken down into sequential phases, often including a change control process in which changes force the project into a prior phase

→ Answers on page 67

reflect regularly

there are no Dumb Questions

Q: Does each principle match up with exactly one practice? Is there a one-to-one correspondence?

A: Not at all. The first three principles behind the Agile Manifesto emphasize early and continuous delivery of software, welcoming changing requirements, and delivering working software frequently. And we used two practices (iteration and backlog) to help you understand the principles on a deeper level. But that doesn't mean there's a one-to-one relationship between the practices and the principles.

In fact, the opposite is true. You can have principles without practices, and you can have practices without principles.

Q: I'm not sure how that works. What exactly do you mean by "practices without principles"?

A: Here's an example of what it looks like when a team puts a practice in place without really understanding or internalizing the agile principles. Scrum teams hold a **retrospective** at the end of each sprint so that they can talk about what went well and what can be improved.

But take another look at the last principle in the list:

> At regular intervals, the team reflects on how to become more effective, then tunes and adjusts its behavior accordingly.

What if the team hasn't really taken this idea to heart? They'll still have the retrospective meeting because the Scrum rules tell them to do it. And they'll probably talk about some of the problems that they had, which can certainly lead to some marginal improvements.

The problem is that while the new meeting did something, it feels like it's somewhat "empty" or superfluous. The people on the team feel like they're taking time away from their "real" jobs to do it. Eventually, they'll start talking about replacing it with something more "efficient," like using an email discussion list or wiki page. A lot of teams have that experience when they add practices without principles.

Q: OK, I think I see how you can have practices without really believing in the principles. But how can you have principles without practices?

A: A lot of people have a little trouble with this idea when they first come in contact with the idea of an "agile mindset" driven by principles.

So what does it look like if the team takes very seriously the agile principle of reflecting on how to become more effective, but doesn't have a specific practice for doing that reflection? That's actually pretty common on *very effective* agile teams. Everyone is in the mindset of reflecting often, so when someone feels like it's time to review how the project has progressed and make necessary corrections, that person usually grabs a few other team members and has an informal retrospective. If something good comes out of it, they talk it over, and make the necessary correction. For a team accustomed to a framework with well-specified rules, such as Scrum, that feels disorganized, chaotic, or "loosey-goosey." That's one reason teams like to standardize on a set of practices—so that everyone has common ground rules.

Q: Why did you capitalize "Product Owner" on the bottom of page 47?

A: Because while many people have the job title "product owner," spelling Product Owner with capital letters refers to a specific role with responsibilities specified by the Scrum rules. You'll learn more about that in the next chapter.

> When the team adopts practices without the right principle-driven mindset, it often feels "empty" or superfluous, like they're just going through the motions, and they'll start looking for alternatives that take less effort.

agile *values* and principles

> **WOW, KATE! I'M FEELING SO MUCH BETTER ABOUT THE PROJECT NOW. EVERY TIME I GET A NEW BUILD, I CAN SEE *EXACTLY HOW MUCH PROGRESS* THE TEAM'S MADE.**

> **SOMETHING'S STILL REALLY BOTHERING ME ABOUT HOW THINGS ARE GOING, THOUGH.**

Ben: And just as I was feeling good about how things were going. I wish you didn't have to be so negative. What's the bad news now?

Kate: I'm not trying to be negative. I'm really happy about the progress we've made just by starting to use iteration.

Ben: Right! I went back to the users with early builds, and they found all sorts of changes that we could fix early without a lot of rework.

Kate: Yeah, and that's great. But we still have problems.

Ben: Such as...?

Kate: Well, like that meeting we had last Wednesday. We spent all afternoon arguing about documentation.

Ben: Why do you want to bring that up again? You and Mike keep asking for specifications with every tiny detail about what to build.

Kate: Yes, because then I can help the team plan out the project, and Mike and the team know exactly what to build.

Ben: But it's not that simple! These specs are really hard to write. And even when we write a spec for one iteration, it still gets really long.

Kate: Look, if you've got a better idea about how to get the team to build the right software, I'd love to hear it.

A lot of teams seem to have trouble writing and reading highly detailed specifications. Can you think of a more effective way for a product owner to help the team to understand exactly what the users need them to build?

you are here ▸ **51**

maybe hang up some motivational posters?

The agile principles help your team communicate and work together

Modern software is built by teams, and while individual people are really important to any team, teams work best when everyone works together—which means developers not just working with each other, but with the users, customers, and stakeholders, too. That's what these next principles are all about.

> - Business people and developers must work together daily throughout the project.
> - Build projects around motivated individuals. Give them the environment and support they need, and trust them to get the job done.

It's really common for developers to dread or resent meeting with users, because those meetings often uncover changes, which leads to rework that can often be difficult and frustrating. But when the team has a better, more agile mindset, they know that **meeting with users more often** keeps them in sync, and actually prevents those changes.

A more agile mindset would help Mike see that working with his users more often actually prevents those changes.

> MEET WITH THE USERS *MORE OFTEN*?! BUT ALL THEY EVER DO IS ASK US FOR CHANGES.

Teams do their best work when the people on them are **motivated**. Unfortunately, most of us have had bosses or coworkers who seemed determined to drain all of that great motivation. When people feel like they aren't allowed to make mistakes without serious consequences, are pressured to work extremely long hours, and generally feel like they aren't trusted to do their jobs, the quantity and quality of their work plummets. Teams with a more agile mindset know that when everyone is trusted and given a good working environment, they flourish.

> I DON'T CARE IF THE TEAM'S BEEN WORKING 70 HOURS A WEEK. FAILURE IS NOT AN OPTION, AND **MISTAKES WILL BE REPORTED.**

*This is a great way to **demotivate** your whole team and cause them to do lousy work. Ben isn't even the boss! But he can still create an environment of fear and distrust for everyone around him.*

agile values and principles

> - The most efficient and effective method of conveying information to and within a development team is face-to-face conversation.

Let's be honest—we don't always read every word in the manual before turning on a new gadget. So why would we expect people to do that with a spec?

Waterfall teams typically build a requirements specification first, and then design the software based on those requirements. The problem is that three people can read the same spec and come away with three very different ideas about what the team is supposed to build. This can be a little surprising—shouldn't specifications be precise enough to give everyone the same idea?

There are two problems with that in the real world: writing technical material is hard, and reading it is even harder. Even if the person writing the spec does a perfect job describing what needs to be built (which rarely happens), the people reading it will very often interpret it differently. So how do you get around this problem?

The answer is surprisingly simple: **face-to-face conversation**. When the team gets together and talks about what they need to build, it *really is* the most efficient and effective way to communicate exactly what needs to be built... and also status, ideas, and any other information.

you are here ▸ **53**

there are no Dumb Questions

Q: Are you saying that when people are demotivated, they do bad work on purpose?

A: No, not on purpose. But it's very difficult to innovate, create, or do the mentally taxing tasks required to work on a software team when you're in a demotivating environment. And it's surprisingly easy to demotivate a team: your motivation gets drained when you're not trusted to do your job, harshly punished or publicly embarrassed if you make a mistake (everyone makes mistakes!), or held to unreasonable deadlines that you have no input into or control over. Those are all things that have been shown repeatedly to drag down software teams and make them a lot less productive.

Q: Wait, go back to what you were saying about mistakes. We've been talking about welcoming changes. But if you're making a change, doesn't that mean someone made a mistake earlier that has to be changed now?

A: It's dangerous to think of changes as mistakes, especially when you're using iteration. A lot of times, everyone on the team and the users and stakeholders all agree that the software should be built to do something, but when the users get their hands on the working software at the end of the iteration, they realize that it needs to change—not because they made a mistake earlier, but because they now have information they didn't have at the start of the iteration. That's actually a really effective way to build software. But it only works if people feel comfortable making changes, and if they don't call it a mistake or "blame" anyone for finding the change.

Q: Don't we need specifications for more than just communication? What if you need to refer back to the spec in the future? Or if it needs to be distributed to a lot of people?

A: Sure, and those are good reasons to write things down. And that's why agile teams value comprehensive documentation—they just value working software more.

One thing to keep in mind, though, is that if you're writing documentation to refer back to, or to distribute to a wide audience beyond the software team, then a software specification may not be the right kind of document for the job. Documentation is a tool to get a job done, and you always want to use the right tool for the job. The information that teams need in order to build software is usually different than the information a user or manager might need after the software is built, so trying to create a document that serves both purposes might do neither particularly well.

Q: Hey, it looks like the chapter is almost over, and you haven't covered all twelve principles! Why not?

A: Because the agile principles aren't just an isolated topic that teams learn once and then move on from. They're important because they help you understand how agile teams think about the way they work together to build software. That's why the values and principles of the Agile Manifesto are important.

We're not going to stop talking about the agile mindset, values, or principles, even though we're moving on to methodologies in the next chapter. We'll keep coming back to them, because they help you understand the methodologies (for example, Scrum teams are self-organizing, and XP teams value simplicity).

BULLET POINTS

- Software is **valuable** when it does what the users, customers, or stakeholders need it to do.
- To ensure that software is valuable, teams should deliver an **early** version to the users, and keep delivering **continuously**.
- Agile teams **welcome changing requirements**, and finding those changes early helps prevent rework.
- The best way to find those changes early is to get **working software to the users frequently**.
- Documents are helpful, but the *most effective* way to convey information is **face-to-face conversation**.
- Developers on agile teams **work with business people every day**, including users and stakeholders.
- **Iteration** is a practice in which teams break the software down into frequent timeboxed deliveries.
- A **backlog** is a practice in which teams maintain a list of features that will be built in future iterations.

Scrum teams actually maintain two backlogs: one for the current sprint, and one for the whole product.

You'll learn more about that in the next chapter.

agile *values* and principles

JUDGMENT CALL

Getting into a more agile mindset isn't always easy! Sometimes we get it, but sometimes we need a little work. Here are some things we overheard Mike, Kate, and Ben saying. Draw a line from each speech bubble to either **COMPATIBLE** or **INCOMPATIBLE**, and then to the agile principle that it's either compatible or incompatible with.

COMPATIBLE

> WHY ARE YOU ASKING ME QUESTIONS? I ALREADY WROTE DOWN EVERYTHING THE USERS ASKED FOR IN THE SPECIFICATION.

INCOMPATIBLE

Working software is the primary measure of progress.

COMPATIBLE

> I JUST FOUND OUT THAT THE AUDIENCE SIZE CALCULATION ALGORITHM WE'RE USING DOESN'T WORK. WE NEED TO PUSH THIS FEATURE TO THE NEXT ITERATION.

INCOMPATIBLE

Welcome changing requirements, even late in development. Agile processes harness change for the customer's competitive advantage.

COMPATIBLE

> OK, WHICH OF YOU IDIOTS WROTE THIS BUGGY BLOCK OF SPAGHETTI CODE? IT'S YOUR FAULT THAT WE'RE BEHIND.

INCOMPATIBLE

Deliver working software frequently, from a couple of weeks to a couple of months, with a preference to the shorter timescale.

COMPATIBLE

> I'M RUNNING THE LATEST BUILD, BUT I THOUGHT WE'D BE A LOT FURTHER ALONG ON THAT ANALYTICS FEATURE. IS THERE A PROBLEM I DON'T KNOW ABOUT?

INCOMPATIBLE

Build projects around motivated individuals. Give them the environment and support they need, and trust them to get the job done.

➡ Answers on page 68

nice work team

The new product is a hit!

Kate and Mike delivered a great product, and it's been extremely successful.

"DID YOU SEE THAT EMAIL FROM THE CEO? SALES ARE THROUGH THE ROOF, AND IT'S ALL BECAUSE OF THE NEW FEATURES WE ADDED."

"AND THE TEAM IS REALLY CLICKING! IT'S BEEN A LONG TIME SINCE I'VE ENJOYED WORK SO MUCH."

In fact, it's gone so well that Ben has some fantastic news that everyone will want to hear. Nice work, team!

"THANKS TO THE LATEST SALES, WE JUST GOT A NEW ROUND OF VENTURE CAPITAL FUNDING... AND THAT MEANS **BONUSES FOR EVERYONE!**"

agile values and principles

Mindsetcross

See how well you understand the agile values and principles. Can you solve this without flipping back to the rest of the chapter?

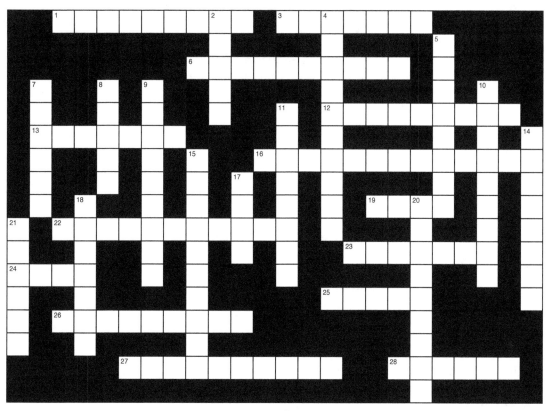

Across

1. When a deadline's been set, and the scope is adjusted to meet it
3. A great way to manage changing requirements
6. How often to deliver
12. When teams repeatedly perform all of the project activities in small chunks
13. What the team does to its behavior after a retrospective
16. An effective way to communicate complex requirements and ideas
19. There's no single "_____" way to build software
22. What we do with customers
23. Something that shouldn't be punished if you want a motivated team
24. What business people and developers must do together daily
25. Very useful for agile teams because they help get the work done
26. At regular _____ the team reflects on how to become more effective
27. The most effective and efficient method of conveying information
28. Teams work best when you pay attention to them

Down

2. When to deliver software
4. The kind of delivery agile teams try to achieve
5. Attitude toward customers or other teams that requires strict agreements before any work can start
7. What agile teams respond to
8. You need to _____ the team to get the job done
9. Traditional but often less-than-effective way to build software
10. The kind of individuals to build projects around
11. Working software is the primary measure of _____
14. Where the original authors of the Agile Manifesto got together
15. What happens to your team if you create a culture of fear
17. Agile teams still follow one
18. The kind of software delivered at the end of every iteration
20. The kind of users that are the highest priority for agile teams
21. Avoid this if you can

⟶ Answers on page 69

Exam Questions

> These practice exam questions will help you review the material in this chapter. You should still try answering them even if you're not using this book to prepare for the PMI-ACP certification. It's a great way to figure out what you do and don't know, which helps get the material into your brain more quickly.

1. You're a project manager on a team building network firmware for embedded systems. You've called a meeting to give a demo of the latest version of code the team has been working on for a control panel interface to a very technical group of business users and customers. This is the fifth time that you've called a meeting to do a demo like this. And for the fifth time, the users and customers asked for specific changes. The team will now go back and work on a sixth version, and you'll repeat the process again.

    Which of the following BEST describes this situation?

    A. The team does not understand the requirements

    B. The users and customers don't know what they want

    C. The project needs better change control and requirements management practices

    D. The team is delivering value early and continuously

2. Which of the following is NOT a Scrum role?

    A. Scrum Master

    B. Team Member

    C. Project Manager

    D. Product Owner

3. Joaquin is a developer, and his software team is in the process of adopting agile. One of the project's users wrote a brief specification that describes exactly what she wants for a new feature, and Joaquin's manager assigned him to work on that feature. What should Joaquin do next?

    A. Demand a meeting with the user, because agile teams recognize that face-to-face conversation is the most efficient and effective method of conveying information

    B. Read the specification

    C. Ignore the specification, because agile teams value customer collaboration over comprehensive documentation

    D. Start writing code immediately, because the team's highest priority is to satisfy the customer through early delivery of valuable software

4. Which of the following is TRUE about working software?

    A. It does what the users need it to do

    B. It meets the requirements in its specification

    C. Both A and B

    D. Neither A nor B

Exam Questions

5. Which of the following statements BEST describes the Agile Manifesto?

A. It outlines the most effective way to build software
B. It contains practices that many agile teams use
C. It contains values that establish an agile mindset
D. It defines rules for building software

6. Scrum projects are divided into:

A. Phases
B. Sprints
C. Milestones
D. Rolling wave planning

7. You are a developer at a social media company working on a project to build a new feature to create a private site for a corporate client. You need to work with your company's network engineers to determine a hosting strategy, and come up with a set of services and tools that the engineers will use to manage the site. The network engineers want to host all of the services internally on your network, but you and your teammates disagree and feel that the services should be hosted on the client's network. Work on the project has come to a halt while everyone tries to come to an agreement. Which agile value BEST applies to this situation?

A. Individuals and interactions over processes and tools
B. Working software over comprehensive documentation
C. Customer collaboration over contract negotiation
D. Responding to change over following a plan

8. Donald is a project manager on a team that follows separate phases for each project, starting with a requirements phase followed by a design phase. Some work can begin on the code before the requirements and design are finished, but the team typically doesn't consider any work to be complete until those phases are finished. Which term BEST describes Donald's projects?

A. Iterative
B. Rolling wave planning
C. Waterfall
D. Scrum

Exam Questions

9. Keith is the manager of a software team. He's made it clear that mistakes are not to be tolerated. A developer spent several hours building "proof of concept" code to test a possible approach to a complex problem. When he eventually discovered from the experiment that the approach wouldn't work, Keith yelled at him in front of the whole team and threatened to fire him if he did it again.

   Which agile principle BEST applies to this situation?

   A. The most efficient and effective method of conveying information to and within a development team is face-to-face conversation.

   B. Build projects around motivated individuals. Give them the environment and support they need, and trust them to get the job done.

   C. Our highest priority is to satisfy the customer through early and continuous delivery of valuable software.

   D. Continuous attention to technical excellence and good design enhances agility.

10. What's the highest priority of an agile team?

    A. Maximizing the work not done

    B. Satisfying the customer by delivering valuable software early and often

    C. Welcoming changing requirements, even late in development

    D. Using iteration to effectively plan the project

11. Which of the following statements is NOT true about the daily standup?

    A. The length is kept short by having everyone stand for the duration of the meeting

    B. It's the same thing as a status meeting

    C. It is most effective when everyone listens to each other.

    D. It's an opportunity for every team member to get involved in planning the project.

12. Which of the following BEST describes the agile mindset with respect to simplicity?

    A. Maximizing the work not done

    B. Satisfying the customer by delivering valuable software early and often

    C. Welcoming changing requirements, even late in development

    D. Using iteration to effectively plan the project

13. A'ja is a project manager on a team that is just starting their agile adoption. The first change they made to the way they work was to start holding daily standup meetings. Several team members have approached her to say that they don't like attending. And despite the fact that she's getting some valuable information from the team at each standup, A'ja is concerned that the extra lines of communication might not be worth damaging the team cohesion.

Exam Questions

What is the BEST thing for A'ja to do?

- A. Stop holding the daily standup and find another way to adopt agile.
- B. Make and enforce a rule that every attendee must put away his or her phone and pay attention.
- C. Follow up with people individually after the meeting to get more detailed status.
- D. Work with the team on changing their mindset.

14. You're a developer on a software team. A user has approached your team about building a new feature, and has provided requirements for it in the form of a specification. She is very certain of exactly how the feature will work, and promises there will be no changes. Which agile value BEST applies to this situation?

- A. Individuals and interactions over processes and tools
- B. Working software over comprehensive documentation
- C. Customer collaboration over contract negotiation
- D. Responding to change over following a plan

15. Which of the following is NOT a benefit of welcoming changing requirements?

- A. It gives the team a way to explain a missed deadline
- B. The team builds more valuable software when customers aren't pressured not to change their minds
- C. There's more time and less pressure so the team can make better decisions
- D. Less code is written before changes happen, which minimizes unnecessary rework

16. Which of the following is NOT part of an agile team's mindset toward working software?

- A. It contains the final version of all features
- B. It is the primary measure of progress
- C. It is delivered frequently
- D. It is an effective way to get feedback

17. Which of the following is NOT true about iteration?

- A. The team must finish all planned work by the end of an iteration
- B. Iterations have a fixed deadline
- C. The scope of work performed during an iteration may change by the time it ends
- D. Projects typically have multiple sequential iterations

Exam ~~Questions~~ Answers

> Here are the answers to this chapter's practice exam questions. How many did you get right? If you got one wrong, that's OK—it's worth taking the time to flip back and re-read the relevant part of the chapter so that you understand what's going on.

1. Answer: D

Did this situation sound negative, like something was going drastically wrong? If it did, you may want to think about your own mindset! This was actually a pretty accurate description of a very successful agile project that uses an iterative methodology. It only sounds like the project is running into problems if you approach it with a mindset that considers change and iteration to be a mistake rather than a healthy activity. If you see the project this way, then you'll be tempted to "blame" the team for not understanding the requirements, or the users for not knowing what they want, or the process for not having adequate controls to prevent and manage changes. Agile teams don't think about things like that. They know that the best way to figure out what the users need is to deliver working software early and frequently.

2. Answer: C

Project managers are very important, but there's no specific role in Scrum called "project manager." Scrum has three roles: Scrum Master, Product Owner, and Team Member. The project manager will fill one of those roles on a project that uses Scrum, but will often still have the "Project Manager" job title.

When your team follows an agile methodology that has specific roles, the role that you fill doesn't always match the title on your business card, especially when your team is just starting to adopt the methodology.

3. Answer: B

It's true that agile teams value customer collaboration, believe face-to-face conversation is the most effective method of conveying information, and place the highest priority on delivering software. However, the user took the time to write the specification, and the information in it could be very helpful in either writing code or having a face-to-face conversation.

When someone takes the time to write down information they think is important, it's very UN-collaborative to ignore it.

4. Answer: D

When agile teams talk about working software, they mean software that they consider "done" and ready to demonstrate to the users. But there's no guarantee that it fulfills the users' needs or that it meets the specific requirements in a specification. In fact, the most effective way to build software that genuinely helps users is to deliver working software frequently. The reason is because the early versions of working software typically **don't fully meet the users' needs**, and the only way for everyone to figure that out is to get it into the hands of the users so they can give feedback about it.

This is why agile teams value early and continuous delivery of working software.

agile values and principles

Exam Questions

5. Answer: C

The Agile Manifesto contains the core values shared by effective agile teams. It doesn't define a "best" way to build software or a set of rules that all teams should follow, because people on agile teams know that there's no "one-size-fits-all" approach that works for all teams.

6. Answer: B

Scrum teams work in sprints, typically (but not always) 30 days long. They plan the next 30 days of work (assuming the length is 30 days) at the start of the sprint. At the end of the sprint, they demonstrate working software to the users, and also hold a retrospective to review what went well and find ways to improve.

7. Answer: C

The project is suffering because the team is having trouble collaborating with their customer. In this case, the network engineers are the customer, because they're the ones who will be using the software. This is a situation where it would be easy to take a contract negotiation approach, laying out specific terms and documents to describe what will be built so that software development work can begin. But it's more effective to genuinely collaborate with them and work together to discover the best technical solution.

8. Answer: C

A waterfall project is divided into phases, typically starting with requirements and design phases. Many waterfall teams will begin "pre-work" on code once the requirements and design have reached a stable point, even if they're not yet complete. However, this is definitely not the same thing as iteration, because the team doesn't change the plan based on what they learned building and demonstrating working software.

9. Answer: B

Agile projects are built around motivated team members. Keith is taking actions that undermine the whole team's motivation by undercutting a team member who's taking a good risk and genuinely trying to make the project better.

Exam Questions

10. Answer: B

Flip back and reread the first principle of agile: "Our highest priority is to satisfy the customer through early and continuous delivery of valuable software." The reason this is the highest priority is because agile teams are focused first and foremost on delivering software that's valuable. All of the other things we do on projects—planning, design, testing, meetings, discussion, documentation—are really, really important, but it's all in service of delivering that valuable software to our customers.

11. Answer: B

While some teams treat the Daily Standup as a status meeting where each team member gives an update to a boss or project manager, that's not really its purpose. It works best when everyone listens to each other, and uses it to plan the project together as a team.

12. Answer: A

Agile teams value simplicity, because simple designs and code are much easier to work with, maintain, and change than complex ones. Simplicity is often called "the art of maximizing the work not done" because—and this is especially true of software—the most effective way to keep something simple is often to simply do less.

13. Answer: D

The reason that the team isn't paying attention during the daily standup is because they don't really care about it or buy into it as an effective tool, and mainly want it to end as quickly as possible so they can get back to their "real" jobs. When teams have this mindset, it's likely that they will eventually stop attending the meeting altogether, and the agile adoption is much less likely to be successful. The daily standup practice will be more effective if the team understands how it helps each of them, both individually and as a team. That mindset shift can only be accomplished through open and honest discussion about what's working and what isn't. That's why working with the team on changing their mindset is the best approach to this situation.

14. Answer: B

It certainly makes sense to read and understand the specification. But the most effective way to truly ascertain whether or not the team really understands what she intended is to deliver working software to her, so that she can see how the requirements she documented were interpreted and work with the team to determine what works well and what needs to change.

Exam Questions

15. Answer: A

There are a lot of great reasons that agile teams welcome changing requirements. When customers are encouraged to change their minds (rather than discouraged from it), they give better information to the team, and that leads to better software. And even when people keep their mouths shut about changes, they almost always eventually get exposed in the end, so when the team gets them early it gives them more time to respond—and the earlier the changes arise, the less code has to be reworked.

However, changes are never an excuse for poor planning or missed deadlines. Effective agile teams generally have an agreement with their users: the teams welcome changing requirements from users, customers, and managers, and in return they aren't blamed for the time it takes to respond to those changes, because everyone recognizes it's still the fastest and most effective way to build software. So nobody really sees welcoming changing requirements as giving the team a way to explain a missed deadline, because the deadlines should already be adjusted to account for the changes.

16. Answer: A

Working software is delivered frequently so that the team can get frequent feedback and make changes early. That's why working software should never be assumed to contain the final version of any requirement. That's why it's "working" software, not "finished" software.

17. Answer: A

Iterations are timeboxed, which means that the deadline is fixed and the scope varies to fit it. The team starts each iteration with a planning meeting to decide what work will be accomplished. But if it turns out that they didn't get the plan right and work takes longer than expected, then any work that didn't get done is returned to the backlog and reprioritized (and often ends up in the next iteration).

exercise solutions

Manifesto Magnets Solution

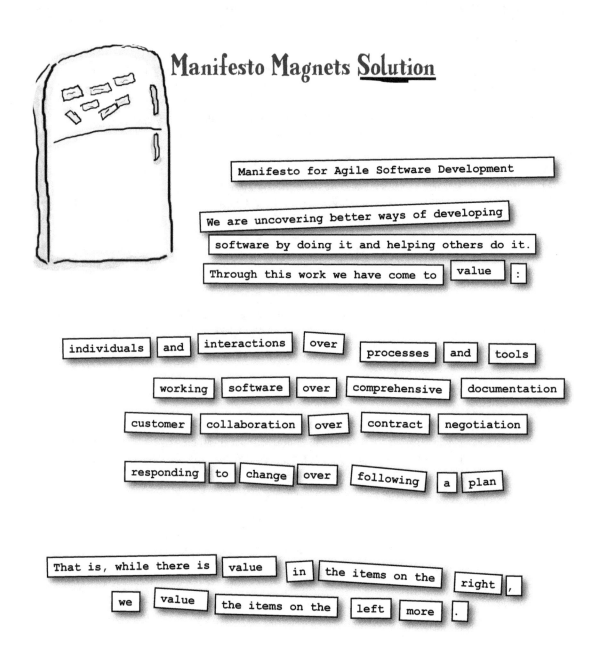

Manifesto for Agile Software Development

We are uncovering better ways of developing software by doing it and helping others do it. Through this work we have come to `value`:

individuals and interactions over processes and tools

working software over comprehensive documentation

customer collaboration over contract negotiation

responding to change over following a plan

That is, while there is value in the items on the right, we value the items on the left more.

Dictionary drill SOLUTION

Here are a bunch of definitions for words that you've seen in this chapter. Can you fill in the words that each definition belongs to?

✓ *We used "rework" as a noun earlier in the chapter: "a major source of bugs is rework."*

rework, noun

work done by the team to change previously written code to make it function differently or serve a different purpose, often considered risky by teams due to its increased likelihood of introducing bugs

Rework can also be a verb: "We had to rework this bit of code to adapt it to a new purpose."

timeboxed, adjective

setting a hard deadline for an activity to be completed, and adjusting the scope of that activity to meet the deadline

This can also be used as a verb: "Let's timebox the work we're doing on this feature to six hours."

backlog, noun

a practice used by teams in which the team works with users, customers, and/or stakeholders to maintain a list of features that will be built in the future, often prioritized with the most valuable features at the top

progressive elaboration, noun (two words)

developing a project artifact (like a plan) in steps, using knowledge gained from the previous step to improve it

iterative, adjective

a kind of methodology in which teams break the project down into smaller parts, delivering working software at the end of each one, and possibly changing direction based on the feedback from the working software

Normally "waterfall" is a noun. But in this case it's actually an adjective that describe a type of process.

waterfall, adjective

a type of model, process, or method for building software in which the entire project is broken down into sequential phases, often including a change control process in which changes force the project into a prior phase

Here's how it's used in a sentence: "Brian used to work at a company that followed a waterfall process, so he's really excited to try an agile process like Scrum."

compatibility test

Solution

Getting into a more agile mindset isn't always easy! Sometimes we get it, but sometimes we need a little work. Here are some things we overheard Mike, Kate, and Ben saying. Draw a line from each speech bubble to either **COMPATIBLE** or **INCOMPATIBLE**, and then to the agile principle that it's either compatible or incompatible with.

WHY ARE YOU ASKING ME QUESTIONS? I ALREADY WROTE DOWN EVERYTHING THE USERS ASKED FOR IN THE SPECIFICATION. → INCOMPATIBLE

> Working software is the primary measure of progress.

When Kate discovered this change, they could have delivered software that didn't work, or delayed the delivery to fix it. But pushing the feature to the next iteration is a better choice, because they'll still deliver working software with the other features.

> It's not fair to ask users for requirements at the start of the project and then refuse to let them change their minds (as long as they understand that the team needs time to make the changes).

I JUST FOUND OUT THAT THE AUDIENCE SIZE CALCULATION ALGORITHM WE'RE USING DOESN'T WORK. WE NEED TO PUSH THIS FEATURE TO THE NEXT ITERATION. → COMPATIBLE

> Welcome changing requirements, even late in development. Agile processes harness change for the customer's competitive advantage.

OK, WHICH OF YOU IDIOTS WROTE THIS BUGGY BLOCK OF SPAGHETTI CODE? IT'S YOUR FAULT THAT WE'RE BEHIND. → INCOMPATIBLE

> Deliver working software frequently, from a couple of weeks to a couple of months, with a preference to the shorter timescale.

If Kate was only relying on her schedule to measure progress, she might think the project was going just fine. Relying on working software as her primary measure of progress helps her spot (and hopefully fix!) problems early.

Technical people like Mike are often really blunt. But even if the team has a culture where it's OK to challenge and even insult people, blaming one person for delays or quality issues is really demotivating.

I'M RUNNING THE LATEST BUILD, BUT I THOUGHT WE'D BE A LOT FURTHER ALONG ON THAT ANALYTICS FEATURE. IS THERE A PROBLEM I DON'T KNOW ABOUT? → COMPATIBLE

> Build projects around motivated individuals. Give them the environment and support they need, and trust them to get the job done.

agile *values* **and principles**

Mindsetcross
SOLUTION

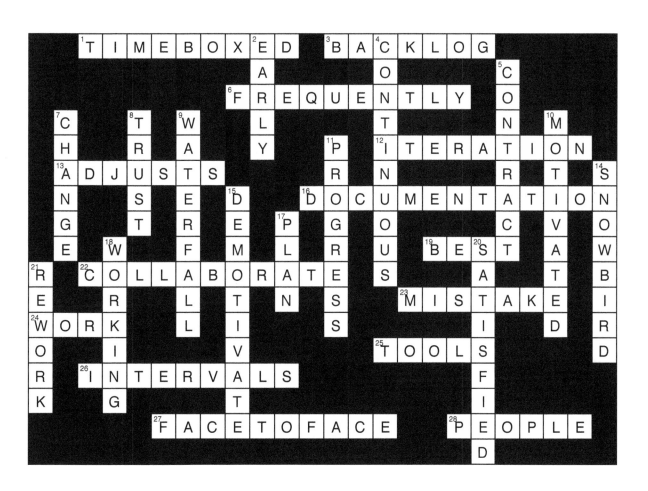

this page *intentionally left blank*

3 managing projects with Scrum

The Rules of Scrum

The rules of Scrum are simple. Using it effectively is not so simple.

Scrum is the most common approach to agile, and for good reason: the **rules of Scrum** are straightforward and easy to learn. Most teams don't need a lot of time to pick up **the events, roles, and artifacts** that make up the rules of Scrum. But for Scrum to be most effective, they need to <u>really</u> understand **the values of Scrum** and the Agile Manifesto principles, which help them get into an effective mindset. Because while Scrum seems simple, the way a Scrum team constantly **inspects and adapts** is a whole new way of thinking about projects.

first person moo-ter

COMING SOON FROM RANCH HAND GAMES

THE TEAM THAT BROUGHT YOU THE BLOCKBUSTER INTERNATIONAL HIT COWS GONE WILD VIDEO GAMES

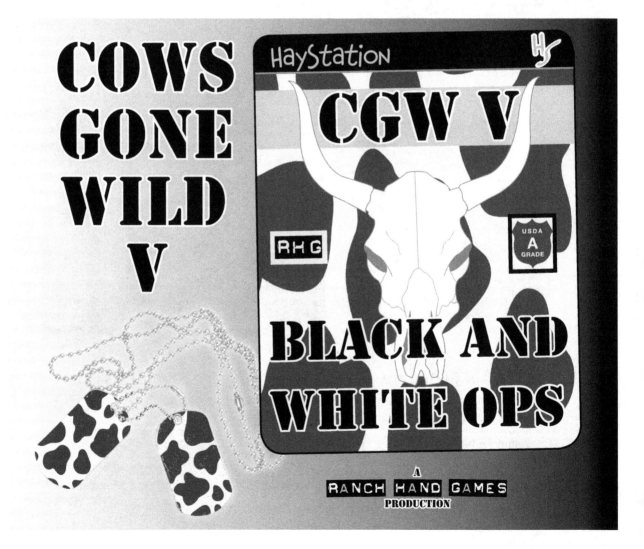

managing projects **with scrum**

Meet the Ranch Hand Games team

They're hot off the successful release of *Cows Gone Wild IV: The Milk Man Cometh*, and about to tackle their most ambitious project yet! But while *CGW4* may have sold well, it was far from perfect as a project, and Amy, Brian, and Rick want *CGW5* to be an improvement. Agile to the rescue!

Amy: Wow, I'm so glad you said that. I can't take another project like that. Just staying on top of the last-minute artwork changes was practically impossible.

Brian: Oh man, we can't have that argument again. You know they were necessary because the levels kept changing. I had my whole team working nights and weekends just to keep up.

Amy: I know, I know. We all got overwhelmed trying to tackle too many things at the same time.

Brian: No matter how much planning we did, our schedules never seemed long enough.

Rick: Yeah, that was definitely a problem—and I've been doing a lot of research about how to fix that. What do you guys think about using **Scrum**?

Amy: I've been reading about it, too. I think it could help.

Brian: Look, anything that can help make things less insane around here gets my vote.

Amy: But don't the Scrum rules say that we need a Product Owner and a Scrum Master?

Rick: Well, I'm looking over what a Scrum Master does, and I think I'm up to that task. Amy, you work with the business side a lot, right? What do you think about being the Product Owner?

Amy: I'll try anything once. I'll start letting the business and PR teams know that I'm the Product Owner.

Brian: Let's do this!

you are here ▶ **73**

the rules of *scrum*

The Scrum **events** help you get your projects done

Scrum is the most popular approach to agile, and for good reason. The rules of Scrum are simple, and teams all around the world have been able to adopt them and improve their ability to deliver projects. Every Scrum project follows the same pattern of behavior, which is defined by a **series of timeboxed events** that always happen in the same order. Here's what the Scrum pattern looks like:

> **The Scrum events:**
> The Sprint
> The Sprint Planning session
> The Daily Scrum
> The Sprint Review
> The Sprint Retrospective

This is the single source for requirements and any changes that will be made to the product throughout the project.

Every Scrum project is organized into timeboxed iterations called Sprints. Many teams use 30-day Sprints, but it's pretty common to see two-week Sprints too.

The team uses the Product Backlog to keep track of the features that they'll build for the whole project.

At the beginning of each Sprint, the Scrum Team gets together for a Sprint Planning session where they choose what items to include in the Sprint.

The items for the Sprint are pulled off of the **Product** Backlog and added to the Sprint Backlog. During the Sprint, all of the development work is focused on building the items in the Sprint Backlog.

	30 days	30 days	30 days	30 days
Backlog:	21 features	17 features	14 features	12 features
	Planning	Planning	Planning	Planning
	Daily Scrum	Daily Scrum	Daily Scrum	Daily Scrum
	Daily Scrum	Daily Scrum	Daily Scrum	Daily Scrum
	Daily Scrum	Daily Scrum	Daily Scrum	Daily Scrum
	Development	Development	Development	Development
	Daily Scrum	Daily Scrum	Daily Scrum	Daily Scrum
	Daily Scrum	Daily Scrum	Daily Scrum	Daily Scrum
	Daily Scrum	Daily Scrum	Daily Scrum	Daily Scrum
	Sprint Review	Sprint Review	Sprint Review	Sprint Review
	Retrospective	Retrospective	Retrospective	Retrospective
New backlog:	17 features	14 features	12 features	9 features

Every day, the team holds a Daily Scrum, a short meeting where each person gives an update on their progress, what they'll work on next, and any roadblocks they've hit.

When Sprint is done, the team holds a meeting called the Sprint Review, where they meet with the users and give a demo of the working software that they built.

The very last thing the team does in the Sprint is meet and hold a Sprint Retrospective, where they talk about what happened during the Sprint so they can reproduce the things that went well and learn from any problems.

If you've read our book *Learning Agile* then you'll recognize this illustration of the basic Scrum pattern!

managing projects **with scrum**

The Scrum roles help you understand who does what

There are three roles that must be filled on every Scrum team. The first role is the one that's most familiar to us: the **Development Team**. People on the team may have different areas of expertise, and maybe even different job titles in the company, but they all participate in the Scrum events the same way. There are also two very important roles filled by team members: the **Product Owner** and the **Scrum Master**. And when you add them to the Development Team, you get the entire **Scrum Team**.

The Product Owner helps the team understand the users' needs so they can build the most valuable product.

The Product Owner works with the team every day to help them understand the features in the Product Backlog: what items are on it and why the users need them. This is a really important job because it helps the team **build the most valuable software they can**.

> Flip back to the last chapter. Can you find an agile principle that talks about delivering valuable software?

The Scrum rules are clear that the Product Owner and Scrum Master roles are each filled by a person and not a committee.

The Scrum Master helps the team understand and execute Scrum.

Scrum may be simple to describe, but it's not always easy to get right. That's why there's one person on the team, the Scrum Master, whose whole job is to help the Development Team, the Product Owner, and the rest of the company to do exactly that—get Scrum right.

The Scrum Master is a leader (which is why the word "master" is right there in the name). But he or she demonstrates a very *particular* kind of leadership: the Scrum Master is a **servant leader**. This means that the person in this role spends all of his or her time helping (or "serving") the Product Owner, the Development Team, and people throughout the organization:

★ Helping the product owner find effective ways to manage the backlog

★ Helping the Development Team understand the Scrum events, and facilitating them if needed

★ Helping the rest of the organization to understand Scrum and work with the team

★ Helping everyone do the best job they can to deliver the most valuable software possible

---

The Scrum Guide lays out the rules for how teams use the Scrum framework.

Before you move on to the next page, go to https://www.scrum.org and download a copy of the Scrum Guide by Ken Schwaber and Jeff Sutherland, the originators of Scrum. It contains the definition of Scrum, and is updated regularly with the latest ideas about how teams use Scrum: when new thinking is incorporated into Scrum, that's where you'll find it. Also, did you notice how Scrum, Daily Scrum, Sprint, Sprint Planning, Sprint Review, Product Backlog, and other terms are ***capitalized*** in this chapter? That's done to follow the standard in the Scrum Guide.

the art of *artifacts*

The Scrum artifacts keep the team informed

Software projects run on information. The team needs to know about the product that they're working on, what they're building in the current Sprint, and how they'll get it built. Scrum teams use three **artifacts** to manage all of this information: the **Product Backlog**, the **Sprint Backlog**, and the **Increment**.

> Here's an example of a Product Backlog—but there's no rule that says that it has to look like this. Many teams keep spreadsheets, or add entries to a database, or use software tools to manage their backlogs.

Cows Gone Wild 5 Product Backlog

Item #1: The stealth sheep barn level needs to be designed and play tested

Estimated effort: 27 person-days

Value: Builds on the most popular level from *CGW4*, gamers will be excited about this

Item #2: The milk gun physics need to be updated to include squirting action

Estimated effort: 4 person-days

Value: Improved gameplay will make the game more fun

Item #3: The mad cow zombie survival level design needs to be finished, including zombie AI and zombie horde attacks

Estimated effort: 16 person-days

Value: Zombies are really popular right now, will be a good selling point

Item #4: The strategic silo assault will have an overhead view of the battlefield, giving the player placement and command/control of turrets, barnyard troops, tractors, feed lots, and snipers

Estimated effort: 19 person-days

Value: Similar to a feature in *CGW4*, we can add a whole level to the game by reusing code

Page 1 of 7

> Every item in the Product Backlog has four attributes: order, description, estimate, and value.

> There's no rule that says that the estimate has to be in person-days. It just needs to be in a unit that everyone on the team understands.

> The value can be a description like this, or it can be a relative number, expected dollar value, or another way to measure or express value.

> The Product Owner keeps the project on track by continuously refining the Product Backlog: staying up to date on what the company needs, and adding, removing, and reordering backlog items.

> The Product Backlog is never complete as long as the project is running. The Product Owner will constantly work with users and stakeholders throughout the company to add, remove, change, and reorder the items in the Product Backlog.

managing projects **with scrum**

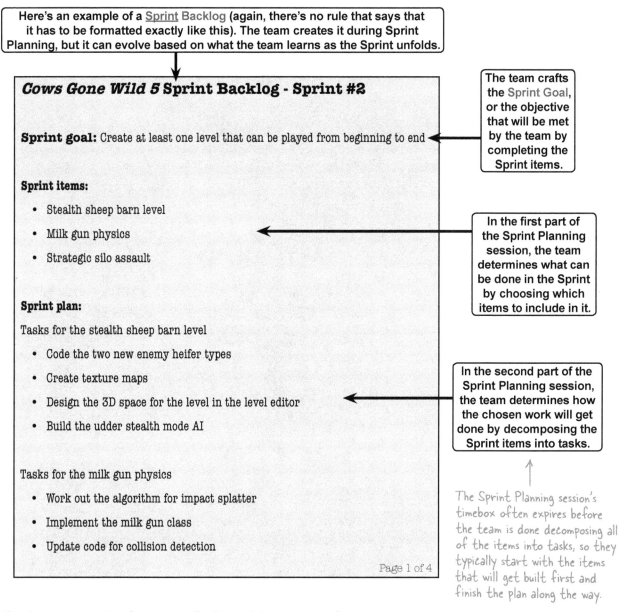

Here's an example of a Sprint Backlog (again, there's no rule that says that it has to be formatted exactly like this). The team creates it during Sprint Planning, but it can evolve based on what the team learns as the Sprint unfolds.

The team crafts the Sprint Goal, or the objective that will be met by the team by completing the Sprint items.

In the first part of the Sprint Planning session, the team determines what can be done in the Sprint by choosing which items to include in it.

In the second part of the Sprint Planning session, the team determines how the chosen work will get done by decomposing the Sprint items into tasks.

The Sprint Planning session's timebox often expires before the team is done decomposing all of the items into tasks, so they typically start with the items that will get built first and finish the plan along the way.

The Increment is the sum of all backlog items that are actually completed and delivered at the end of the Sprint

Scrum is incremental, which means the project is broken into "chunks" that are delivered one after another. Each of those "chunks" is called an **Increment**, and each Increment represents the result of one complete Sprint: the working software that the team demonstrates to the users in the Sprint Review. That working software typically includes all of the features that they previously delivered —which makes sense, because they're not going to delete them!—so the *product Increment is the sum of all of the backlog items completed during this Sprint and all of the previous Sprints*.

page goal *header*

The Scrum Sprint Up Close

|—— 30 days ——|

The **Sprint** is a *timeboxed* iteration. Most teams use a two-week Sprint, but it's common to see 30-day Sprints as well.

Planning

Daily Scrum

Daily Scrum

Daily Scrum

Development

Daily Scrum

Daily Scrum

Daily Scrum

Sprint Review

Retrospective

The **Sprint Planning** session is a meeting with the whole team, including the Scrum Master and Product Owner. For a 30-day Sprint it's timeboxed to 8 hours, for 2-week Sprints it's 4 hours, and other Sprint lengths have proportionally sized timeboxes. It's divided into parts, each timeboxed to half of the meeting length:

★ In the first half, the team figures out *what* can be done in the Sprint. First the team writes down the **Sprint Goal**, a one- or two-sentence statement that says what they'll accomplish in the Sprint. Then they work together to pull items from the Product Backlog to create the **Sprint Backlog**, which has everything they'll build during the Sprint.

★ In the second half, they figure out *how* the work will get done. They break down (or **decompose**) each item on the Sprint Backlog into **tasks** that will take one day or less. This is how they create a *plan* for the Sprint.

All of the work is planned, but not all of it is decomposed. The meeting timebox can expire befores the team's done decomposing every Sprint Backlog item, so they concentrate on decomposing work for the first days of the Sprint.

The **Daily Scrum** is a 15-minute timeboxed meeting that's held at the same time every day, In it, the Development Team and Scrum Master meet and the Product Owner is strongly encouraged to participate. Each person answers three questions:

★ What have I done since the last Daily Scrum to meet the Sprint Goal?

★ What will I do between now and the next Daily Scrum?

★ What roadblocks are in my way?

In the **Sprint Review** the whole team meets with key users and stakeholders who have been invited by the Product Owner. The team demonstrates what they built during the Sprint, and gets feedback from the stakeholders. They'll also discuss the Product Backlog, so that everyone knows what will *probably* be on it for the next Sprint. For 30-day Sprints, this meeting is timeboxed to four hours.

The **Sprint Retrospective** is a meeting that the team uses to figure out what went well and what can be improved. Everyone on the team participates, including the Scrum Master and Product Owner. By the end of the meeting they'll have written down specific improvements that they can make. It's timeboxed to three hours for a 30-day Sprint.

The Sprint is over **when its timebox expires**.

there are no Dumb Questions

Q: Wait, is that all there is to Scrum?

A: That's all of the Scrum rules, but that's definitely not all there is to Scrum. Scrum was designed to be lightweight and simple to understand. But mastering Scrum takes a lot more than just following the rules. You learned in the last chapter how the values of the Agile Manifesto can make a big difference in how a team uses a practice. That same idea applies to Scrum: **mindset and experience** make the difference between a team with an empty "just following the rules" approach and an effective Scrum team that really "gets" it.

Q: I already break my projects into phases. Aren't Sprints the same thing?

A: Not at all. Traditional waterfall projects are often broken into phases, with a complete deliverable at the end of each phase. But those phases are still planned at the beginning of the project. And if there's a change discovered that will affect the next phase, the project needs to be replanned, which usually involves a separate change control process. In other words, the team basically assumes the project plan is mostly correct, and that it's the project manager's job to deal with the relatively few changes that happen along the way.

Scrum is different because it's **iterative**, which is a lot more than just breaking a project down into phases. The team doesn't even plan the next iteration until the current one is complete. The team might discover changes partway through the Sprint that affect the next one, or which may even affect the current one. That's why the Product Owner is such an important member of the team: he or she has the authority to make decisions on behalf of the business or customers about what features the team will build, which gives them the power to make those changes immediately.

Q: What's the difference between the Product Backlog and the Sprint Backlog?

A: The Product Backlog contains a list of everything that might be needed in the product. Scrum teams are usually working on ongoing releases of a single product, so the Product Backlog is never complete. The first version they release typically contains the requirements that are best understood by everyone. As the project goes on, the Product Owner will add and remove items based on what's valuable to the company.

The Sprint Backlog contains the specific items that the team will build during the Sprint, which were moved from the Product Backlog during Sprint Planning. The software that the team finishes during the Sprint is the **Increment**: the team delivers and reviews a complete Increment in each Sprint. The Sprint Backlog *also* contains a **plan for delivering the Increment**. The team builds that plan during Sprint Planning by decomposing backlog items into tasks.

Q: What exactly *is* a backlog item?

A: Every item in the Product Backlog consists of a brief description, a (usually rough) estimate of how long it will take, the business value, and an order. The Product Owner will routinely **refine the Product Backlog**. This means going through the items in the backlog, removing ones that are no longer useful, re-evaluating the value of each item, and updating the order so the most valuable ones are first.

Q: Wait a minute—in the story, Brian is a team lead. But there are only three roles in Scrum, and "team lead" isn't one of them. What's going on there?

A: Roles and jobs aren't the same thing. As far as Scrum is concerned, Brian is just another member of the Development Team, even though his job in the company is team lead, giving him more authority than the other the developers, and he has his own skills and expertise. Scrum may not have a distinct role for him to play, but *he still plays an important and unique role on the team.* It's just that there are no Scrum-related events or artifacts that are specific to him.

Q: What do you mean by "artifacts"?

A: An artifact is just a by-product specific to a process or methodology. Scrum has **three artifacts**: the Product Backlog, the Sprint Backlog, and the Increment.

Q: What happens if we get partway through a Sprint and discover there's some sort of emergency situation? Do we still have to wait for the timebox to expire before the Sprint can end?

A: On **very rare occasions** the Product Owner has the authority to cancel the Sprint before the timebox is over. When he or she does this, a Sprint Review is held for any Sprint Backlog items that are done, and all other items are put back on the Product Backlog and left for the next Sprint Planning session. *Be really, really careful about cancelling a Sprint.* It's an in-case-of-emergency "break glass" action because it can waste a lot of the team's energy—and more importantly, it causes people in the company to distrust the team, and to lose confidence in the effectiveness of Scrum.

> **Scrum was designed to be lightweight and simple to understand, but mastering Scrum takes a lot more than just following the rules.**

they followed the rules but something went wrong

Sharpen your pencil

Write down the name of each of the Scrum events, when the event happens, and the length of the event's timebox. <u>We filled in the first event</u>. Then write down the three Scrum roles and three Scrum artifacts.

Event names in the order they occur	When the events happen	Length of the event's timebox
Sprint		*Assume that the Sprint is timeboxed to 30 days*

Write down the Scrum roles

Write down the Scrum artifacts

➡ Answers on page 112

80   Chapter 3

managing projects with scrum

FOUR MONTHS LATER AT A SPRINT RETROSPECTIVE...

> I DON'T THINK THIS SPRINT WENT SO WELL.

> WHAT TIPPED YOU OFF, SHERLOCK? WAS IT THE FACT THAT WE **BARELY GOT ANYTHING DONE**?

Rick: Hey, there's no need for insults. We're all doing our best here!

Amy: Sorry. Things have just been really tense with the business lately, and it's got me on edge. I spend so much time with Scrum that I don't have time to do my job.

Rick: What do you mean?

Amy: I mean, you guys constantly ask me to make decisions. Like when we were planning this Sprint, and I had to decide whether Brian's team should start the strategic silo assault level or improve the milk grenade mechanics.

Rick: Yeah. You had us get started on the strategic silo assault. Was that wrong?

Amy: Oh man, you have no idea! After the demo I got hauled into the CEO's office and screamed at for an hour. Gamers hated the strategic levels in *CGW4*, and the last thing our stakeholders want is to see bad reviews for them in *CGW5*.

Rick: Wait, what? We worked super hard on those. I thought they came out well!

Amy: Me too. But they said that I had no business making the decision to include any strategic levels in the Sprint—or any other decision like that.

Rick: But you're the Product Owner! Making those decisions is part of your job now.

Amy: Exactly. So I have no idea what to do. Plus, I spend so much time answering the team's questions about features that I barely have time for my real job.

Rick: Well, it's no easier for me. It's getting harder and harder to drag Brian's team members into my Daily Scrums—you know how programmers hate meetings.

Amy: You know what? We're following the Scrum rules, it's kind of helping. I think. Right? Maybe? Um... are we still sure this Scrum stuff is actually worth the trouble?

The team followed the Scrum rules to the letter, but the project still ran into trouble. What went wrong?

just following rules isn't enough

The Scrum values make the team more effective

We already saw how agile teams are much more effective when everyone on the team has a mindset that's driven by the values in the Agile Manifesto. Scrum comes with its own set of five **values** that do exactly the same thing for Scrum teams.

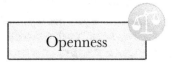

Each of the five Scrum values is represented by one word. In this case, the value is "openness".

You always know what your team members are working on, and you're comfortable that they know what you're working on. If you run into a problem or an obstacle, you can bring it up with the team.

It's not always easy to talk to your teammates when you run into problems. None of us like making mistakes, especially at work. That's why everyone on the team needs to share this value: it's a lot easier to be open with teammates about problems and roadblocks that you run into if they're just as open with you about theirs—and that helps the whole team.

I'M REALLY SORRY, I KNOW I PROMISED I'D BE DONE WITH THE CODE FOR THE SILO SNIPER LEVEL BY NOW, BUT I JUST CAN'T GET THE LOGIC RIGHT.

It was uncomfortable and a little embarrassing for Brian to be open about this roadblock, but it's what was best for the project.

I WON'T LIE, THAT'S GOING TO SET US BACK. BUT WE'LL FIND A WAY TO GET YOU THE TIME YOU NEED.

Respect

You and your teammates have mutual respect for each other, and every person on your team trusts everyone else to do their jobs.

Trust and respect go hand in hand. People on Scrum teams listen to each other, and when they disagree they take the time to understand each other's ideas. It's natural for people on teams to disagree on an approach. On an effective Scrum team, you'll listen when your teammates disagree with the approach you're taking, but in the end they'll respect your decisions, and give you the benefit of the doubt if you take an approach they disagree with.

Trust doesn't always come easy to a traditional waterfall team where the project manager does the planning by demanding estimates from the team members. If things go wrong, the project manager can blame the team for underestimating, and the team can blame the project manager for a bad plan.

> Scrum teams need at least three members, and they usually max out at nine people. A Scrum team needs at least three people in order to get a significant amount of work done in a Sprint. But once the team size gets up to ten or more people, it becomes really hard for them to coordinate with each other—the Daily Scrum gets too chaotic, and it gets difficult to plan effectively.

82 Chapter 3

managing projects **with scrum**

Courage

Scrum teams have the courage to take on challenges. Individual people on the team have the courage to stand up for their project.

What do you do when the boss demands that you and your team meet an impossible goal? What if he demands that they meet a two-week deadline for a project that can't possibly be done in less than two months? Effective Scrum teams have the courage to stand up and say "no" to impossible goals because they have the planning tools to show what is possible, and users and stakeholders who trust them to deliver the most valuable software that they can.

Focus

This is why the team writes a one- or two-sentence Sprint Goal at the beginning of the Sprint Planning meeting. It helps keep them focused during the Sprint.

Every team member is focused on the Sprint Goal, and everything they do in the Sprint helps move them toward it. During the Sprint, everyone works only on Sprint tasks, and does <u>one task at a time</u> until the Sprint is done.

During a Scrum Sprint, every single person on a Scrum team is focused exclusively on items in the Sprint Backlog and tasks that they decomposed them into during the planning meeting. Each person works on one item in the Sprint Backlog at a time, focuses exclusively on one task in the plan, and moves on to the next task only when the current one is done.

WHAT ABOUT **MULTITASKING**? AREN'T TEAMS MORE EFFECTIVE WHEN PEOPLE DO MULTIPLE TASKS AT THE SAME TIME?

People on Scrum teams know that focusing on one task at a time is more effective than trying to multitask.

There's a myth that switching back and forth between multiple tasks many times a day is more effective than concentrating on one at a time. Think about it this way: let's say you have two tasks that will each take one week. If you try to multitask and do them at the same time, the best you'll do is deliver them both at the end of two weeks. But if you don't start the second task until the first one is done, you'll finish the first task a week earlier.

There's one more Scrum value. Flip the page... ➡

pigs and chickens

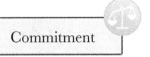

Commitment

Every person on the team is committed to delivering the most valuable product that they can—and the rest of the company is committed, too.

When the team is committed to the project, it means that the tasks that they planned out to meet the Sprint Goal are the most important tasks they have to work on. Each team member feels that his or her personal success at the company is tied to the success of the project. Not only that, but every single person on the team then *feels committed to each item in the Increment*, not just the items that they're working on. This is called **collective commitment**.

But what happens if something comes up that's important to the company, but not part of the project? For Scrum to be effective, the boss needs to keep that from happening. In other words, **the company needs to be fully committed to the project,** to respecting the team's collective commitment to the Sprint Goal, and to following the rules of Scrum.

So how does a company express that kind of commitment?

By **giving the team the authority** to determine what features are going to be developed during each Sprint, and trusting that's how the team will deliver the most valuable software possible. The company does that by assigning a full-time Product Owner to the team with the authority (and willingness!) to decide what features will be built and accept them as done.

<u>Collective commitment</u> means that every team member feels committed to delivering the complete Increment, and not just the items he or she is working on.

THIS IS THE THIRD TIME I ACCEPTED A FEATURE, ONLY TO GET YELLED AT BY MY BOSS BECAUSE I HAD "NO BUSINESS" DOING THAT.

YEAH, IT'S LIKE YOU DON'T REALLY HAVE THE AUTHORITY TO **MAKE A REAL COMMITMENT** ON BEHALF OF THE COMPANY.

YOU KNOW WHAT? I THINK I'M THE **WRONG PERSON** TO BE THE PRODUCT OWNER!

LUCKILY, I'M THE **RIGHT** PERSON TO BE THE SCRUM MASTER. LET ME TALK TO SOME SENIOR MANAGERS AND SEE IF I CAN FIX THIS.

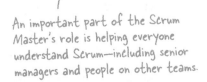

An important part of the Scrum Master's role is helping everyone understand Scrum—including senior managers and people on other teams.

84   Chapter 3

Story Time

Once upon a time there lived a pig and a chicken who were the very best of friends.

One day the chicken said to the pig, "I have a great idea. Let's open a restaurant together!"

The pig said, "That <u>is</u> a great idea! What should we call it?"

The chicken replied, "How about **Bacon 'n Eggs**?"

The pig thought about it for a minute. Then he said, "You know what? Never mind. <u>You're just involved</u>, but **I'm <u>committed</u>.**"

find the right product owner

there are no Dumb Questions

Q: Uh... why are you talking about farm animals?

A: Because the fable of the pig and the chicken is a good way to understand commitment (because the chicken just lays eggs, but the pig has to let himself get eaten—a much greater commitment). In fact, *some* teams even use these terms in practice this way, calling people who are committed to the project "pigs," and those who have an interest but not a true commitment "chickens." (Occasionally, they'll even call themselves pigs in cultures where "pig" is used as an insult!)

The Scrum value of commitment means that when you're on a Scrum team, you genuinely believe that your professional success or failure rests on your ability to deliver valuable software, and that makes you a committed "pig." Your users and stakeholders may have a big interest in the project getting done, which makes them "chickens"—they're very important to the project, but they don't feel the same sense of strong personal commitment as "pigs" do.

Q: What if the team doesn't really "get" some of the Scrum values?

A: An important part of the Scrum Master's job is to help everyone on the team to understand the Scrum values, and to eventually internalize them so that each person has the right mindset for Scrum. Very few teams <u>start out</u> Scrum genuinely believing in all of these values at first. They <u>grow and evolve together</u> over time. The Scrum Master helps by using the Scrum values to help the team understand and deal with project obstacles that come up.

Q: What if my team can't find a Product Owner with enough authority?

A: If the Product Owner doesn't have the authority to decide what the team will build, or to accept each item in the Sprint Backlog as done, then the company hasn't given the team the authority to do their jobs, and they haven't truly committed to the project—or to Scrum. That can be trouble.

A lot of projects fail because the team **did a <u>great</u> job building the <u>wrong</u> software**. The Product Owner keeps that from happening because his or her **full-time job** is to work with the rest of the company and understand what the business needs, decide what features will be built, and help the team to understand those features.

If the Product Owner doesn't have the authority to decide what features the team builds, then the company isn't <u>really</u> <u>committed</u> to delivering the project with Scrum.

BULLET POINTS

- **Scrum** is the most popular approach to agile, and is most successful when used by software teams of at least three and up to nine people.

- Scrum includes five timeboxed **events**: the Sprint, Sprint Planning, the Daily Scrum, the Sprint Review, and the Sprint Retrospective.

- Scrum has three **roles**: the Product Owner, the Scrum Master, and the Development Team.

- Scrum uses three **artifacts**: the Product Backlog, the Sprint Backlog, and the Increment.

- Projects are divided into **Sprints**, iterations that are typically timeboxed to 30 days (but can be shorter).

- Each Sprint starts with **Sprint Planning**, a timeboxed meeting in which the team decides which items (such as features) to include in the Sprint Backlog, and decomposes them into tasks for at least the first week.

- The **Daily Scrum** is a meeting where each team member talks about what they've done, what they'll work on next, and if they see any roadblocks ahead.

- The Product Owner invites key stakeholders to the **Sprint Review** where the team demonstrates working software and discusses the next Sprint Backlog.

- At the **Sprint Retrospective** the team talks about what went well and identifies specific things that can be improved.

- Scrum teams have five **values** that help them get into a more effective mindset: openness, respect, courage, focus, and commitment.

86   Chapter 3

90% done 90% left to go

Here are some things we overheard Amy, Rick, and Brian saying. Draw a line from each speech bubble to either COMPATIBLE or INCOMPATIBLE, and then to the Scrum value that it's either compatible or incompatible with.

IT'S MY TURN TO TALK? OK. SINCE THE LAST DAILY SCRUM I KEPT WORKING ON THE SAME FEATURE, AND I'LL DO THE SAME UNTIL THE NEXT SPRINT. NO ROADBLOCKS. WHO'S NEXT?

COMPATIBLE / INCOMPATIBLE

Courage

LOOK, ALEX, I KNOW REDOING ALL OF THE GRAPHICS WILL IMPRESS REVIEWERS, BUT THERE'S ABSOLUTELY NO WAY THE TEAM CAN DO THAT WITHOUT MISSING THE RELEASE DATE.

COMPATIBLE / INCOMPATIBLE

Focus

THE ONLY THING I'M IMPRESSED WITH IS TECHNICAL ABILITY. IF YOU CAN'T CODE, I HAVE NO USE FOR YOU.

COMPATIBLE / INCOMPATIBLE

Openness

I CAN'T WORK ON ANY SPRINT TASKS TODAY. ANOTHER TEAM HAS A REALLY BIG DEADLINE AND THEY NEED TO BORROW ME.

COMPATIBLE / INCOMPATIBLE

Respect

Answers on page 113

managing projects with scrum

> I THOUGHT GOING TO THE DAILY SCRUM EVERY DAY WOULD TELL ME HOW THINGS ARE GOING, BUT **I HAVE NO IDEA WHAT'S GOING ON** WITH THE BLACK ANGUS BOSS BATTLE.

> WHAT DO YOU MEAN? THE DEVELOPER WORKING ON THAT SAID IT'S 90% DONE.

> YEAH, I KNOW! BUT IT WAS 90% DONE **YESTERDAY**, AND 90% DONE **A WEEK AGO**. WHAT AM I SUPPOSED TO DO WITH THAT?

Rick: I... uh...

Alex: Yeah, just what I thought.

Rick: Hey, that's not fair. I know he's been working really hard on that feature. It's just taking a lot longer than any of us thought it would.

Alex: So what are you going to do about it?

Rick: Hey, we're all on the same team here. And that includes you, Alex. So I think you meant to ask what are **we** going to do about it?

Alex: OK, fine. Well, since I'm on the team, let me give you an answer. Other teams build contingency and padding into their schedules so *they never have to tell a senior manager like me* that they're running late.

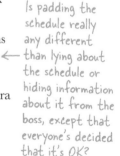

Is padding the schedule really any different than lying about the schedule or hiding information about it from the boss, except that everyone's decided that it's OK?

Rick: Sure. Yeah. I've done that before on other projects I've managed. I'd add extra tasks to my schedule to account for things taking a lot longer than we anticipated.

Alex: But you're not doing that now?

Rick: Well, no. The Scrum rules don't really give me a way to add contingency, padding, or extra tasks. I don't really see any way to pad the schedule without breaking the Scrum rules.

Alex: Then maybe there's something wrong with Scrum.

What do you do when one of the tasks in the Sprint takes a lot longer than the team planned for it to take?

you are here ▶ **89**

Question Clinic: The "which-comes-next" question

> A LOT OF PRACTICES OR EVENTS HAPPEN IN A SPECIFIC ORDER, AND YOU'LL BE ASKED ABOUT THAT ORDER IN A "WHICH-COMES-NEXT" QUESTION. THESE ARE QUESTIONS THAT QUIZ YOU ON HOW PRACTICES FIT TOGETHER ON A REAL PROJECT. THESE QUESTIONS AREN'T USUALLY VERY DIFFICULT, BUT THEY CAN BE A LITTLE MISLEADING.

Most which-comes-next questions describe a situation and ask you what you'd do next. Sometimes it's not immediately obvious that it's asking about the order of events. Watch for any question that lay out a scenario and then asks what comes next, what happens afterward, or how the team should proceed.

Don't be thrown if the question asks about an industry you don't know much about.

The key to a "which-comes-next" question is figuring out what the team is currently doing. So what's another word for "objective that will be met"? That's the definition of the Sprint Goal! So this must be happening at the beginning of the Sprint Planning session.

27. You're the Scrum Master for an (automotive) industry software team working on firmware for an antilock brake system. Your team just finished writing down the objective that will be met when the team completes the Sprint items. What is the next thing your team should do?

A. Decompose the Sprint Backlog items into tasks

B. Review the working software with the users

C. Meet with the business users

(D.) Decide what items to include in the Sprint Backlog

This also happens during the Sprint Planning session. But when you compare this with the other answers, there's another one that happens before it.

Neither of these answers are things that happen during the Sprint Planning session.

Aha! If the team is holding a Sprint Planning session and just finished writing the Sprint Goal, the next thing they'll do is decide what items from the Product Backlog will be included in the Sprint Backlog. This is what comes next!

HEAD 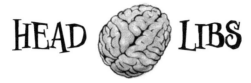 LIBS

Fill in the blanks to come up with your own "which-comes-next" question! **Start** by thinking of a Scrum event or activity to be the correct answer, and **then** figure out exactly what the team just finished doing before it—that's what you'll describe in the question!

You're a Product Owner on a _____ project. Your team just finished
                                    (an industry)

_____. A _____ just informed you that
(description of a Scrum activity)   (a type of user)

your project _____.
              (a problem that came up on the project)

The last part of this question doesn't change the answer at all. A lot of questions are like that.

Which of the following does the team do next?

A. _____
   (the correct answer—description of the Scrum activity, tool, or practice that comes next)

B. _____
   (description of a different Scrum activity, tool, or practice)

C. _____
   (the name of activity, tool, or practice that's part of a different methodology)

D. _____
   (description of one of the Scrum values or roles)

THE WHICH-COMES-NEXT QUESTION DOESN'T ALWAYS LOOK LIKE IT'S ASKING ABOUT THE ORDER OF EVENTS, TOOLS, OR PRACTICES! KEEP AN EYE OUT FOR QUESTIONS THAT DESCRIBE SPECIFIC ARTIFACTS THAT GET CREATED OR ACTIONS YOU'LL PERFORM AND THEN ASK YOU WHAT YOU'RE SUPPOSED TO DO NEXT.

it ain't over *'til it's over*

A task isn't done until it's "Done" done

If you look in the Scrum guide, you'll actually see it refer to a "Done" product increment.

When you're on a Scrum team, you work on one backlog item or task at a time—that's what the value of Focus is all about—and you work on it until it's done. But when, exactly, is it done? Is there still some testing left to do? It's easy to think that you're done... except for this one teeny little thing. That's why Scrum teams have a **definition of "Done"** for every item or feature that they add to the backlog. Before an item can go into the Sprint Backlog, everyone on the team needs to understand and agree about exactly what it means for it to be not just done, but "Done" done. And since every item in the backlog has a definition of "Done," the **entire Increment has a definition of "Done"** and the team is committed to delivering a "Done" Increment at the end of the Sprint.

Under the Hood: Sprint Planning

Sprint Planning relies on the definition of "Done"

During the first part of the Sprint Planning meeting, the team decides which items to include in the Sprint Backlog. But what happens if there's an item that doesn't *really* have a definition of "Done" that everyone on the team understands? Say one person thinks that it means all the code is done, but someone else assumes that it also includes documentation or testing. Even if they don't realize that there's a disagreement about what it means for that item to be done, they'll discover that they can't come to a real agreement on how many other items they can deliver in the Sprint. That's why Sprint Planning only works when every item has a clear definition of "Done" that the whole team can agree on.

Scrum teams adapt to changes throughout the Sprint

Project teams have to make decisions every day: Which features will we build in this Sprint? What order will we build them in? How will users interact with this feature? What technical approach will we take? Traditional waterfall teams have one answer: all of the planning is done at the beginning of the project. The problem is that at the time the plan is being built, most of those questions don't have answers yet. So the project manager works with the team to make assumptions, and relies on a change control process to change the plan when they guessed incorrectly.

Scrum teams *reject the idea* that every project question can be answered at the beginning of the project, or even at the beginning of a Sprint. Instead, they make decisions based on real information as soon as it's discovered. They do this using the **three pillars** of Scrum: a cycle of **transparency**, **inspection**, and **adaptation**:

* The cycle starts with **transparency**, where the team decides together what items to include in the Sprint, and what it means for each of them to be done. All work done by everyone is visible to the team all the time.
* The whole team meets every day in the Daily Scrum to **inspect** each item that's being worked on.
* If they discover changes, they **adapt** to them (like adding or removing items from the Sprint Backlog when roadblocks happen).
* The next day **the cycle begins again**, with team members giving complete transparency into what they're doing at the Daily Scrum. This cycle continues every day until the timebox expires and the Sprint is complete.
* The team **also inspects and adapts at the other Scrum events**—Sprint Planning, the Sprint Review, and the Sprint Retrospective—by reviewing and modifying the Sprint Goal, items, tasks, and the way they work.

---

there are no Dumb Questions

Q: This is starting to sound *really* theoretical. Can you tie it back to the real world?

A: Sure. "Transparency" just means that every single person understands all of the features they're building in the Sprint, and are open about the work they're doing, what they plan to do next, and what problems they've run into. "Inspection" means that they constantly make sure that knowledge is up to date using the Scrum events (*especially* the Daily Scrum). And "adaptation" means that they're constantly finding ways to change what they plan to do next based on this new information.

Q: If the team plans some of the tasks but not all of them, won't that just cause chaos later on in the project?

A: No. Making all of the decisions at the beginning of the project lets you feel like everything's under control. But very often they aren't, leaving you surprised that somehow your project—which seemed fine yesterday—is suddenly late, and now everyone's in crunch mode. That's why so many teams turn to agile: because their traditionally planned projects keep blowing their schedules, and *something* needs to change. Scrum avoids those pitfalls by recognizing that many important decisions depend on information that **won't be known** until partway through the project.

Q: Can you give me an example of how that would work in real life?

A: Here's one that our *Cows Gone Wild* team might face. Brian needs to program the behavior for a new tractor vehicle for the player to use, but he can't start until Amy finishes the basic behavior for it. If Rick used traditional project management, he would either have to assume that Amy will finish first (and add padding to the schedule in case she takes too long), or he'd have Brian work on something else first, even if the other task is less important.

Now that they're using Scrum, the team can give themselves more options. They know Brian can't start the coding for the tractor until Amy finishes designing its behavior. But they also know they'll **constantly inspect their progress and adapt the plan along the way**. So instead of deciding *today* whether Brian work on the tractor or on a less important task, they can add **both** of them to the Sprint Backlog—and they'll **put off the decision** until Brian is ready to start the code. If Amy is done with her work, he can start coding. If she's not, he'll pull the other task off of the Sprint Backlog and come back to the tractor coding when he's done (but <u>only</u> when he's *"Done"* done!).

never do today what you can put off 'til tomorrow

> I STILL DON'T SEE HOW SPRINT PLANNING CAN POSSIBLY WORK. HOW CAN YOU HAVE A TIMEBOXED PLANNING SESSION? YOU SAID THAT THE TIMEBOX OFTEN EXPIRES BEFORE ALL OF THE WORK IS PLANNED. DOESN'T A HALF-BAKED PLAN LEAD TO HALF-FINISHED PROJECTS?

No—because agile teams make decisions as late as possible.

When a lot of people learn about Scrum, they're surprised that the Sprint Planning session is timeboxed to four hours for a two-week Sprint (sized proportionally for different Sprint lengths), because they're accustomed to having every single task in the project completely planned out before any of the work can begin. But we've already seen that teams following a traditional waterfall process often have trouble when their plans change partway through the project. In fact, changes that would deliver more value often get rejected during the change control process simply because replanning a months-long schedule is too much work.

Scrum teams rarely (if ever!) run into this problem because they don't plan every single task at the very beginning of the project. In fact, usually they don't even plan out all of the tasks at the beginning of the Sprint. Instead of planning everything up front, they make decisions at the **last responsible moment**. What that means is that they only need to do the planning that's absolutely necessary to get started with the Sprint. If more planning needs to be done, it can be done later in the Sprint.

This is a new way of thinking about planning for a lot of teams. Luckily, we have the Agile Manifesto to help our teams get into a mindset that's really effective.

Sharpen your pencil — MINI

The Agile Manifesto is really useful because it helps get into a mindset where concepts like the last responsible moment *really make sense*. One of the twelve Agile Manifesto principles is especially helpful for understanding the last responsible moment. Write down which one you think it is—you can see our answer on page 98.

_____

_____

managing projects with scrum

The Daily Scrum Way Up Close

The "ceremony"

Even though a lot of teams refer to answering the Daily Scrum questions as a "ceremony," every single person is engaged and paying attention.

The whole team gets together at the same time every day, and most teams start with a different person each time. Everyone (including the Product Owner and Scrum Master) answers three questions:

★ What did I do **yesterday** to move us toward the Sprint Goal?

★ What will I do **today** to help meet the Sprint Goal?

★ Are there any **impediments** that prevent me from meeting the Sprint Goal?

Each person's answers are brief and to the point because the meeting is timeboxed to 15 minutes.

Inspect and adapt

The reason that each person answers those three questions is to give complete **transparency** into what he or she is doing. But this only works if every other person on the team is listening carefully (which is why it's important to keep the updates brief!). One of the most common things that happens during a Daily Scrum is that one team member realizes his or her teammate is going to do something that doesn't quite make sense: maybe they're starting one Sprint Backlog item when a different one is more valuable, or taking one approach when there's a better alternative, or has hit a roadblock that other team members can help with.

When this happens, they'll set up a meeting later in the day to talk about it. More often than not, they'll have a discussion that causes them to change the plan: they may take a different approach, or choose another item from the Sprint Backlog, or have to add more work (and possibly remove a different item from the Sprint Backlog) to get around the roadblock. This is how the team **adapts** to the change.

> THIS IS THE THREE PILLARS AGAN, RIGHT? ANSWERING THE QUESTIONS IS THE *INSPECTION* PART, AND WHEN THEY MAKE CHANGES THAT'S THE *ADAPTATION* PART.

That's right. And it all works because of transparency.

Each team member answers the questions every day, so everyone has current, accurate information about the project. Not to get too theoretical, but this is actually a really good example of **empirical process control theory**, which tells us that when a process (in this case, an agile approach) is based on empiricism, it lets the team optimize the way they work to reduce risk and give them sustainable, predictable results.

DICTIONARY DEFINITION

em-pir-i-cism, noun

the theory that knowledge comes from experience, and that decisions are to be made based on information that is known

The team was driven by **empiricism**, *and rejected the project plan based on guesswork.*

trust the team to self-organize

The Agile Manifesto helps you really "get" Scrum

We learned back in Chapter 2 that the most effective way to approach any agile methodology, framework, method, or approach—like Scrum, for instance—is with a mindset that's driven by the values and principles of the Agile Manifesto. Let's take a closer look at **three of those principles** that are especially helpful for Scrum teams.

> THERE'S NO WAY I CAN *ACCEPT* THE HAY THRESHER BATTLE AS *DONE* UNTIL YOU GUYS GET THAT MILK GUN WORKING.

Our highest priority is to satisfy the customer through early and continuous delivery of valuable software.

A really important word in this principle is "valuable." Everyone on the team takes it really seriously, and does their best to deliver the most value that they can.

The Product Owner makes sure the team delivers value

That's the whole reason that the team needs a Product Owner with the authority to accept items in the Sprint Backlog as Done—or, if they're not "Done" done, refuse to accept them. If everyone on the *Cows Gone Wild 5* team really "gets" this principle, then they won't be mad at Alex for not accepting this feature as done yet, because they genuinely want to deliver the most **value** that they can. They want to deliver it **early** and **continuously**, and the best way to do that is to finish the current item (which means it's "Done" and ready to demo to the users in the Sprint Review) so that they can move on to the next one and get as many backlog items done in the Sprint as they can.

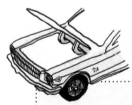

Under the Hood: The Sprint Review

The Sprint Review is all about maximizing value

During the Sprint Review, the team works with key users and stakeholders invited by the Product Owner to inspect the Increment and the Product Backlog, and everyone understands the goal is to maximize value by reviewing the value delivered during the Sprint, and maximizing the value of the items for the next Sprint. Here's how they do it:

- The Product Owner goes over what was "Done" in the Sprint and the team demonstrates the working software.

- The team talks about what went well and what could be improved, and answers user and stakeholder questions.

- The Product Owner walks everyone through the current Product Backlog, and everyone collaborates on what they think should probably go into the next Sprint, so the team learns directly from the users and stakeholders **what they think is going to be most valuable**. This is really important for planning the next Sprint.

- They'll have an open and honest discussion of anything that's changed in the marketplace (in case it changes the most valuable thing to do next), the company's timeline and budget, and anything else that's relevant.

managing projects with scrum

Here's Amy's update at the Daily Serum.

> I FINISHED THE CHICKEN COOP AIR ASSAULT ARTWORK, SO THAT'S ANOTHER SPRINT BACKLOG ITEM DONE. THERE ARE A FEW ITEMS I CAN CHOOSE TO DO NEXT. I WAS THINKING I'D WORK ON THE HORSE CORRAL BATTLE LEVEL DESIGN.

> I'M NOT SURE THAT'S THE RIGHT THING TO WORK ON NEXT. CAN WE MEET RIGHT AFTER THIS DAILY SCRUM IS OVER AND TALK IT THROUGH?

Sounds like Amy may be going down the wrong track. If Brian hadn't been paying attention, he might not have caught this the potential problem. Now they can figure it out together.

> The best architectures, requirements, and designs emerge from self-organizing teams.

Self-organizing means <u>deciding as a team</u> what to work on next

If you've worked on a traditional waterfall team with a project manager, then you've probably worked on tasks that were part of a project plan created by a project manager and assigned to you by either the project manager or your boss. But that's <u>not</u> how Scrum teams work. Scrum teams are **self-organizing**, which is a new way to work for a lot of people.

The whole Scrum team works together to plan the project. There isn't a single person building a plan and telling them what to do. The Development Team members decide what will be delivered, adding new work to the Sprint Backlog as needed. The whole team decides together how they'll meet those goals.

But self-organization doesn't just happen during Sprint Planning. They ***constantly adapt their plan*** by checking in with each other at the Daily Scrum: each person tells the whole team what they're planning on doing next to meet the goals they decided on during Sprint Planning. If a teammate sees a problem with the approach, they'll work together that day to fix it.

> The Daily Scrum is timeboxed, so when two team members discover an issue like this, they'll meet afterward to work it out (inviting other team members to join if they have input). Brian and Amy will meet and talk about how to handle this situation. Once they've got a new plan, they'll review it at the next Daily Scrum to make sure everyone is on board, and to see if anyone has anything to add.

What did you get for your **MINI Sharpen your pencil** answer on page 94? Flip the page to find out ours... ➡

decomposing zombies?

MINI Sharpen your pencil Solution

Did you come up with a different answer? This is the agile principle we think is most helpful for understanding the last responsible moment, but other principles can shed light on it too!

> Welcome changing requirements, even late in development. Agile processes harness change for the customer's competitive advantage.

Making decisions at the last responsible moment probably sounds really weird if you've never done it before. But to someone who <u>genuinely</u> *welcomes changing requirements*, it really does feel as normal as riding a bike.

This is a new way of thinking about planning for a lot of teams.

That's part of the mindset shift that we talked about in the last chapter. Here's how it works on a Scrum project in the real world:

★ There's **just enough** written down about each item in the Product Backlog to be able to start Sprint Planning.

★ During Sprint Planning, the team decomposes enough of the Sprint Backlog to get everyone started, but they **don't feel like they need to decompose everything**. (Have a look at the Sprint Planning section in the Scrum Guide, which says the team decomposes *work planned for the first days of the Sprint* by the end of the meeting.)

★ Self-organizing teams don't need to decide the exact order of every single task in painstaking detail when they're planning a Sprint. They **trust themselves** to make good decisions when the time comes.

★ As the team works through the Sprint Backlog over the course of the Sprint, they discover new tasks and changes and bring them up in the Daily Scrum, and they use that information to work together to *create a plan* **for the next 24 hours**.

★ They're **constantly inspecting and adapting**, so they trust themselves to make good decisions in the future.

Scrum teams make decisions at the last responsible moment. They only make project decisions that need to be made now, and leave the rest for later.

> THE MAD COW ZOMBIE LEVEL WE BUILT JUST ISN'T AS FUN AS WE THOUGHT IT WOULD BE, AND GAMERS WON'T LIKE IT. LET'S PUSH IT BACK TO THE PRODUCT BACKLOG. MAYBE WE'LL PUT IT OUT IT AS DOWNLOADABLE CONTENT AFTER THE GAME'S RELEASED.

It's a good thing Alex is comfortable making decisions at the last responsible moment! In this case, the only way the team could find out the feature wouldn't be fun (and deliver the value the project needs from it) was to build it. Notice how he's staying positive, not blaming anyone for wasting effort, and keeping the option open to release it later.

managing projects **with scrum**

Watch it! **The Daily Scrum is timeboxed to 15 minutes, so make sure everyone keeps their updates focused and to the point.**

That's easier said than done. Once you and your team start holding a Daily Scrum every day, you'll discover that some people are really uncomfortable talking about their work in front of everyone else on the team, while others just can't seem to stop talking, and will take up the entire 15 minute timebox if you let them.

This is why it's really important that the Scrum Master takes his or her role seriously—especially the responsibility of making sure that the team understands and adheres to the Scrum rules:

- *If someone is reluctant to speak up, the Scrum Master can help that team member to accept the Scrum value of openness and understand how transparency is critical for making Scrum work.*
- *When team members give updates that take up too much of the Daily Scrum, the Scrum Master can show them which facts are relevant and help them manage their Daily Scrum time better.*
- *If a team member gives an update that goes beyond just answering the three questions and raises issues for discussion, the Scrum Master can remind the team to set up a separate meeting for some of the team members to discuss the issue and report back to the whole group.*

BULLET POINTS

- The team agrees on the **definition of "Done"** for every item in the Sprint Backlog and for the entire increment.
- The Product Owner doesn't accept an item to include in the Sprint Review until it's **"Done"** (i.e., it meets the definition of "Done" that the team decided on for it).
- Scrum uses **empirical process control**, based on the **three pillars** of *transparency, inspection, and adaptation* to make decisions based on real facts.
- **Transparency** (or visibility) means everyone on the team understands what their teammates are doing.
- **Inspection** means they frequently check with each other on what they're building and how they're building it at the Daily Scrum and the other Sprint events.

- The team constantly **adapts** their plan based on what they learn from that inspection.
- Agile teams make decisions at the **last responsible moment**, and plan only those tasks that absolutely need to be planned right now.
- The team delivers value **early** (by working on each backlog item until it's done) and **continuously** (by delivering a complete Increment that's "Done" at each Sprint Review).
- Scrum teams are **self-organizing**: they decide as a team how to meet their goals and who will do the work.
- Scrum teams **welcome changing requirements** more easily than traditional waterfall teams because they're self-organizing and make decisions as late as possible.

"customizing" scrum rarely ends well

Who am I?

A bunch of Scrum artifacts, events, and roles are playing a party game, "Who am I?" They'll give you a clue, and you try to guess who they are based on what they say. Write down its name, and what kind of thing it is (like whether it's an event, role, etc.).

And watch out—another Scrum concept that's not an event, artifact, or role might just show up and crash the party!

Name   **Kind of thing**

I'm a servant leader who guides the team in understanding and implementing Scrum, and helps people outside of the team grasp it.

_____   _____

I'm held at the end of the Sprint to do an inspection of each item that the team built with users and stakeholders who were invited.

_____   _____

I'm how the team inspects itself, where they look for the things that went well, and make a plan to improve things that didn't.

_____   _____

I'm the sum of all items that the team delivers to the users at the end of the Sprint, and I can only be delivered if every item in me is "Done."

_____   _____

I'm the group of professionals who actually do all of the work that's needed to deliver the software to the users and stakeholders.

_____   _____

I'm responsible for deciding what items will go into the product, and I have the authority to accept them as "Done" on behalf of the company.

_____   _____

I'm a 15-minute timeboxed meeting held every day, where team members create a plan for the next 24 hours.

_____   _____

I'm what the Product Owner helps the team optimize and maximize, and the team tries to prioritize the items that have the most of me.

_____   _____

I'm the set of items the team will build during the Sprint, along with a plan to build them (usually a set of tasks the items are decomposed into).

_____   _____

I'm a timeboxed meeting where the team comes up with a Sprint Goal, decides which items they'll deliver, and decomposes them into tasks.

_____   _____

I'm an ordered list of all of the items (with descriptions, estimates, and values) that might be needed in the product at some time in the future.

_____   _____

Answers on page 114

there are no Dumb Questions

Q: Is the *Cows Gone Wild* team's story realistic? Can you really use Scrum for something like a video game, which requires a lot of creativity and on-the-fly—and sometimes last-minute—changes with heavy deadline pressure?

A: Not only can Scrum be used for a complex and dynamic project that constantly changes, it's actually better suited to an environment like that than a traditional waterfall process. Scrum teams constantly look for changes and find ways to adapt to them, which makes them much better at dealing with complexity and even chaos. And the timeboxed nature of Sprints helps the team meet their deadlines. Alex just showed us a good example of the kind of decision that video game teams make in real life. The team built a feature but it turned out not to be fun enough in its current form, so they shelved it for now and might revisit it to sell later as downloadable content. Scrum gave the team the flexibility to handle this on the fly, while a traditional waterfall team would probably have to go through a lengthy change control process. More importantly, a Scrum team *sees this change as a victory* because they'll welcome any change that delivers more value to the product. A traditional waterfall team is likely to see it as a defeat because it "wasted" effort and required a change to the plan.

Q: Does "self-organizing team" mean that there's no boss?

A: Of course there's a boss. If you work in a company and you're not the CEO, then you have a boss. But an effective self-organizing team typically has a manager who doesn't micromanage, and who trusts all of his or her employees to deliver the most valuable software they can. Self-organizing teams are given the authority to decide which features to include in the software, usually by having a Product Owner assigned to the team who's senior enough to make those decisions. They're given the freedom to plan the work so that they can build those features in the way they feel is most effective. And they're given the flexibility to make decisions at the last responsible moment, because that's the most effective time to make important project decisions.

Q: What exactly happens during the Sprint Retrospective?

A: The Sprint Retrospective is how the team inspects the Sprint that just ended and tries to find ways that they can improve. They look at all sorts of things: the people on the team can improve the processes and tools that they use to do the job, find ways to improve the quality of the software they're building, work on their relationships with others in the organization, and do anything else that might have an impact on the work—especially anything that can make their work more enjoyable or effective. By the time the Sprint Retrospective ends, the team has put together a plan for improvement. This plan typically consists of a small number of discrete and specific tasks that individual people on the team will carry out. Before the meeting, the Scrum Master helps everyone understand how it works, and makes sure that they all respect the timebox. This happens *before* the meeting because the Scrum Master and Product Owner have to participate in the retrospective as team members, offering their own opinions and ideas.

Q: Hold on—the Product Owner goes to the retrospective too? Does the Product Owner really need to be at *all* of the Scrum events?

A: Absolutely. The Product Owner is a real member of the team, and participates in the Scrum events just like everyone else. In fact, on a lot of Scrum teams the Product Owner does his or her share of the development work. But even in that case, that person still has the authority to make decisions about what items will go into the backlog and how to maximize the value of what the team delivers, and the company still respects his or her decisions.

Q: I'm having trouble finding anyone who has the clout to be the Product Owner but also has enough time to attend all of the Scrum events. Can the Product Owner be a committee instead?

A: Absolutely not. The Product Owner definitely **must** be a person. That person **must** have the authority to make the decisions about what goes into the software and what doesn't, and *the Product Owner role must be his or her top priority*.

Modifying Scrum like this almost always renders it *much less effective*, usually by removing the parts that make its empirical process control work. Teams often try to "customize" Scrum by bending or breaking any rules that highlight a serious flaw in the way their team is structured. In this case, Scrum makes it really obvious that the team doesn't have the authority to decide what features go into the software. It's **scary** for a manager to say, "The team should build this feature, but not that one." The wrong decision will cost the company a lot of money, and *there will be a lot of blame* if it goes wrong. That's why Scrum requires the team to have a Product Owner who has the authority to make that call.

> **Teams often try to "customize" Scrum when they don't have a true Product Owner who can make tough decisions about what to build.**

demo went great we're halfway there

Things are looking good for the team

Cows Gone Wild 5 is this year's most anticipated video game release! Now they just have to get it out the door. (Easier said than done?)

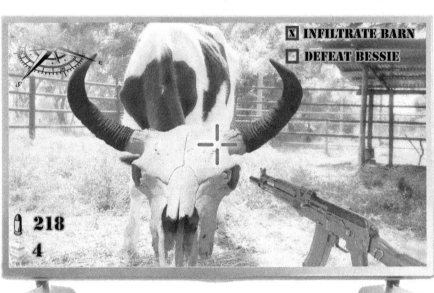

GREAT NEWS! I DEMOED THE HAY THRESHER LEVEL AT THE WISCONSIN GAME CONFERENCE LAST WEEK AND **WE TOTALLY KILLED IT!**

YEAH! I WAS JUST READING SOME GREAT REVIEWS ON THE TOP TWO GAMER BLOGS FROM PEOPLE WHO ATTENDED. LOOKS LIKE CGW5 IS THE **MOST ANTICIPATED GAME OF THE YEAR!**

managing projects with scrum

Scrumcross

Here's a great opportunity to seal the Scrum concepts, values, and ideas into your brain. See how many answers you can get without flipping back to the rest of the chapter.

Across

1. The Sprint, Sprint Planning, the Daily Scrum, the Sprint Review, and the Sprint Retrospective
4. The Product Owner is allowed to _____ the Sprint, but it wastes the team's energy and damages the company's trust in the team
8. The most effective time to make a decision is at the last _____ moment
10. The Product Owner makes sure the team maximizes this
12. The Product Backlog, the Sprint Backlog, and the Increment
13. When this value isn't part of your mindset, you don't trust your teammates, and tend to blame them when things go wrong
16. What the team does to turn the Sprint Backlog items into tasks
17. Product Owner, Scrum Master, and Development Team
18. When this is part of your mindset, you don't even consider trying to do two things at once
20. What Product Owners routinely do to keep the Product Backlog current
21. Responsible for maximizing the value of the product and managing the Product Backlog
23. During Sprint _____ the team chooses what items to include in the Sprint and builds a plan
24. The three Daily Scrum questions give the team complete _____ into what each team member is doing
25. The _____ Backlog is the single source of requirements and changes for the product
26. The timeboxed iteration used in Scrum
27. What's missing from the mindset when a team member is uncomfortable bringing up a roadblock or problem during the Daily Scrum

Down

2. The Sprint is over when the timebox _____
3. The team decides what can be done in the Sprin during the _____ part of the Sprint Planning session
5. Each item in the backlog has a description, the business value, an order, and a rough _____
6. The Sprint _____ contains items the team will build during the Sprint
7. The Sprint _____ is an objective crafted by the team when they plan the Sprint
8. The Sprint _____ is how the team inspects itself and creates a plan to improve
9. The theory that knowledge comes from experience
11. A self _____ team decides as a team how they'll meet their goals
13. The Sprint _____ is how the team inspects what they built and adapts the Product Backlog
14. What pigs have, chickens don't, and everyone on a Scrum team feels collectively toward the whole project
15. The Sprint Review is timeboxed to _____ hours for a 30-day sprint
16. What teams do with working software at the Sprint Review
19. The team decides how the work will be done during the _____ part of the Sprint Planning session
22. The team inspects and _____ at each Scrum event by reviewing and modifying artifacts and the way they work

→ Answers on page 115

Exam Questions

> These practice exam questions will help you review the material in this chapter. You should still try answering them even if you're not using this book to prepare for the PMI-ACP certification. It's a great way to figure out what you do and don't know, which helps get the material into your brain more quickly.

1. The Scrum Master is responsible for all of the following except:

    A. Helping the team understand what goes on during the Daily Scrum
    B. Giving the Product Owner guidance in effectively managing the Product Backlog
    C. Helping the team understand customer requirements
    D. Giving the rest of the organization guidance on understanding Scrum and working with the team

2. Which of the following is NOT an attribute of a Product Backlog item?

    A. Status
    B. Value
    C. Estimate
    D. Order

3. Juliette is a Product Owner on a Scrum project in a healthcare organization. She was called into a meeting with a steering committee made up of her company's senior managers because she decided to include a planned health privacy feature in the most recent Sprint. At the meeting, the senior managers told her in the future that she must consult with the whole committee before making business decisions like that.

    Which of the following BEST describes Juliette's role?

    A. She is in a servant-leadership role
    B. She is not committed to the project
    C. She needs to concentrate on focus and courage
    D. She does not have the authority to adequately fill her Product Owner role

4. When is the Increment considered done?

    A. When the timebox expires
    B. When every item to be delivered meets its definition of "Done" and the Product Owner accepts it
    C. When the team holds the Sprint Review and demonstrates it to the users and stakeholders
    D. When the team holds the Sprint Retrospective

Exam Questions

5. Which of the following is an example of collective commitment?

A. Everyone on the team feels personally responsible for delivering the entire Increment, not just their individual parts of it

B. Everyone on the team always stays late and often works weekends

C. Everyone on the team is responsible for delivering an important part of the project

D. Everyone on the team participates in the Sprint Planning and retrospective meetings

6. Which of the following is NOT a Scrum event?

A. Sprint Review

B. Product Backlog

C. Retrospective

D. Daily Scrum

7. Amina is a Scrum Master on a team that is working on adopting Scrum. She wants to make a change to help her team get better at self-organizing. Which of the following is the best area to focus their improvement effort?

A. Daily Scrum

B. Sprint Planning

C. Sprint Retrospective

D. Product Backlog

8. When is a Scrum Sprint over?

A. When the team finishes the work

B. When the team completes the Sprint Retrospective

C. When the timebox expires

D. When the team completes the Sprint Review

Exam Questions

9. Each person on the team answers all of the following questions during the Daily Scrum except:

    A. What roadblocks are in my way?
    B. What planned work did I fail to accomplish?
    C. What will I do between now and the next Daily Scrum to meet the Sprint Goal?
    D. What have I done since the last Daily Scrum?

10. Barry is a developer at an online retailer. His project manager told him the deadline for the current feature that he's working on is three weeks from now, even though Barry made it clear that he would need four weeks, and there were no specific deadlines or external pressures that require it to be done earlier than that. Barry's team is starting to adopt Scrum. Which of the Scrum values will make the team's Scrum adoption difficult or less effective?

    A. Openness
    B. Respect
    C. Courage
    D. Focus

11. Sandeep is a product owner on a Scrum team working on a telecommunications project. The business users let him know about a major regulatory change in one of his regular meetings with them. Handling this regulatory change is now a very high priority for the team, and will need to be the main objective of the next Sprint.

    Which of the following is used to describe the main objective of the next Sprint?

    A. The Increment
    B. The Sprint Backlog
    C. The Sprint Goal
    D. The Sprint plan

12. What aspect of empirical process control theory involves frequently examining the different Scrum artifacts and making sure the team is still on track to meet the current goal?

    A. Examination
    B. Adaptation
    C. Transparency
    D. Inspection

Exam Questions

13. What is an Increment in Scrum?

    A. The items from the Sprint Backlog that the team actually completes during the Sprint

    B. The items from the Product Backlog that the team plans to complete during the Sprint

    C. The result of decomposing the Sprint Backlog items

    D. A statement that describes the objective of the Sprint

14. Which of the following helps Scrum teams focus?

    A. Multitasking

    B. Holding a Daily Scrum

    C. Writing a Sprint Goal

    D. Holding a retrospective

15. Danielle is a Product Owner on a Scrum team. She's talking to one of her business users, who gives her a new requirement. Which of the following should Danielle do next?

    A. Update the Product Backlog

    B. Hold a Sprint Planning session

    C. Update the Sprint Backlog

    D. Bring up the new requirement at the next Daily Scrum

16. Which of the following BEST describes how the team determines what specific work will be needed to complete Sprint Backlog items?

    A. The Product Owner works with the business users to determine which items go into the Product Backlog

    B. The team decomposes Sprint Backlog items into tasks

    C. The team chooses which Product Backlog items to include in the Sprint Backlog

    D. The team decides on each Sprint Backlog item's definition of "Done"

17. Which of the following does NOT take place during a Sprint Review?

    A. The Product Backlog is updated to reflect what will probably be in the next Sprint

    B. The team collaborates with business users on what they will work on next

    C. The working software the team built during the Sprint is demonstrated

    D. The team looks back at the Sprint and creates a plan to improve

exam answers

~~Exam Questions~~ Answers

> Here are the answers to this chapter's practice exam questions. How many did you get right? If you got one wrong, that's OK—it's worth taking the time to flip back and re-read the relevant part of the chapter so that you understand what's going on.

1. Answer: C

It's the Product Owner's job to help the team understand customer requirements, not the Scrum Master's. The other three answers are good examples of the Scrum Master's servant leadership role.

2. Answer: A

The Product Backlog does not contain any information about the status of a task. This makes sense—none of the items on in the Product Backlog are currently being worked on, so they all have the same status of not having been started yet.

Take a minute and read through the description of the Product Backlog in the Scrum Guide. It says that Product Backlog items have the attributes of a description, order, estimate, and value.

3. Answer: D

One of the most common problems that Scrum teams run into is that the person in the Product Owner role does not have the authority to decide on behalf of the company what features the team will build during the Sprint, or accept them as done on behalf of the company.

4. Answer: B

The Increment is done when every item to be delivered by the team meets its definition of "Done" and the Product Owner accepts it. Every item in the Sprint Backlog has a definition of "Done" that the team uses to determine when it's ready to release to the users. The Product Owner can only accept an item on behalf of the company if it meets its definition of "Done"—any item that isn't "Done" done when the timebox expires must be pushed to the next Sprint.

5. Answer: A

Collective commitment means that everyone on the team feels a personal sense of responsibility to deliver not just the piece that he or she is working on, but to do what it takes to help the team deliver the whole Increment at each Sprint.

Just because everyone on a team works long hours, that doesn't mean they feel a genuine commitment to it. In fact, they might resent the project and the organization for interfering with their lives, and only work the extra hours out of pressure to keep their jobs.

Exam Questions

6. Answer: B

The Product Backlog is a Scrum artifact, not an event.

7. Answer: A

Self-organizing teams take responsibility for their own planning to meet their objectives, assign work themselves (rather than depending on a single manager or project manager to make those assignments), and fix problems with the plan as they come up. Of all of the practlices listed as answers, the Daily Scrum is the only one that impacts the way the team plans their work and executes that plan.

One reason the Daily Scrum is so important is that it's part of the transparency-inspection-adaptation cycle. The team inspects the plan every day, and adjusts it as they uncover new information about the project.

8. Answer: C

The Sprint is over when the timebox expires. The same is true of any timeboxed event. Answer D seems correct because the Sprint Retrospective is typically the last thing that the team does during the Sprint. But if the timebox expires before the team has a chance to hold their retrospective, the Sprint still ends. (And that's a good opportunity for the Scrum Master to help them understand how to plan better next time.)

The Product Owner is allowed to cancel the Sprint before the timebox is over. But this can waste a lot of the team's energy, and cause people in the company to lose trust in the team, so it should be extremely rare.

9. Answer: B

The purpose of the Daily Scrum questions is to give everyone a good idea of how each person on the team is progressing, so they can help identify problems with the current plan that need to be fixed. But none of the questions are about failures—that could create a negative and possibly embarrassing environment, and detract from the environment of openness.

10. Answer: B

The project manager is having trouble with the Scrum value of respect. Barry gave an honest assessment of the work that needed to be done, but the project manager ignored it and demanded a shorter deadline even though there was no business need to apply the extra pressure. That's disrespectful.

It's also extremely demotivating!

11. Answer: C

When the team holds their Sprint Planning meeting at the start of the Sprint, the first thing they do is to decide on the Sprint Goal, a brief description of the objective of the Sprint that will be met by completing backlog items.

Exam ~~Questions~~ Answers

12. Answer: D

The core of empirical process control theory—the theoretical underpinning for Scrum—is the "three pillars" cycle of transparency, inspection, and adaptation. In the inspection step, the Scrum team members frequently examine the Scrum artifacts, as well as their current progress toward the Sprint Goal. They try to detect any differences between where they are versus where they expected to be, so that they can take action (which is what adaptation is all about).

13. Answer: A

The Increment is what the team actually delivered during the Sprint. The items that the team intended to complete at the beginning of the Sprint often don't exactly match up with the work they actually did. That's a good thing—it means the team used the information they learned along the way to change direction. The Increment is the product of what actually happened, and the team doesn't know exactly what the current Sprint's Increment will contain until they've delivered it.

We learned earlier that Scrum takes an incremental approach. It's the delivery of successive Increments that makes it incremental.

14. Answer: B

The Sprint Goal helps the team focus on the specific objective that they planned on accomplishing during the Sprint.

Retrospectives and Daily Scrums can be very useful, but holding meetings is not typically a tool that teams use to help focus.

15. Answer: A

The Product Backlog is the single source for product requirements, and it's maintained by the Product Owner. When the Product Owner discovers a new requirement, she adds it to the Product Backlog.

Exam Questions Answers

16. Answer: B

Scrum teams determine what work will be done during the Sprint to complete Sprint Backlog items by decomposing them into tasks. The other answers are also things that the team does during Sprint Planning, but it's not how the team determines what work they're going to do.

17. Answer: D

During the Sprint Review, the team meets with the business users and customers to review what they've done and collaborate on what the next Sprint will accomplish. They'll review the Increment, which typically involves demonstrating the working software that they built. They'll also discuss the backlog and update it to show the items that they'll probably work on in the next Sprint. The Sprint Review isn't for looking back at what happened and making improvements—that's what the Sprint Retrospective is for.

The updated backlog only reflects the probable items that will be worked on during the next Sprint. That's not the same thing as committing to build certain items—the team will come up with the Sprint Backlog during Sprint Planning, and the Product Owner can make changes to it during the Sprint.

> DID YOU GET SOME OF THE QUESTIONS WRONG? THAT'S **ABSOLUTELY OK!** JUST KEEP TRACK OF THEM, AND MAKE SURE TO GO BACK AND RE-READ THE PARTS OF THE CHAPTER THAT COVERED THEM.

When you get a question wrong now, that actually makes it more likely that you'll get a question on the same topic right when you take the exam!

exercise solutions

Sharpen your pencil Solution

Here are the five Scrum events, three Scrum roles, and three Scrum artifacts. Remember, each event is timeboxed, but the length of the timebox changes proportionally if the team uses a shorter Sprint.

Event names in the order they occur	When the events happen	Length of the event's timebox
Sprint	throughout the project	Assume that the Sprint is timeboxed to 30 days
Sprint Planning	beginning of the Sprint	8 hours
Daily Scrum	every day	15 minutes
Sprint Review	end of the Sprint	4 hours
retrospective	after the Sprint Review	3 hours

Write down the Scrum roles

- Scrum Master
- Product Owner
- Development Team

Write down the Scrum artifacts

- Sprint Backlog
- Product Backlog
- Increment

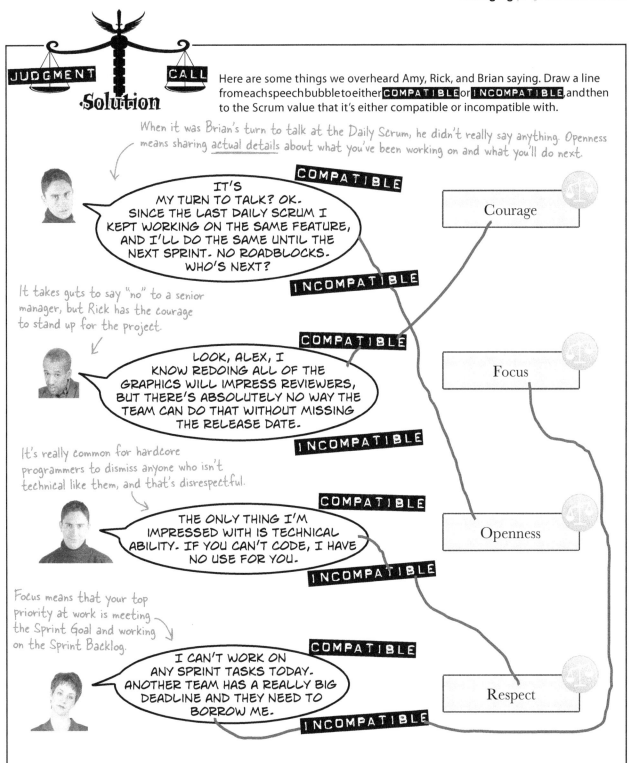

exercise solutions

Who am I?
Solution

A bunch of Scrum artifacts, events, and roles are playing a party game, "Who am I?" They'll give you a clue, and you try to guess who they are based on what they say. Write down its name, and what kind of thing it is (like whether it's an event, role, etc.).

And watch out—another Scrum concept that's not an event, artifact, or role might just show up and crash the party!

Clue	Name	Kind of thing
I'm a servant leader who guides the team in understanding and implementing Scrum, and helps people outside of the team grasp it.	Scrum Master	role
I'm held at the end of the Sprint to do an inspection with users and stakeholders who were invited of each item that the team built.	Sprint Review	event
I'm how the team inspects itself, where they look for the things that went well, and make a plan to improve things that didn't.	Sprint Retrospective	event
I'm the sum of all items that the team delivers to the users at the end of the Sprint, and I can only be delivered if every item in me is "Done."	Increment	artifact
I'm the group of professionals who actually do all of the work that's needed to deliver the software to the users and stakeholders.	Development Team	role
I'm responsible for deciding what items will go into the product, and I have the authority to accept them as "Done" on behalf of the company.	Product Owner	role
I'm a 15-minute timeboxed meeting held every day, where team members create a plan for the next 24 hours.	Daily Scrum	event
I'm what the Product Owner helps the team optimize and maximize, and the team tries to prioritize the items that have the most of me.	value	concept
I'm the set of items the team will build during the Sprint, along with a plan to build them (usually a set of tasks the items are decomposed into).	Sprint Backlog	artifact
I'm a timeboxed meeting where the team comes up with a Sprint Goal, decides which items they'll deliver, and decomposes them into tasks.	Sprint Planning	event
I'm an ordered list of all of the items (with descriptions, estimates, and value) that might be needed in the product at some time in the future.	Product Backlog	artifact

managing projects with scrum

Scrumcross
SOLUTION

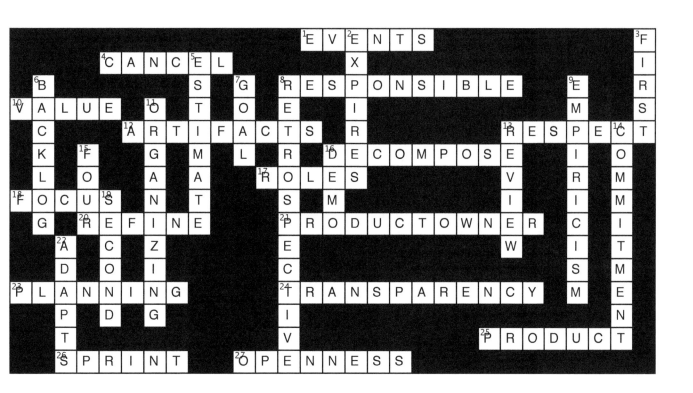

this page *intentionally left blank*

4 Agile Planning and Estimation
Generally Accepted Scrum Practices

Agile teams use straightforward planning tools to get a handle on their projects. Scrum teams plan their projects together so that everybody on the team commits to each sprint's goal. To maintain the team's **collective commitment**, planning, estimating, and tracking need to be simple and easy for the whole team to do as a group. From **user stories** and **planning poker** to **velocity** and **burndown charts**, Scrum teams always know what they've done and what's left to do. Get ready to learn the tools that keep Scrum teams informed and in control of what they build!

victims of their own success

Meanwhile, back at the ranch...

The demo of *CGW5* was the most exciting thing at the Wisconsin Game conference. But is the team a victim of their own success? Now it seems like everyone in the gaming world expects it to be the most innovative and fun game of the year. That's a lot of pressure on the team!

THIS MORNING, WHEN I WENT TO GET COFFEE, A TOTAL STRANGER SAW THE COMPANY LOGO ON MY MUG AND STARTED ASKING ME WHEN THIS GAME WAS COMING OUT. ALL WE HAVE IS A 15-MINUTE DEMO!

LOOK, THIS IS *A GOOD PROBLEM TO HAVE.* WE'RE DEFINITELY ON THE RIGHT TRACK.

I'M GLAD THEY'RE EXCITED TOO. BUT HOW ARE WE GOING TO DO THIS?

generally accepted scrum practices

Exercise

How would you solve these problems that have the *CGW5* team concerned about meeting their users' expectations? Get creative with this!

Just write down a short sentence for each of these.

1. When they started out, the team thought they'd market the product as a kids' game, but some of the content is kind of violent and the gamers at the conference like the more mature focus.

....

The demo looks really cool, but it's not exactly integrated with the CGW5 code base just yet.

2. A lot of the features in the demo are limited. To make them work for a full-length game, the team will have to go back and make changes to some pretty old parts of the code.

....

3. A developer had an idea to add a mini-game as a downloadable add-on. But that feature looks like it's a lot more work than the team initially thought. Is it worth building that downloadable content if it means losing one of the team's best coders for most of the project?

....

4. The biggest complaint from gamers about the demo is that the player had to stop running in order to change weapons while they were fighting a big battle. That caused them to die all the time and made the game less fun.

....

you are here ▶ **119**

looks like the team has some work to do

Exercise Solution

Here are a few ways that we came up with for the team to handle these situations. Did you come up with different answers? Changes happen all the time in agile teams—the way that you and your team handle them can make the difference between success and failure.

1. When they started out, the team thought they'd market the product as a kids' game, but some of the content is kind of violent and the gamers at the conference like the more mature focus.

   Brainstorm real-world users who will use the game and target features to them

   > You can't build a product that suits your users' needs if you don't know who they are.

2. A lot of the features in the demo are limited. To make them work for a full-length game, the team will have to go back and make changes to some pretty old parts of the code.

   Add this work to the product backlog and try to do it early in the project

   > If the team knows there's work to do that will affect everything else, it's best to do it early in the project.

3. A developer had an idea to add a mini-game as a downloadable add-on. But that feature looks like it's a lot more work than the team initially thought. Is it worth building that downloadable content if it means losing one of the team's best coders for most of the project?

   The Product Owner figures out if this feature is valuable and prioritizes it in the backlog

   > Since the Product Owner knows what the customer wants, he needs to make sure that features like this get the right priority.

4. The biggest complaint from gamers about the demo is that the player had to stop running in order to change weapons while they were fighting a big battle. That caused them to die all the time and made the game less fun.

   Meet with the team and talk about how users will want to play the game

   > This is the kind of discussion that happens when the Scrum Team meets with the important stakeholders at the end of each sprint during the sprint review.

   > Writing down the features they need to build from a user perspective will help the team get it right the first time.

So... what's next?

The team had enough backlogged requests to build out a crowd-pleasing demo at the beginning of the project. But now they've got to make sure that the full-length game will make gamers as happy as the demo did.

WE'RE FOLLOWING ALL OF THE RULES OF SCRUM, BUT I'M NOT SURE WE'VE GOT A HANDLE ON THIS PROJECT JUST YET. THERE HAVE TO BE TOOLS TO HELP US MAKE SURE WE'RE PLANNING AND BUILDING **THE BEST PRODUCT WE CAN.**

> The *CGW* team needs to find a way to plan and track their project. Can you think of tools you've used on your own projects that might help? How effective would they be for working on an agile team?

gasps and scrum *perfect together*

Introducing GASPs!

When Scrum teams start working on a sprint, they use tools that include the whole team in setting goals and tracking them. While these practices are not part of the core Scrum rules, they've been used by many Scrum teams to plan work and keep everybody on the same page. That's where the **Generally Accepted Scrum Practices**—or "GASPs"—come in. They're not technically part of the Scrum framework, but they're so common among Scrum teams that they're found on almost every Scrum project.

All of these tools help teams share all of the information they gather for planning so that the whole team can plan and track the project together.

❶ User stories and story points

User stories help you to capture what your users need from the software so you can build it out in chunks that they can use. Story Points are a way of saying how much effort will be needed to build a user story.

```
SWITCH WEAPONS WHILE      / 2 POINTS
      SPRINTING

AS A PLAYER,

I WANT TO SWITCH BETWEEN MY EQUIPPED
WEAPONS WHILE SPRINTING

SO THAT I DON'T HAVE TO STOP TO SEE MY
INVENTORY
```

❸ Task boards

Task boards keep everybody in the team on the same page about the current sprint's progress. It's a quick, visual way to see what everybody is doing.

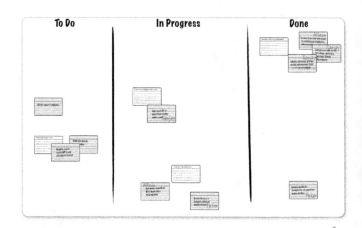

Task boards keep all the status of stories transparent to the team.

generally accepted scrum practices

> The term GASP was created by leading agile thinker Mike Cohn in a widely read 2012 blog post called "The Rules vs. Generally Accepted Scrum Practices"
> https://www.mountaingoatsoftware.com/blog/rules-versus-generally-accepted-practices-scrum

❸ Planning poker

Teams use **planning poker** to get everybody thinking about how big each story is and how they'll build it.

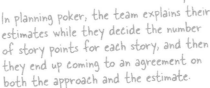

In planning poker, the team explains their estimates while they decide the number of story points for each story, and then they end up coming to an agreement on both the approach and the estimate.

❹ Burndown chart

Everyone on the team can see much work they've done and how much is left to do using **burndown charts**.

The burndown chart is a great tool for helping you to figure out how much work is left to do, and get a pretty good sense of whether or not you'll finish all of the work planned for the Sprint.

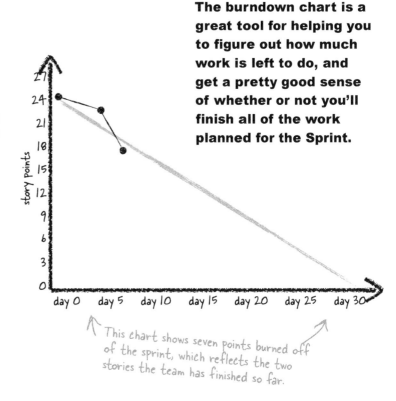

This chart shows seven points burned off of the sprint, which reflects the two stories the team has finished so far.

> The Y-axis of this chart is labeled story points. How do you think the team will use them?

it's easy to get *overwhelmed with documentation*

No more 300-page specs... please?

The team used to create detailed specifications, because writing up all of the requirements for the game seemed like the most efficient way to communicate everything the users needed. But a lot gets lost in translation when one person tries to write down everything and communicate it to a development team. Is there a better, more agile way to write down requirements?

I'VE GOT **150 PAGES OF REPORTS** ON THE FEATURES OUR USERS HAVE REQUESTED. DO I JUST WRITE THOSE UP AS REQUIREMENTS AND TURN THEM OVER TO THE TEAM TO BUILD?

I THINK WE'RE GOING TO NEED TO **TALK ABOUT THOSE FEATURES** A LOT AS WE BUILD THEM. LET'S TRY BREAKING THEM UP INTO USER STORIES AND PLANNING THEM THAT WAY.

Alex: I was thinking that too. But I need more than just a report of requested features to do that, don't I?

Rick: Huh, I guess you're right. It looks like we need to figure out who our users are first, and then divide the list of features into a list of actions and benefits. Do you feel like you have enough information to do that?

Alex: Actually, when I think about it, I do have all of that. There are three kinds of gamers we target with the *CGW* series: novices, casual gamers, and experts.

Rick: OK, let's use those roles to write up our stories.

Alex: Yeah, figuring out the actions the users will take is pretty easy. That's usually the meat of the feature request. The benefit they receive is easy to understand too. Actually, writing our requirements this way won't be very hard.

Rick: Great! Let's just get them in the backlog and get started on our first sprint.

Alex: Not so fast, Rick. We still need to figure out how to estimate this stuff. And we don't even know what "this stuff" really is! Seriously... what, exactly, are we going to build?

124   Chapter 4

generally accepted scrum practices

User stories help teams understand what users need

Software helps people do things. When a user asks a team to build a feature, it's because they need to be able to do something in the future with it that they can't do today. The most efficient way to make sure the team builds the right thing is to keep those needs in mind throughout the development process. A **user story** is a very short description of a specific thing that users need. A lot of teams write them on index cards or sticky notes. By organizing all of their work around user stories, Scrum teams make sure they keep user needs front and center in their planning and prioritization process. That way, they stay focused on building what their users need and there are no surprises when the team demonstrates the stories at the end of the sprint.

User Stories

User stories describe how the user will use the software in just a few sentences. Many teams write user stories on note cards following a fill-in-the-blank format:

As a <*type of user*>, I want to <*specific action I'm taking*> so that <*what I want to happen as a result*>.

Because stories are short and modular, they're a good reminder for the team to constantly confirm that they're building the right features. You can think of each story card as a symbol for a conversation that the team is having with users to make sure that they're building features that are useful to them.

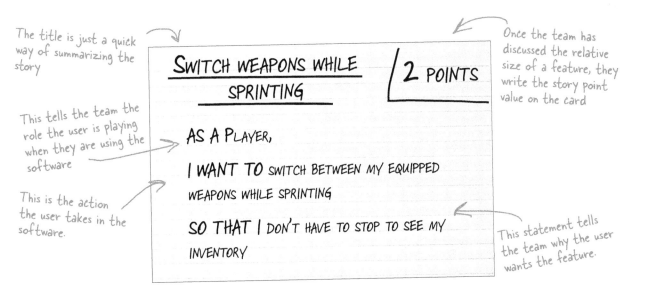

you are here ▸ **125**

all things *being relative*

Story points let the team focus on the relative size of each story

The goal of planning isn't to predict the order features will be done in or their exact completion dates. Instead, the team assigns a point value to each story based on how big it is. That's why most Scrum teams plan their projects using **story points**, which let them compare stories with each other. By focusing on the relative size of features rather than the exact amount of time it will take to develop them, Scrum teams keep the team engaged in planning together and allow for uncertainty in their plans.

How story points work

Story points are simple: the team just picks a number of points that represents the amount of work required for each story, and assigns that number to every story in the sprint backlog. Instead of trying to predict exactly how long it will take to build a feature, the team assigns a point value to each story based on its size relative to other features they've built before. At first, the estimates vary a lot from story to story. But after a while, the team gets used to the scale they're using to estimate and it gets easier to figure out how big each story is.

One way that teams start using story points is to divide stories up into **T-shirt sizes**, and assign a point value to each size. For example, they might decide to use 1 point for extra small features, 2 points for small features, 3 points for medium features, 4 for large, and 5 for extra large. Once they decide on a scale, they just need to decide which category each story fits into. Some teams use the Fibonacci sequence (1, 2, 3, 5, 8, 13, 21...) for story point scales because they think it provides a more realistic weight for bigger features. As long as your team uses a scale consistently, it doesn't matter which one they use.

Whatever doesn't get done in a sprint is moved from that sprint to the next and the total number of story points that are completed in each sprint is tracked as the project's **velocity.** If a team finishes 15 stories totalling 55 story points in a sprint, they track the 55 points as the sprint velocity and that gives them a general idea of roughly how much they can do in the next sprint.

Over time the team gets better and better at assigning story points and more and more consistent in the number of points they deliver in each sprint. That way the team gets a feel for how much they can do in a sprint and takes control of planning together.

Extra Small	Small	Medium	Large	Extra Large	Extra Extra Large
1 point	2 points	3 points	5 points	8 points	13 points

> Why could it be better for teams to assign a general size value to each story than an exact date?

126   Chapter 4

generally accepted scrum practices

THESE USER STORIES ARE GREAT! WE'RE REALLY GETTING A HANDLE ON WHAT THE USERS WANT FROM THE GAME. BUT HOW DO WE FIGURE OUT HOW MUCH WE CAN BUILD IN A SPRINT?

This story applies to all of the novice players, casual players, and expert players.

> **FIGHT MAD COW ZOMBIES**
>
> AS A PLAYER,
>
> I WANT TO SWITCH TO ZOMBIE MODE WHERE THERE ARE TRIPLE THE NUMBER OF ENEMIES AND THEY ARE EASIER TO KILL
>
> SO THAT I CAN REPLAY THE GAME WITH A DIFFERENT MODE

The team needs to agree on how big this story is before we can assign a story point number to it.

Making the game replayable is one of the major goals of this release.

User stories are really simple, which is why Scrum teams find them so valuable. But the most important part of a user story is the discussion that's sparked among the team and with their users and stakeholders.

One reason user stories are so effective is because of this agile principle.

> The most efficient and effective method of conveying information to and within a development team is face-to-face conversation.

get some practice writing stories

Re-write the items in this backlog as user stories

Cows Gone Wild 5.2 Product Backlog

Item #1: Stealth chicken coop level

Value: Adds a different play mode for expert users who want to be able to replay levels.

Item #2: Big Bessie's fighting sequence needs to anticipate the hero's attacks and react faster in expert play mode

Value: This will make Bessie harder to defeat.

Item #3: Novice gamers wanted the haymaker gun to include a super-baler that doubles damage when a bale is fired

Value: Will help novice gamers get through harder battles in Easy mode.

Item #4: Users need to be able to switch weapons while sprinting

Value: Users can see their inventory without stopping

Page 1 of 7

generally accepted scrum practices

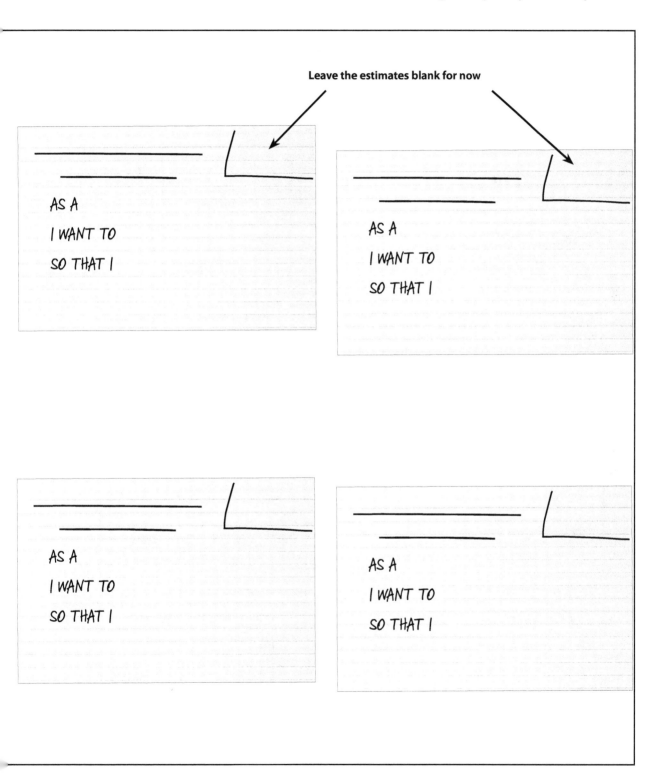

invest in stories

Sharpen your pencil Solution

Re-write the items in the product backlog as user stories.

STEALTH CHICKEN COOP LEVEL

AS AN EXPERT PLAYER

I WANT TO PLAY THE CHICKEN COOP LEVEL IN STEALTH MODE

SO THAT I CAN REPLAY THE GAME IN A DIFFERENT MODE

BIG BESSIE FIGHT MOVES

AS AN EXPERT PLAYER

I WANT TO HAVE BESSIE ANTICIPATE MY MOVES AND REACT FASTER

SO THAT I WILL HAVE MORE FUN FIGHTING BESSIE

Here are the stories we came up with. It's OK if the wording that you used is different, as long as you got practice writing stories.

HAYMAKER — SUPER BALER

AS A NOVICE PLAYER

I WANT TO GET DOUBLE DAMAGE BY SHOOTING THE HAYMAKER IN SUPER-BALER MODE

SO THAT I CAN DEFEAT ENEMIES MORE EASILY

SWITCH WEAPONS WHILE SPRINTING

AS A PLAYER

I WANT TO SWITCH BETWEEN MY EQUIPPED WEAPONS WHILE SPRINTING

SO THAT I DON'T HAVE TO STOP TO SEE MY INVENTORY

Under the Hood: User Stories

User Stories are all about delivering testable software

During the planning for an agile project, the Product Owner will work with end users to identify user stories. Those stories identify the user's needs and their rationale for asking for the feature. But just writing down the user story is only the beginning of the team's understanding of the user's need. Even though the team doesn't know all of the details of what's needed when a user story is first written, they use the card as a reminder that they need to figure out the details and plan the work to develop it. By not delving into the details of each story up front, Scrum teams keep their options open and allow themselves to make decisions about each story at the last responsible moment.

User stories were originally developed as an XP practice (we'll learn more about that in the next chapter), but they're used by many, many Scrum teams. Even though they're much shorter and less-detailed than traditional software requirements, they manage to serve the same purpose while offering teams the flexibility to plan their approach to development as late as possible. Here's how they do it:

- **Card:** First the Product Owner writes down the user story (often using the "As a... I want to... So that..." template we've been talking about), and that card reminds them to understand the details of what needs to be built.

- **Conversation:** When it's time to estimate the story, the team has a conversation with the Product Owner, and sometimes the users, to figure out the details they need to know to estimate the card. Sometimes the Product Owner will work with designers and users to produce mock-ups, or sometimes the team will produce technical designs that help them flesh out the approach to building out a story.

- **Confirmation:** Next, the team turns their attention to the tests they'll write to make sure the user story has been built. This confirmation is an important feedback loop and the fact that user stories are small and self-contained helps both the team and users agree on the tests to run.

Some teams will write the tests that confirm each user story on the back of the user story card. That helps the team to remember how the story should work when it's done. It also helps the users and team come to an agreement on how the story will behave when the software's ready. These tests are alternately called conditions of satisfaction and acceptance criteria.

The guidelines for writing a good user story can be summed up with the acronym **INVEST**:

I - Independent: user stories should be able to be described apart from one another.

N - Negotiable: all of the features in a product are the product of negotiation

V - Valuable: there's no reason to spend time writing a card that isn't valuable to your users

E - Estimatable: each user story needs to convey a feature that the team can assign as size or effort number to

S - Small: user stories should describe independent interactions, not huge categories of functionality

T - Testable: being able to test each user story is what makes it such an effective feedback loop for Scrum teams

> INVEST and the three C's originated with XP, which you'll learn about in the next chapter. You can read more about them in the original 2003 post by Bill Wake:
> http://xp123.com/articles/invest-in-good-stories-and-smart-tasks/
> (as he mentions, the three C's were originated by XP pioneer Ron Jeffries)

put on your poker face

The whole team estimates together

Once the team has a prioritized list of user stories to get started with, they need to figure out how much effort it will take to build them. They usually estimate the story points needed to build each story out as part of the Scrum planning meeting at the start of each sprint. Most often the team knows which stories are the highest priority by looking at the backlog, so they try to commit to as many high priority stories as they can in each sprint. One way the team does this is planning poker.

❶ The setup

Each team member has a deck of cards with valid estimation numbers on each card. Usually the Scrum Master moderates the session.

When the team can't be in the same room to use cards, the team will agree on the point scale they're going to use up front and a method for communicating estimates. Many distributed teams will have everybody give their estimates over an instant message system to the moderator instead of using physical cards.

❷ Understanding each story

The team goes through each story in the sprint backlog in priority order with the Product Owner and asks questions about the story to figure out what the users need.

> HAYMAKER – SUPER BALER
>
> AS A NOVICE PLAYER
>
> I WANT TO GET DOUBLE DAMAGE BY SHOOTING THE HAYMAKER IN SUPER-BALER MODE
>
> SO THAT I CAN DEFEAT ENEMIES MORE EASILY

❸ Assigning a story point value

Once the team has discussed the feature, each person assigns a story point value by choosing a card from the deck and shares that value with the group.

❹ Explaining the high and low numbers

If the estimates differ between team members, the high number and the low number explain their estimates.

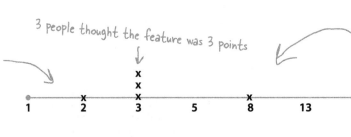

2 is the low estimate. Maybe the person who estimated it knows a way to develop the feature faster than the rest of the team is assuming.

3 people thought the feature was 3 points

The person who estimated 8 points might know of some complexity to the feature that the rest of the team isn't thinking of.

generally accepted scrum **practices**

❺ Adjusting the estimates

Once the team has heard the explanations, they have a chance to choose an estimation card again. If the team can't be in the same room, they communicate their estimate using e-mail or IM to the moderator without sharing it out loud to the team.

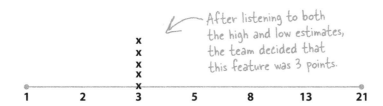

After listening to both the high and low estimates, the team decided that this feature was 3 points.

❻ Converging on an estimate

Usually, teams start with a significant range of estimates but that range narrows over the course of explanation and adjustment. After a few iterations through the process, the estimates converge on a number that the team is comfortable with. It usually only takes 2-3 iterations of discussion until the team can unanimously agree on a story point value.

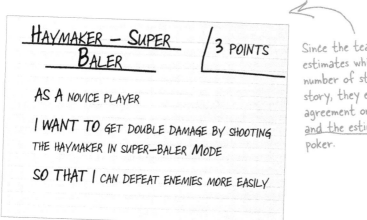

Since the team explains their estimates while they decide the number of story points for each story, they end up coming to an agreement on both the approach and the estimate using planning poker.

Planning poker is very effective, in part because it's collaborative. When the team estimates effort for an item, each team member estimates the whole effort, not just his or her part of it. So even if you're not doing the work, you're estimating it... and that helps everyone on the team get a better understanding of the whole project.

perfect accuracy *isn't possible*

No more detailed project plans

It feels like you've got a really good handle on your project if you create a plan to map out all of the dependencies and figure out who will be doing what from the time you start until the end. Traditional project plans make everybody feel like the there's a guarantee of success because everything has been thought through. More often than not, the information you have at the beginning of the project isn't enough to make a completely accurate detailed plan. But some of the decisions traditional project plans ask you to make in the beginning turn out to be different than the ones you'd make if you were in the middle of the project.

Scrum teams try to make decisions at the last responsible moment and allow for change, because they realize that detailed project plans can lead a team to focus on following the plan rather than responding to the changes that come up naturally. That's why Scrum teams work on prioritizing the backlog and doing the highest priority work first. That way, they're always working on the most important tasks, even when things change.

OK. WE'VE PLAYED PLANNING POKER, AND I KNOW HOW MANY POINTS EACH OF THESE STORIES WILL BE. BUT **HOW DO I BUILD A SCHEDULE** FROM STORY POINTS? I STILL DON'T KNOW WHO WILL DO WHICH TASK OR HOW LONG THEY'LL EACH TAKE.

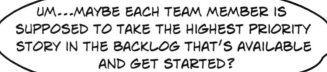

UM...MAYBE EACH TEAM MEMBER IS SUPPOSED TO TAKE THE HIGHEST PRIORITY STORY IN THE BACKLOG THAT'S AVAILABLE AND GET STARTED?

Rick: Wait, no! That's definitely not right. How do I know when we'll be done? Or who's doing what?

Brian: You're right. Assigning a story point value to a user story is a lot different than telling you how many days it will take to do the work.

Rick: So I'm just supposed to tell upper management that we'll do as much work as we can over the next two-week sprint?

Brian: Well, as long as we're all working on the highest priority features, we're doing the most valuable work we can be doing. I think it's a combination of us doing high priority work and demoing what we've done in every sprint review that keeps everyone in the loop.

Rick: OK. But without a project plan, how do I even know if everybody is busy working on the right tasks?

generally accepted scrum practices

Estimation Way Up Close

Here are a few concepts used in estimating software in general that will help you understand how Scrum teams estimate:

Elapsed time

Estimates done in elapsed time predict the date a task will be completed. Estimates like this often require buffers and contingency to set expectations. If a member of the team is going to be on vacation during a project, that person can't be assigned work during that time, and the overall project estimate needs to be adjusted to deal with it. Some projects go as far as trying to predict the total number of hours in a day that each person will be able to work on project work versus the time they spend in meetings or other overhead.

Traditional project management practices attempt to account for all of the possible interruptions and schedule adjustments from the beginning of the project. Projects that are planned in this way attempt to forecast a hard-and-fast end date based on the scope of work and effort estimate.

Ideal time

This is the amount of time necessary to accomplish a task if the person who is assigned to it is able to work without interruption. When an estimate is made in ideal time, it assumes that there is no overhead, no sick days, and no competing priorities that would take the person away from the project work and affect your delivery date. Agile teams estimate in ideal time and they use empirical measurements of how much work each team has delivered in past sprints to set expectations of what can be done in a given time frame.

Story Point

A numeric indicator of the relative size of a feature. Features that require roughly the same amount of effort are given the same story point value. Estimates that are done using story points don't require buffers. When you assign a story point value to a feature, you assume it's a measure of relative size given all of the normal disruptions and uncertainty your team deals with. Because they're relative size values, they do not translate to specific time values. You can't say a story point is equal to one hour of work, for example. You can, however, say that a story point is equal to the effort required to create a button and link it to an action.

Velocity

The number of story points completed in a sprint. This number is averaged over time to predict the amount of work that can be accomplished across multiple sprints. Velocity values normally vary at the start of a project and stabilize as team members gets more comfortable with each other and the work they're doing. When a team has been working with a consistent velocity for some time you can forecast that they will continue to deliver at that rate. Rather than predicting the delivery date for each feature in an agile project, the team focuses on delivering the highest value work possible in each sprint and maintaining a sustainable velocity of work.

> **Velocity is a historical measure.** Teams use it to understand their capacity based on past performance. But that capacity can change over time—and when it does, your team's velocity will, too.

keep track *of tasks*

Taskboards keep the team informed

Once your team has planned the sprint, they need to get started building it. But Scrum teams generally don't sit down and figure out who will do each task at the beginning of each sprint. They try to make it easy for team members to make decisions at the last responsible moment by giving the whole team constantly updated information about how the sprint is progressing.

Most teams start by marking a whiteboard with three columns: To Do, In Progress, and Done. As a team member starts working on a story, he or she will move it from the To Do column to the In Progress column and to the Done column when it's complete.

> This task board has just the stories, and a lot of agile teams do this. But many teams add cards or stickies to their board for the tasks that the stories are decomposed into, grouped under each story card. When the first task moves into the In Progress column, the story goes with it. It stays there until the last task is Done.

 Sprint Begins
All of the user stories are in the To Do column because no one has started working on them yet.

Task boards keep the status of all stories transparent to the team.

Chapter 4

generally accepted scrum **practices**

② Mid Sprint

Team members move their tickets to In Progress when they start working on them and to the Done column as they are completed. The team usually agrees to a definition of done up front so that everyone is clear on what it means to be done with a story.

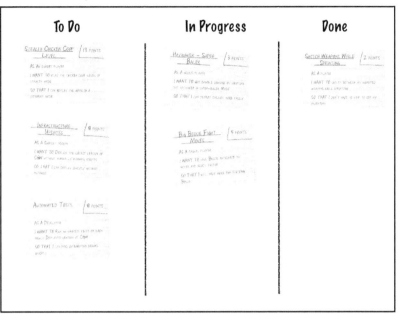

Because the team knows which stories are being worked on, they know what they can take on next to help out.

Now everyone can pick what they should work on next instead of waiting for someone to assign it to them.

③ Sprint Ends

If the team estimated well, all of the user stories they put in the backlog have been moved to the Done column. If there are leftover user stories, they're added to the next sprint backlog.

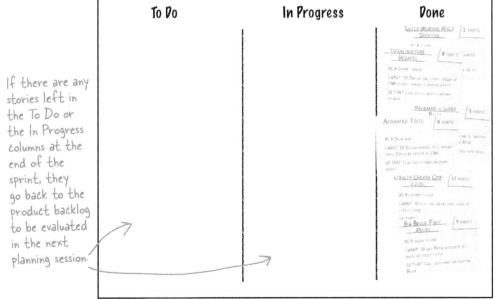

If there are any stories left in the To Do or the In Progress columns at the end of the sprint, they go back to the product backlog to be evaluated in the next planning session.

The team finished all of the user stories in the backlog for this sprint.

you are here ▶ **137**

projects work best *when teams plan together*

there are no Dumb Questions

Q: So what's the difference between a user story and requirement? Is it just that it's written on an index card?

A: No. In fact a lot of teams don't write their user stories on cards at all. Often, they're created as tickets in an issue tracking system. They can be rows in a spreadsheet, or bullets in a document. The biggest difference between user stories and traditional software requirements is that user stories don't attempt to nail down the specific details about the feature that's being described.

The goal when you're writing a user story is to capture just enough information that everybody can remember who's going to use the feature, what it is, and why users want it. The story itself is a way to make sure that the team talks about the feature and understands it well enough to do the work. Sometimes teams will need to write more documentation once they've confirmed the story with the users. Sometimes just having the conversation is enough, and the team can build the feature without any more documentation. However they do it, the most important thing is for them to understand what the user needs, and his or her perspective.

Q: How do I know how many story points to assign to a story?

A: When a team is starting to use story points for the first time, the first thing they typically do is get together and decide what type of work is worth one story point—usually a simple task that everyone on the team can understand. (For example, a team working on a web application might decide that one story point is equivalent to the effort it takes to add a button with some simple, specific functionality to a web page.) Depending on the kind of work they usually do, they'll choose a scale that makes sense to the team. But once they decide the value of one point, it helps the team understand the rest of the possible point ranges.

Some teams use a practice called **T-shirt sizing** to assign all of the stories they're estimating into small, medium, or large categories and assign points that way (1 point for small, 3 points for medium, 5 points for large). Other teams use broader scales (XS, S, M, L, and XL) with corresponding point values. Other teams assign values using the Fibonacci sequence (1, 2, 3, 5, 8, 13, 21...). As long as the team is consistent in how they assign points to stories, it doesn't matter which approach they use.

Q: What's the point of planning poker? Can't developers just estimate their work on their own?

A: Like most of the other GASPs, planning poker focuses on involving the whole team in planning and tracking progress on your project. Planning poker is all about getting the team to discuss their estimates and agree on the right approach for development. By being transparent about estimates and approach, the team can help each other avoid mistakes and think through the most efficient method for developing each feature together. Planning poker helps teams make their estimation reasoning transparent. When the team decides on the approach and the estimate together, they have a much better chance of catching flaws in their approach earlier and they have a better handle on the work that needs to be done.

Q: What happens if you get the estimate wrong?

A: That happens—and it's OK. You might end up thinking that a feature is worth 3 story points at the beginning of the sprint but realize by the end of it that it should've been a 5. But since story points are used to measure your overall velocity over time, you'll find that the whole team gets better and better at figuring out which features are which as they work with their scale. What's great about planning poker and story points is that they don't expect you to be able to predict the future. Once you assign story points to your sprint backlog, you build your sprint backlog and then track your velocity numbers over time. If you have more story points in the backlog than you can do in the sprint, they go back to the product backlog so they can be reprioritized. As the team estimates and keeps track of what they're delivering, they get better and better at it.

In the beginning, you'll see that the number of story points your team completes per sprint varies a lot. But as the team gets more and more comfortable working together, the number of story points they can accomplish in a sprint becomes more and more predictable.

Rather than focusing on getting each estimate right, GASPs help your team to get a handle on how much work you can actually do. That way you can take on the right amount in each sprint and keep your team working as efficiently as possible.

> **Planning poker, story points, and velocity get the whole team planning and tracking work together. All of these tools are about the whole team staying accountable for the project's vision and plan.**

138 Chapter 4

generally accepted scrum practices

MY STAKEHOLDERS WANT TO KNOW WHEN THE PROJECT WILL BE DONE, NOT HOW MANY POINTS WE'VE BURNED DOWN TODAY.

That's true. Stakeholders who are used to traditional status reports and project plans will need to adjust to these practices.

Traditional project management methods create a plan for delivery up front and then track delivery to that plan. Getting everybody in your organization to give up that feeling that they know exactly how a project will be developed from the very beginning can be a major challenge for agile teams.

Instead of planning each detail up front and holding the team to that plan, Scrum teams promise transparency, the ability to change, and a team focus on building the best product possible with the time and resources available. By building incrementally and delivering frequently, they often have much happier stakeholders once they get accustomed to working differently.

Watch it!

Your team can put more than just stories in a backlog.

The Ranch Hand Games team only has stories in their Product Backlog and Sprint Backlog right now. But it's very common to see other types of backlog items, too. A lot of teams will add backlog items for important bug fixes, performance improvements (or other nonfunctional requirements), dealing with risks, or other kinds of work. So you should feel comfortable doing that with your own projects!

you are here ▸ **139**

Question Clinic: The red herring

> SOMETIMES A QUESTION WILL GIVE YOU A LOT OF EXTRA INFORMATION THAT YOU DON'T NEED. IT'LL INCLUDE A RAMBLING STORY OR A BUNCH OF EXTRA NUMBERS THAT ARE IRRELEVANT.

> 104. You are managing an advertising software project. You have to build an interface for buying space in online publications at an average cost of $75,000 per placement. Your project team consists of an advertising analyst as a Product Owner, and a team of experienced software engineers. Your business case document is complete, and you have met with your stakeholders and sponsor. Your senior managers are now asking you to plan your first sprint. Your team has done four other projects very similar to this one, and you have decided to make your estimate by having the team provide story point values in a group session and converge on an agreed estimate and approach together.
>
> What estimation practice involves having the team provide individual estimates and discuss them until they converge on an agreed value?
>
> A. **Planning Poker**
> B. Planning method
> C. Bottom-up
> D. Rough order of magnitude

Did you read that whole paragraph, only to find out the question had nothing to do with it?

> WHEN YOU SEE A RED HERRING QUESTION, YOUR JOB IS TO FIGURE OUT WHAT PART OF IT IS RELEVANT AND WHAT'S INCLUDED JUST TO DISTRACT YOU. IT SEEMS TRICKY, BUT IT'S ACTUALLY PRETTY EASY ONCE YOU GET THE HANG OF IT.

Red Herring

HEAD 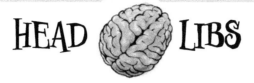 LIBS

Fill in the blanks to come up with your own red herring question!

You are managing a _____ project.
                              (kind of project)
You have _____ at your disposal, with _____ . Your
            (describe a resource)   (how that resource is restricted)
_____ contains _____ . The _____
  (a project document)   (something that document would contain)   (a team member)
alerts you that _____ , and suggests _____ .
              (a problem that affected your project)   (a suggested solution)
_____ ?
        (a question vaguely related to one of the things in the paragraph above)

A. _____
                         (wrong answer)

B. _____
                      (trickly wrong answer)

C. _____
                        (correct answer)

D. _____
                    (ridiculously wrong answer)

keep an eye on your progress

Rick: It feels like we've almost got it down. Alex keeps the product backlog prioritized. At the start of each sprint we go through the highest priority stories, play planning poker, and assign them story point values. Then we add them into the sprint backlog and get started.

Brian: That all works really well. The team likes getting a chance to talk through the work before we do it. It helps everybody stay on the same page about what needs to get done in a sprint, too.

Rick: It does seem like it's working, but then we get to the end of each sprint and we've always got stories that were bigger than we thought they were. There's always stuff in the sprint backlog that needs to get carried over the next sprint. That's making Alex nervous and making our sprint reviews with the users more tense.

Brian: We need to know whether or not we're on track during the sprint, so we can make adjustments if we need to. We shouldn't commit to stories at the beginning of the sprint if we don't think we can get them done.

Rick: I think it's time for us to start tracking our progress more closely. So... um, how exactly do we do that with Scrum?

Burndown charts help the team see how much work is left

Once a team has assigned a story point value to all of the user stories in the sprint backlog, they can use **burndown charts** to get a handle on how the project is progressing. A burndown chart is a simple line chart that shows how many story points are completed each day during the sprint. The burndown chart gives everybody a clear sense of how much work is left to be done at any time. Using a burndown chart, it's clear to everyone on the team how close they are to achieving their sprint goals.

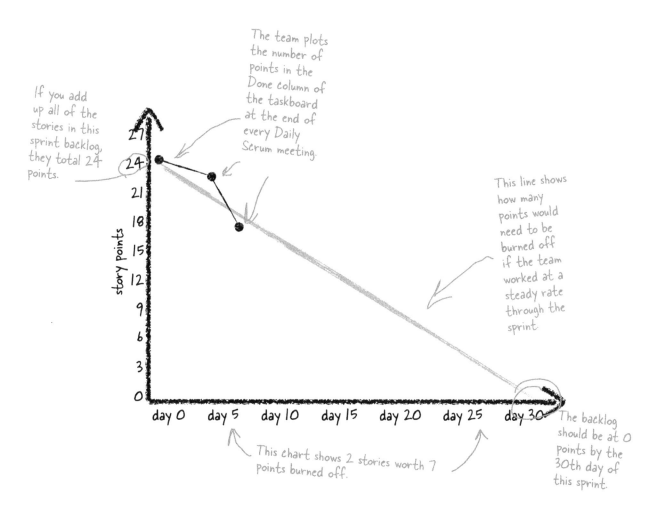

Burndown charts and velocity help the whole team stay in control of the sprint.

quick and steady *wins the race*

Velocity tells you how much your team can do in a sprint

At the end of each sprint, you can count the total number of story points that have been accepted by the Product Owner. The number of points per sprint is called the **velocity**, and it's a great way to gauge how consistently the team is delivering work. Many teams plot their velocity per sprint as a bar chart so they can see how they did across multiple sprints. Since each team's scale for estimating story points is different, **you can't use velocity to compare teams to one another**. But you can use it to help figure out how much work your team should commit to based on their past performance.

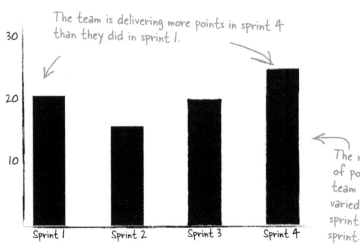

Sprint velocity

This is a bar chart of the total number of story points completed in each of four sprints. If the team is using the same scale for estimation in each sprint, you can use this number to compare how much work has been done from one sprint to the next. To create this chart, the team just adds up the number of story points in the Done column of the task board at the end of each sprint.

Sprint velocity with committed points

This is a bar chart of the total number of story points the team put into the sprint backlog in gray and the number they actually completed in black. To create this chart, the team just adds up the number of story points in the sprint backlog after the planning session and marks that as the committed number. At the end of the sprint they track the velocity number by adding up all of the story points in the Done column of the taskboard.

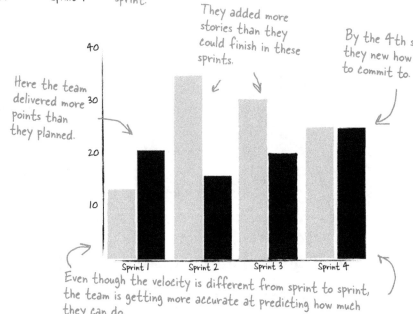

144   Chapter 4

Sharpen your pencil

Here are Rick's notes from looking at the task board after each day's Daily Scrum. The total estimate for this sprint's backlog is 40 points. Draw the burndown chart.

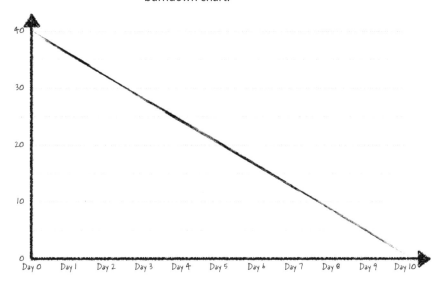

Day 1: We finished the cleanup on the super-baler feature, that's 2 points. We can mark the build script as complete too, that's another 2 points.

Day 2: We can't mark anything complete today.

Day 3: We completed the new finishing move for the big Bessie fight, 3 points.

Day 4: Add two points, we found a refactoring work that has be done to get the haymaker working again.

Day 5: Finished the haymaker refactoring, 2 points.

Day 6: The stealth chicken coop is complete, 8 points.

Day 7: Completed the distribution package script, 5 points.

Day 8: Updated Bessie's AI to react faster, 10 points.

Day 9: Added animations for the super-baler reload, 2 points.

Day 10: Finished the chicken coop refactor, 7 points.

burn down burn up

Sharpen your pencil

Here are Rick's notes from looking at the task board after each day's Daily Scrum. The total estimate for this sprint's backlog is 40 points. Draw the burndown chart.

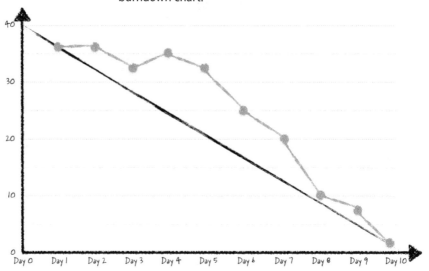

Day 1: We finished the cleanup on the super-baler feature, that's 2 points. We can mark the build script as complete too, that's another 2 points.

Day 2: We can't mark anything complete today.

Day 3: We completed the new finishing move for the big Bessie fight, 3 points.

Day 4: Add two points, we found a refactoring work that has be done to get the haymaker working again.

Day 5: Finished the haymaker refactoring, 2 points.

Day 6: The stealth chicken coop is complete, 8 points.

Day 7: Completed the distribution package script, 5 points.

Day 8: Updated Bessie's AI to react faster, 10 points.

Day 9: Added animations for the super-baler reload, 2 points.

Day 10: Finished the chicken coop refactor, 7 points.

generally accepted scrum practices

Burn-ups keep your progress and your scope separate from each other

Another way to track your progress during a sprint is to use a burn-up chart. Instead of subtracting the number you've completed from the number you committed to, burn-ups track a cumulative total throughout the sprint and show the total committed scope on a separate line. When stories are added or deleted from the scope it's obvious by looking at the scope line. When stories are put into the "Done" column on the task board, that's easy to see too, by looking at the total number of points burned up in the sprint. Because the scope is tracked on a different line from the number of points accomplished, it's clearer when the scope is changing.

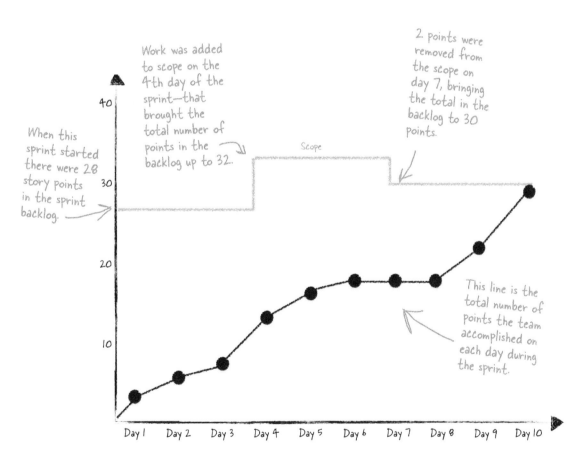

you are here ▸ 147

plan your releases with story maps

How do we know what to build?

The Product Owner's role on a sprint team is to keep everybody working on the most important thing in every sprint. They're in charge of the order of stories in the sprint backlog and the product backlog. When the team has questions about a user story, the Product Owner is the one who tracks down the answers. Many teams set a time near the end of each sprint to make sure that the backlog is in order before the team starts to plan the next sprint. That meeting is called the **product backlog refinement** meeting.

Backlog refinement is a good way for the Product Owner to prepare for sprint planning.

Product backlog refinement is all about adding detail and estimates to each backlog item, and revising the order. Teams usually rely on the estimation that they do during Sprint Planning, but should feel comfortable re-estimating Product Backlog items any time. This is a collaborative effort between the Product Owner and the Development Team—and it's focused entirely on the Product Backlog (the Development Team is solely responsible for the ordering of the Sprint Backlog).

Once the backlog refinement meeting is over, the Product Owner has a couple of days before the start of the next sprint to follow up on any open questions and make sure that the priorities make sense to business stakeholders as well.

A lot of teams block out time 2 or 3 days before the end of the sprint to do backlog refinement. They use the time to come up with questions that need to be answered before the planning session, and to double-check the priority order of the stories.

Some teams refer to product backlog refinement as PBR. In general, teams typically spend less than 10% of their time doing this.

generally accepted scrum practices

Story maps help you prioritize your backlog

One way to visualize the backlog is to lay it out in a story map. Story maps start by identifying the most core features of your product as its **backbone**. Then that functionality is broken up into the backbone's most important user stories. Those are called the **walking skeleton**. Your first sprints should be focused on delivering as much of the walking skeleton as possible. After that, you can plan your releases to include features in their prioritized order on the map.

Remember, these practices are not required by Scrum, but are generally accepted!

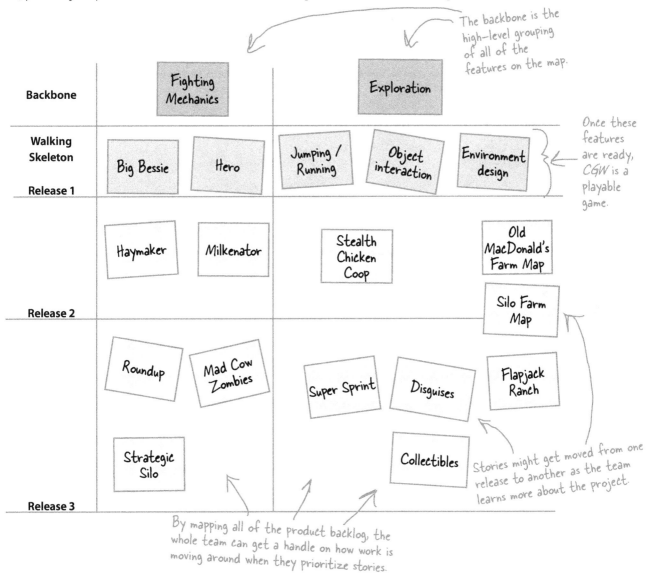

Story maps give teams a way to visualize the release plan

you are here ▶ 149

understand your users

Personas help you get to know your users

A **persona** is a profile of a made-up user that includes personal facts and often a photo, which Scrum teams use to help them understand their users and stakeholders better. Putting a face to each user role and writing down what motivates them will help you make the right choices when you're thinking about how and what to develop. Personas make your user stories more personal. Once the Ranch Hand Games team created them, they started thinking about how each user would react to the features they were building.

Melinda Oglesby

Age: 28
Occupation: IT Consultant
Location: New York, NY
Role: Expert Gamer

Bio: Attends gaming conferences when possible. Owns all of the available consoles. Plays most games more than once. Has built PCs for gaming.

Goals:
- Satisfying story
- Multiple play styles/story options
- Challenging puzzles/fights
- Co-operative game play

Frustrations:
- Poor quality (bugs, getting stuck, crashes)
- Plot holes
- Poor server performance

To create this persona, Alex interviewed 50 gamers at the conference about how they play CGW games.

These goals and frustrations came up in many of the interviews.

Now that they have a face and a name for their expert gamers, the team often thinks about Melinda's opinion when they're deciding how to design a feature.

How do personas and story maps fit into the ideas of transparency, inspection, and adaptation that make Scrum work?

generally accepted scrum practices

BULLET POINTS

- By involving the whole team in planning, GASPs help Scrum teams **adapt** their plan based on what they learn from sprint to sprint.

- **User stories** capture a user need that describes the role of the user, the action they want to accomplish, and the benefit they want to achieve.

- User stories are often written using the following template: *As a <role>, I want to <action>, so that <benefit>*.

- **T-shirt sizing** is a method used by many teams to group features into sizes (S, M, L, XL, XXL) based on the amount of effort necessary to build them.

- **Story points** are a way of measuring the size of the effort necessary to build a story. They do not correspond to hours, or calendar time.

- **Velocity** is the total number of story points a team has accomplished on average during a past sprint.

- **Planning poker** is a collaborative estimation technique used by Scrum teams to determine the story point values for each story in a sprint by gathering estimates and adjusting them after listening to the reasoning behind the low and high values team members have given.

- Teams use **burn-down charts** to track how many story points they've accomplished on each day during a sprint.

- Product owners hold **product backlog refinement** (PBR) meetings near the end of a sprint to get the backlog ready for the next planning session.

there are no Dumb Questions

Q: So how do I use points to make sure I'm on track for the whole project? Do I estimate the whole backlog?

A: Some teams do. Some teams estimate the entire backlog to come up with a **release burndown** chart where they track how far they've come in burning down all of the features initially put into the product backlog. This is how some teams predict their overall release dates for major projects.

But that only works if you're relatively sure that everything that's in the Product Backlog actually needs to be delivered as part of your project. In many cases, there are features in a product's backlog that are so low priority that they will probably never be built. When that's true, it doesn't really make sense to estimate everything in the product backlog and use it to gauge a release date. Instead, many teams focus on building the highest priority functionality possible in every release and releasing software frequently.

That way the most important features are always available as soon as possible.

Q: How do I know how many stories to put into a sprint?

A: When your team sits down to plan a sprint, they always use the Sprint Goal as the starting point to understand the priority of each feature in the Sprint Backlog. That should be enough for everyone to have a good idea of which features are most important. This is the idea behind **commitment-driven** planning (created by Mike Cohn, one of the most influential thinkers in Agile planning): that you deliver something tangible and valuable at the end of each sprint, and that you make trade-offs during the sprint to achieve that.

The other option teams have is **velocity-driven** planning. This means starting with the team's average velocity, and adding the top stories from the backlog until they reach the average velocity. Cohn prefers commitment-driven planning because it relies on the team members' judgement of what's necessary to build a valuable product.

Q: Are story maps and task boards the same thing?

A: No. Both of them can use whiteboards and story cards to show what's going on in your project, but they're displaying very different information.

You can think of the taskboard as an up-to-date look at the sprint backlog. Checking it will always tell you what's going on with each story in the current sprint.

The story map gives you a similar view of the current plan for all of the stories in the product backlog.

The story map helps everyone on your team have the same vision of where the product is going. Story maps give teams a way to visualize the release plan and understand how stories fit together.

you are here ▸ **151**

the team's making progress *but not enough of it*

The news could be better...

Now that the team has simple metrics to track their performance, it's easier for them to see when things don't go as planned. When they take a look across a number of sprints, they can tell that they aren't as predictable as they'd hoped they would be.

generally accepted scrum practices

WE'VE PLANNED THE PROJECT TOGETHER AS A TEAM, BUILT THE PRODUCT AS A TEAM, AND TRACKED OUR PROGRESS TOGETHER AS WELL. SCRUM **HAS TO HAVE PRACTICES** TO HELP EVERYBODY FIX THE PROBLEMS WE SEE WITH THE WAY WE'RE WORKING... RIGHT?

Can you think of ways to involve the whole team in transparency, inspection, and adaptation around the process the team is using in each sprint?

look back at the sprint *find ways to improve*

Retrospectives help your team improve the way they work

At the end of each sprint, the team reflects over the experience they just had and works together to fix any issues that came up. Retrospectives help your team stay aware of how things are going and stay focused on making things better with each sprint. As long as the team is learning from their experience, they'll get better and better at working together as your project progresses. In *Agile Retrospectives: Making Good Teams Great*, Esther Derby and Diana Larsen lay out a simple outline for a retrospective meeting.

❶ Set the stage

At the beginning of the meeting, everybody needs to understand the goal and focus of the retrospective. Derby and Larsen also recommend getting each team member to tell the team their overall mood as an opening activity. If everybody gets a chance to talk when the meeting starts, it's more likely they'll feel comfortable sharing their opinions later on.

A team might focus a retrospective on why they're finding more defects in recent sprints, or how to communicate better about design changes.

❷ Gather data

During this part of the meeting the team takes a look at all of the events of the last sprint using hard facts. They walk through the timeline and discuss the work that was completed and the decisions that were made. Often, team members are asked to vote on these events and decisions to determine whether they were high or low points for them in the sprint.

generally accepted scrum **practices**

③ Generate insights

When the team has gathered the data about the sprint, they can zero in on the events that seemed to be the most problematic for the group. During this part of the meeting, the team identifies the root causes of the problems they encountered and spends time thinking about what they could do differently in the future.

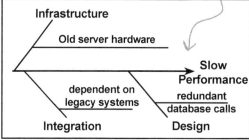

Fishbone or Ishikawa diagram

The vertical "fishbone" lines are categories to help you find and organize the root causes of defects.

Horizontal lines show the root causes you've found for each category.

For this example, a team is looking at causes of defects in a sprint.

Teams use fishbone diagrams to understand the root causes of issues.

④ Decide what to do

Now that they've reviewed what happened during the sprint and spent time thinking about what they might do differently, the next step is deciding which improvements to implement for the next sprint.

get some insight

Some tools to help you get more out of your retrospectives

One way Scrum teams implement the Agile Manifesto principle of periodically reflecting and improving they way they work is to use these tools in their retrospective meetings:

Tools to help you set the stage:

★ **Check-ins** are a way to get your team to engage at the beginning of the retrospective. The retrospective leader will often ask team members to go around the room with each person giving a one- or two-word answer to a question at the start of the meeting.

★ **ESVP** is a technique where each member of the team is asked to categorize themselves into one of four designations: Explorer, Shopper, Vacationer, or Prisoner. **Explorers** want to learn and get as much as they can out of the retrospective. **Shoppers** are looking for one or two improvements out of the retrospectives. **Vacationers** are just happy to be doing something different and away from their desks for the meeting. **Prisoners** would rather be doing something else and feel forced to be part of the retrospective. Asking team members to say which group they think they are a part of helps everybody understand where they're coming from and also helps them to feel more engaged in the meeting.

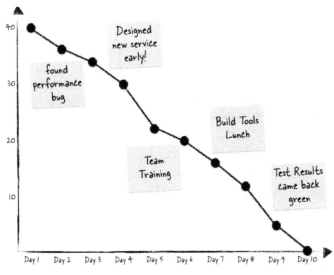

Tools to help you gather data:

★ **Timeline** is a way of displaying all of the meaningful activities that happened in a sprint in chronological order. Each member of the team gets an opportunity to add cards to the timeline with the events that were significant to him or her. Once the team creates the first round of cards to go on the timeline, they review it together and add new cards if they can think of events that should be on timeline.

★ **Color Code Dots** are used to indicate how team members felt about all of the events on the timeline. The moderator might hand out green dots to indicate positive feelings toward an event on the timeline and yellow ones for negative feelings. Then everyone on the team would go through the timeline indicating whether the activities on the timeline were positive or negative to them.

Tools to help you generate insights:

★ **Fishbone** diagrams are also called **cause and effect** and **Ishikawa** diagrams. They are used to figure out what caused a defect. You list all of the categories of the defects that you have identified and then write the possible causes of the defect you are analyzing from each category. Fishbone diagrams help you **see all of the possible causes** in one place so you can think of how you might prevent the defect in the future.

★ **Prioritize with dots** is a technique where each team member is given 10 dots to stick on the issues that they want the team to attempt to address first. Then choose the issues that have the most dots to focus on in the "decide what to do" phase of the retrospective.

Tools to help you decide what to do:

★ **Short Subjects** are a way a categorizing all of the insights the team has come up with into an action plan. Typically the moderator will put short subjects on the top of a whiteboard and the team will work together to put all of their suggestions in the right categories. One common set of short subjects is Stop Doing/Start Doing/Keep Doing. The team takes the time to categorize all of the feedback they've given in the retrospective into the kinds of actions that they can take to make sure they preserve the practices they're doing that are working and change the ones that aren't.

Stop Doing	Start Doing	Keep Doing
Communicating sprint goals to users prior to planning meeting	keeping the scrum to 15 minutes	Think about last sprint's velocity when committing in to stories
	Pair programming for risky features	write tests on user stories

The team had a problem with setting user expectations before they had a planning meeting, so they want to make sure to stop doing that in the next sprint.

making changes *for the better*

Cubicle Conversation

Scrum teams are always focused on improving. At the end of each sprint, everyone on the team takes a look at their burndown chart, their velocity, and the number of stories they've got on the backlog as input to their retrospective. Using the metrics the team has produced through the sprint helps everyone stay on the same page about the root causes of team issues, and that helps the team work together to solve problems that might've come up along the way. Retrospectives are just one more example of how Scrum teams use transparency, inspection, and adaptation to get better and better at building software.

> LET'S FOCUS OUR RETROSPECTIVE ON FIGURING OUT WHY **OUR VELOCITY IS SO UNPREDICTABLE**.

> GREAT IDEA! IF WE COULD TELL STAKEHOLDERS THE RIGHT LIST OF STORIES AFTER OUR PLANNING SESSION, **WE'D HAVE A LOT MORE SUCCESS** IN OUR DEMOS AT THE END OF THE SPRINT.

Rick: Now that we've got velocity numbers for the past four sprints, it looks like we really need to get a better handle on how much we commit to at the beginning of the sprint.

Alex: But how do we do that? I'm only putting the stuff in the sprint backlog that the users need.

Rick: I think we should bring these velocity numbers to the team retrospective, show them the problem, and figure out some solutions together.

generally accepted scrum practices

Here's what the team had to say about their velocity variance at the most recent sprint retrospective. Match the comment with its short subject.

> PROGRAMMERS SHOULDN'T HAVE TO PAY ATTENTION TO VELOCITY NUMBERS. THAT'S THE SCRUM MASTER'S JOB.

Continue Doing

> WE SHOULD PROBABLY NOT COMMIT TO MORE STORY POINTS THAN WE DELIVERED IN THE MOST RECENT SPRINT.

Stop Doing

> I REALLY LIKE PLANNING POKER, IT HELPS THE TEAM TO FIGURE OUT A DESIGN APPROACH REALLY EFFICIENTLY.

Start Doing

> I COULD WAIT TO TALK TO THE STAKEHOLDERS ABOUT OUR GOALS AFTER WE'VE ALL AGREED ON WHAT GOES IN THE SPRINT BACKLOG.

Not Constructive

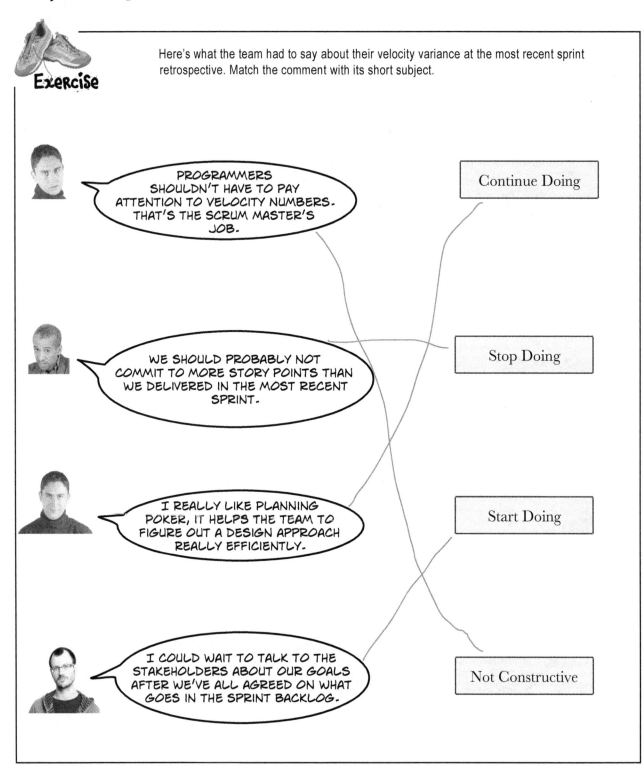

generally accepted scrum practices

GASPcross

Here's a great opportunity to seal the GASPS into your brain. See how many answers you can get without flipping back to the rest of the chapter.

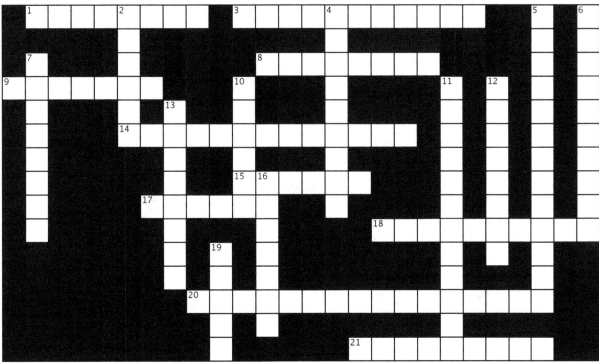

Across

1. _____ diagrams are used to determine the root cause of issues
3. ESVP stands for explorers, shoppers, _____, and prisoners
8. _____ are a way of assigning a name and personal facts to a made-up user of your system
9. The features needed to implement the minimum functionality needed in a product are shown on a story map in the _____ skeleton
14. A Scrum planning technique that depends on the team describing the high and low individual estimates from team members until the group reaches consensus
15. Grouping features in small, medium, large effort categories is called _____ sizing
17. An acronym to help you identify a good user story
18. When planning a sprint, some teams use velocity-driven planning to determine how many story points to include. Others use _____-driven planning to do the same thing
20. Derby and Larsen's basic progression for a retrospective is: set the stage, gather data, _____, decide what to do
21. When estimating effort in story points, some teams use _____ series numbers to appropriately size features

Down

2. _____ charts track scope and completed story points on different lines
4. A tool for tracking work by showing the state all stories are in during a sprint
5. _____ _____ are a way of categorizing follow-up actions from retrospectives
6. When the Product Owner prepares the backlog for a planning session a few days in advance, that's called _____
7. The topmost line in a story map is called the _____
10. The y-axis of a burndown chart is labeled story _____
11. User stories are signifiers for a three-step process that focuses on face-to-face communication between development team members and stakeholders; that process is often described as card, conversation, and _____
12. The number of story points completed in a given sprint is called _____
13. Estimates that tell the date features will be delivered are done in _____ time
16. Stakeholder needs written using a template (*as a <role>, I want to <action>, so that <benefit>*) are called user _____
16. _____ charts track scope and completed story points on different lines
19. Time required to accomplish a task if it's done without interruption

you are here ▸ **161**

GASPcross solution

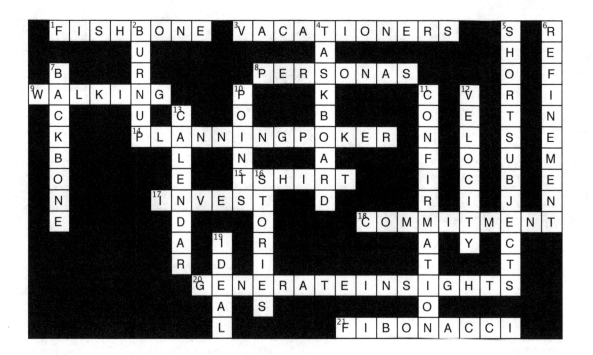

generally accepted scrum practices

Pizza party!

The Ranch Hand Games team used their retrospectives to find and fix issues with planning and delivering *CGW5*. Over time, their sprints went more and more smoothly and they found themselves able to demo great new features in every sprint. By the time the team was ready to ship the game, the whole company knew they had a winner on their hands!

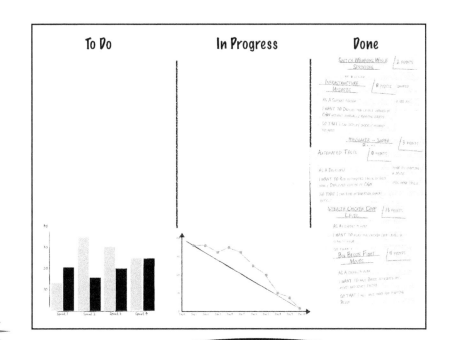

> I FEEL LIKE WE HAVE **REAL CONTROL** OVER WHAT WE'RE BUILDING NOW! WE'RE BUILDING SOFTWARE FASTER AND BETTER THAN WE EVER HAVE BEFORE AND I'M REALLY ENJOYING EVERY DAY AT WORK THESE DAYS.

> THE STAKEHOLDERS ARE REALLY EXCITED TO SEE OUR DEMOS. MORE IMPORTANTLY, **THE PRODUCT LOOKS GREAT!** WE'RE GOING TO MAKE SOME GAMERS VERY HAPPY.

Exam Questions

> These practice exam questions will help you review the material in this chapter. You should still try answering them even if you're not using this book to prepare for the PMI-ACP certification. It's a great way to figure out what you do and don't know, which helps get the material into your brain more quickly.

1. Burndown charts are used for all of the following except:

    A. Helping the team understand how many points have been delivered in a given sprint
    B. Helping the team understand how many points are left to be delivered before the end of a sprint
    C. How many points each team member has delivered
    D. Whether or not the team will deliver everything they committed to in a given sprint

2. The total number of story points delivered in a sprint is called the sprint _____

    A. Increment
    B. Review
    C. Ideal time
    D. Velocity

3. Jim is a Scrum Master on a Scrum project in a media company. His team has been asked to build a new advertising presentation component. They've been working together for 5 sprints and have seen increased velocity over the past two sprints. The team gets together on the first day of the sixth sprint for a planning session. In that session they use a method where the team discusses the features that will be built with the Product Owner, provide estimates on cards, and adjust their estimates as a group until they converge on a number they all agree to.

    Which of the following BEST describes the practice they are using?

    A. Planning poker
    B. Convergence planning
    C. Sprint planning
    D. Analogous estimation

Exam Questions

4. What acronym can be used to describe good user stories?

   A. INSPECT
   B. ADAPT
   C. INVEST
   D. CONFIRM

5. Velocity can be used for all of the following except:

   A. To measure team productivity over multiple sprints
   B. To compare teams to each other and find out who's more productive
   C. To understand how much a team can do when they're estimating a sprint
   D. To understand if the team is committing to too much or too little

6. Which tool is used to visualize scope changes?

   A. Velocity Bar Charts
   B. Burn-up Charts
   C. Cumulative Flow Charts
   D. Scope Histograms

7. How are user stories commonly written?

   A. *As a <persona> I want to <action>, so that <benefit>*
   B. *As a <resource>, I want to <goal>, so that <rationale>*
   C. *As a <role> I want to <action>, so that <benefit>*
   D. None of the above

8. Which of the following BEST describes a taskboard?

   A. The Scrum Master uses it to see if the team is following the plan
   B. It is used to identify new tasks during a sprint
   C. It shows the total number of points accomplished in a sprint
   D. It visualizes the work for the current sprint

Exam Questions

9. Which of the following BEST describes Derby and Larsen's retrospective method:

    A. Set the stage, gather information, decide what to do, document decisions
    B. Check in, create timelines, interpret the data, decide where to focus, measure
    C. Set the stage, gather data, generate insights, decide what to do
    D. ESVP, Color Code Dots, Short Subjects

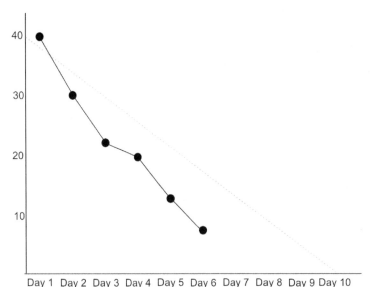

10. What can you tell about this sprint by looking at the burndown chart above?

    A. The sprint is ahead of schedule
    B. The sprint is behind schedule
    C. The project is in trouble
    D. The velocity is too low

11. What is the difference between a burndown and a burn-up chart?

    A. Burndown charts subtract story points form the total number committed while burn-up charts start at 0 and add the story points to the total as they're completed
    B. Burndown charts have a line for scope that tells you how much is added or deleted as you go
    C. Burn-up charts have a trend line to show you the constant rate of completion
    D. Burn-up charts and burndown charts are the same

Exam Questions

12. Which of the following is the BEST tool for determining the root cause of a problem?

- A. Personas
- B. Velocity
- C. Fishbone diagrams
- D. Short Subjects

13. A Scrum team for a medical software company took all of the user stories in their product backlog and arranged them on the wall according to how important the functionality is to a successful product. Then they used that information to determine which features to work on first. What term best describes the practice they were using?

- A. Release planning
- B. A walking skeleton
- C. Velocity planning
- D. Story mapping

14. The process of identifying requirements based on user stories is often referred to as

- A. Card, Call, Confession
- B. Story, Conversation, Product
- C. Card, Conversation, Confirmation
- D. Card, Test, Documentation

15. ESVP stands for

- A. Executive, Student, Vice President
- B. Explorer, Student, Vacationer, Prisoner
- C. Explorer, Shopper, Vacationer, Practitioner
- D. Explorer, Shopper, Vacationer, Prisoner

16. Your Scrum team began measuring velocity over the past three sprints and recorded the following numbers: 30, 42, 23. What can you tell about the team from these measurements?

- A. The team is still determining its story point scale
- B. The team is becoming less productive and actions must be taken to correct this
- C. The velocity is evening out over multiple sprints
- D. The velocity has not been measured correctly

Exam Questions

17. Which of the following BEST describes a tool for identifying a representative software user and describing his or her needs and motivations?

    A. Ishikawa diagrams
    B. User Identification Matrices
    C. Personas
    D. Story Mapping

18. Scrum Planning tools help Scrum teams make project decisions...

    A. As early as possible
    B. Just in time
    C. At the last responsible moment
    D. Responsively

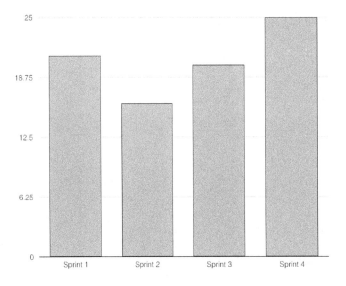

19. What can you learn about this project from looking at the velocity bar chart above?

    A. The project has too much velocity
    B. The team is delivering more story points as the project goes on
    C. Too many scope changes are happening
    D. The project is running behind

Exam Questions

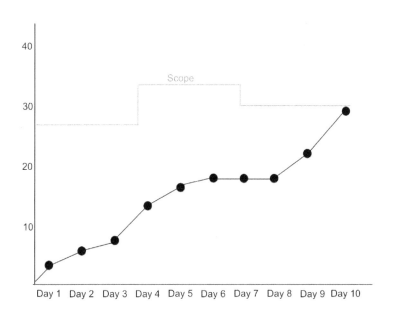

20. What can you learn about this project by reading the burn-up chart above?

    A. Some story points were added to the project scope on day 4 and some were removed on day 7
    B. The team is adding stories to the scope each day of the sprint
    C. The team is not making progress
    D. Stories were added to the project on day 4 and that caused a delay on day 8

Exam Questions

> Here are the answers to this chapter's practice exam questions. How many did you get right? If you got one wrong, that's OK—it's worth taking the time to flip back and re-read the relevant part of the chapter so that you understand what's going on.

1. Answer: C

Burndown charts help the whole team see how much they've accomplished and how much is left to do. They do not show individual team member productivity.

Some people will argue that burndown charts are only about understanding work that's left to do. All three of these things do appear graphically on the chart, so C is technically correct.

2. Answer: D

Teams measure the total number of points they deliver in each sprint as velocity. Velocity can be measured across multiple sprints to help teams get better at estimating and committing to work. Velocity is often used to show the effect of process changes as well.

3. Answer: A

The team is using planning poker. They are doing sprint planning, but since the question was specific about how they were planning, that's not the best answer. They're also doing analogous estimation, but that's not the BEST answer for this question because it is not a generally accepted Scrum pratice. Convergence planning is a made-up name, so don't be confused by counterfeit practice names like these.

4. Answer: C

The acronym INVEST stands for Independent, Negotiable, Valuable, Estimatable, Small, and Testable. A good user story should be all of these things.

5. Answer: B

Since velocity is the sum of all of the story points estimated in a given sprint, it can only be used within one team. Another team would have a different scale since their story point values would come from their sprint planning discussions, or from the estimation that they do during product backlog refinement.

generally accepted scrum practices

Exam ~~Questions~~ Answers

6. Answer: B

Burn-up charts show scope as a separate line on the chart and make it easy to see when scope is added or removed.

7. Answer: C

The correct template for a user story is *As a <role>, I want to <action> so that <benefit>*. Although the other answers were close, there's a big difference between resources, personas, and roles. Roles help you to identify the various perspectives that will need to be accounted for in the application.

8. Answer: D

Taskboards show everyone on the team the status of each task that's in the sprint backlog. It's a visual way to make sure that everyone has the same information about what's available to work on, what's in progress, and what the team has completed.

9. Answer: C

Retrospectives start by setting the stage and making sure that the whole team is included in the conversation. Then the team reviews the information they can gather from the sprint. Once everybody agrees on the facts, they use that inforamation to generate insights on what might be causing problems for the team, Once they've identified the problem, they can figure out what they want to do to try to fix the issue.

10. Answer: A

The dotted line shows a constant burn rate for the sprint. It's normal for the number of points to fluctuate and be to the left of the line at some points and to the right of it at others. In this case, the actual completion line is far to the left of the dotted line and that indicates that the team is burning story points faster than the constant rate necessary for on-time completion.

Some agile practitioners really dislike using the term "on schedule" to describe a burndown chart. However, this terminology may appear on the PMI-ACP exam, and a lot of managers talk in these terms. So it's good to get used to seeing it!

11. Answer: A

Burndown and burn-up charts track the same information, the rate at which the team is completing story points. Burndowns track that rate by subtracting completed points from the total each day. Burn-ups track it by adding the number of points to the total each day.

exam questions

Exam Questions Answers

12. Answer: C

Ishikawa diagrams are tools that are used to categorize common root causes for defects and issues in projects and help you to determine which issues fit into which categories. They're often used to help you find out where you might improve the way you work so that you can fix process problems.

13. Answer: D

The team was mapping their stories so that they could determine the best sequence for delivering them.

14. Answer: C

Card, Conversation, Confirmation is a good way to remember that user stories cards are just reminders to talk to the people who have the information you need for building a story. This is one way Scrum teams value face-to-face communication over comprehensive documentation. They try to write down only what they need out of the conversations they have about each user story card.

15. Answer: D

ESVP is used as a means of checking in with each team member at the start of a retrospective. By asking each team member to tell if they are approaching the retrospective as an explorer, shopper, vacationer, or prisoner, the team can engage each team member in the conversation and let everybody know each person's mindset from the beginning of the discussion.

16. Answer: A

It's very common for teams to have a lot of variance when they're first figuring out the scale they'll use for estimation. It's important not to be alarmed when velocity numbers vary. The goal of measuring velocity is to make the team aware of how much they are doing with each sprint so they get better at figuring out how much to take on in future planning sessions.

17. Answer: C

Personas are fake users that the team creates to help them understand how a user might be feeling when they use the software they're building. (Some teams use actual people and not fake ones, but this is generally frowned upon for privacy reasons.)

Exam Questions

18. Answer: C

Scrum teams know that making too many decisions up front, when you don't know as much about the situations that will come up in a project, can cause more problems than it solves. That's why they focus on making decisions at the last responsible moment.

19. Answer: B

This team is delivering more points with each successive sprint. That's a great trend to observe over a project. It can mean that the team is continuously improving as they work.

20. Answer: A

Since burn-up charts show the scope line separately from the burn-up line, it's easy to see when stories are re-estimated, added, or taken away from scope (as opposed to being completed as part of daily work).

Congratulations! You know enough to be dangerous.

Scrum is by far the most successful and popular approach to agile. It's because it's an empirical method: you work together as a team to understand what's actually going on in your project, you make simple adjustments—again, as a team!—to fix any problems, and then you go back to real, observed behavior to see if those adjustments actually worked.

What's really nice about Scrum is that it gives you a starting point that a huge number of real teams have used on real projects. But the real power of Scrum comes from working with your team to make your own observations, and run your own experiments for improvement. **This is what makes it a framework.**

But there's one extremely important guideline that you should always follow:

USE COMMON SENSE!

The project goals come first

It's possible to abuse the rules of Scrum in a way that damages the project. For example, if there's a huge, critical bug that will cost the company billions of dollars if it's not fixed in the next day, it's not right for a developer to say this:

> I'M WORKING ON SOMETHING ALREADY. THE SCRUM VALUE OF FOCUS TELLS ME TO KEEP WORKING ON WHAT I'M DOING. WE'LL PLAN IT IN THE NEXT SPRINT.

You probably already figured out that Scrum does have a good way to handle this. The Product Owner can add work to fix the bug into the current Sprint and work with the Scrum Team to prioritize it above everything else.

No silver bullet

Scrum Team members often get caught up in the rules of Scrum in ways that can cause harm to the project. It takes time, effort, and experience to get used to Scrum and to really get a deep understanding if its practices, or to genuinely internalize its values. So start experimenting with Scrum, but be careful that you keep the real goals of the project and your organization first! That's what a great Scrum team would always do.

generally accepted scrum practices

Don't just change things because they feel different...

Give the practices a try as written—really, genuinely try as a team—before you try changing or adapting them. It's very common for teams to run into trouble because a Scrum practice feels bad, and then change it:

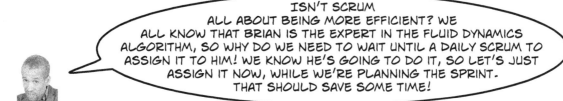

ISN'T SCRUM ALL ABOUT BEING MORE EFFICIENT? WE ALL KNOW THAT BRIAN IS THE EXPERT IN THE FLUID DYNAMICS ALGORITHM, SO WHY DO WE NEED TO WAIT UNTIL A DAILY SCRUM TO ASSIGN IT TO HIM! WE KNOW HE'S GOING TO DO IT, SO LET'S JUST ASSIGN IT NOW, WHILE WE'RE PLANNING THE SPRINT. THAT SHOULD SAVE SOME TIME!

Rick means well! He's trying to find a way to save the team some time. But there's a little more to it than that. **He's also uncomfortable with self-organization.** He would feel a lot better if he knew that Brian was going to do this particular task, and doesn't want to wait for the team to self-organize.

Are you OK with it?

One of the basic rules of Scrum is that the Development Team decides for themselves how they want to meet the Sprint Goal and deliver the Increment. The Scrum values help you really "get" that. There are times when you want to make a change because it will be a genuine improvement to the way that your team works on projects. But there are also times when the values of Scrum **don't quite match the culture of your team.**

That's why it's so important to discuss the values of Scrum, and to discuss collective commitment and self-organization. That's how your whole team can **shift their mindset** and not just their practices.

When everyone on a team shifts their mindset together, they can accomplish <u>astonishing results</u>! The Scrum values are a really important tool for doing that. If you want to take a deeper dive into Scrum, its values, and how teams really learn to self-organize, check out *Learning Agile*.

you are here ▶   175

this page *intentionally left blank*

5 xp (extreme programming)

Embracing change

Software teams are successful when they build great code.
Even really good software teams with very talented developers run into problems with their code. When small code changes "bloom" into a series of **cascading hacks**, or everyday code commits lead to hours of fixing merge conflicts, work that *used to be satisfying* becomes **annoying, tedious, and frustrating**. And that's where **XP** comes in. XP is an agile methodology that's focused on building cohesive teams that **communicate** well, and creating a **relaxed, energized environment**. When teams build code that's **simple**, not complex, they can *embrace change* rather than fear it.

new team new project

Meet the team behind CircuitTrak

CircuitTrak is a fast-growing startup that builds software that gyms, yoga studios, and martial arts dojos use to keep track of classes and attendance.

Gary's the founder and CEO

He's a former college football player who went on to coach high school and, later, college teams. He started the company in his garage two years ago. It's been successful almost since day one, and they just moved into an office downtown. Gary's proud of the company he built, and wants to keep it growing.

Gary always looks for employees who have an athletic background because he knows they're naturally motivated to do whatever it takes to get the job done.

His employees always call him "Coach" because he's as much their coach as he is their boss.

The new downtown office has the latest furnishings, and even exercise equipment so the team doesn't have to leave in order to work out.

178   Chapter 5

xp *extreme programming*

Gyms, yoga studios, and martial arts dojos use CircuitTrak's software to manage their schedules and track their customers' attendance using a website or a mobile app.

Ana and Ryan are the lead engineers

There wouldn't be a CircuitTrak without Ana and Ryan. They were the company's first two employees, and have been around since the beginning, when it was just the three of them working out of Gary's garage. CircuitTrak is up to nine people now—in the last year they hired four other engineers and two sales people—but Ryan and Ana are still the core of the team.

Ana was Gary's first hire. She played lacrosse, softball, and soccer in high school, and got a softball scholarship to put herself through college, where she majored in computer science. Not long after Gary hired her, she recommended that he also hire her college classmate, Ryan. He graduated from the same computer science program a year after she did, and was also a college athlete.

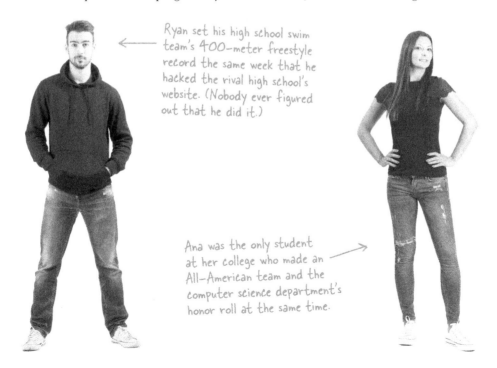

Ryan set his high school swim team's 400-meter freestyle record the same week that he hacked the rival high school's website. (Nobody ever figured out that he did it.)

Ana was the only student at her college who made an All-American team and the computer science department's honor roll at the same time.

we'll clean up the code *later*

Late nights and weekends lead to code problems

Ryan and Ana built the first two versions of CircuitTrak by working 90-hour weeks filled with grit, determination, and lots of caffeine. Now that sales are growing and they're working on version 3.0, they thought they'd be able to relax a little—but they still work plenty of nights and weekends... and Ryan's worried about the effect it's having on the code.

> YOU KNOW I DON'T MIND HARD WORK, BUT I THOUGHT WE'D EVENTUALLY BE ABLE TO STOP WORKING NIGHTS AND WEEKENDS... BUT WE HAVEN'T, AND IT'S CAUSING US TO **BUILD WORSE CODE.**

Ana: What's the problem? And let's make this fast, I need to get back to coding.

Ryan: That's my point! We're always up against some deadline.

Ana: Well, it's a startup. What do you expect?

Ryan: I expected this for the first year, maybe. But we've got customers now. We're growing the team. We shouldn't be scrambling like this.

Ana: That's just how software projects are, right?

Ryan: Maybe. But look at what it does to the code.

Ana: What do you mean?

Ryan: Like, look at what we did here. Remember when we had to change the way we stored group identifiers in the trainer management service?

Ana: Yeah, that was ugly. We still needed the old ID in some parts of the code.

Ryan: Right, so some code uses the old format, and some uses the new format.

Ana: Wait, weren't we supposed to clean that up?

Ryan: Add it to the **long list of things** we were "supposed to" clean up.

Ana: Well, it's basically working, right? If we clean it up now, we'll fall behind.

Ryan: So... what? We'll keep adding lousy code, and never get to clean it up?

Ana: I don't know what to tell you. I think that's just how it is.

It seems like Ryan and Ana don't have enough time to build the code as well as they want, so it's littered with scary TODO comments describing cleanup work they never actually have time to do.

```
public class TrainerContact
{
    // TODO: Need to clean up the ugly hack in getTrainer() when
    // we switched from integer group identifiers to GUIDs

    public Object getTrainerByOldId(String oldId)
        throws TrainerException {
        UUID trainerGroup = GroupManager.convertGuidToId(oldId);
        if (trainerGroup != null) {
```

xp *extreme* **programming**

XP brings a mindset that helps the team <u>and</u> the code

XP (or E**x**treme **P**rogramming) is an agile methodology that's been popular with software teams since the 1990s. XP is focused not just on project management (like Scrum is), but also on how teams actually build code. Like Scrum, XP has **practices** and **values** that help teams get into an effective mindset. XP's mindset helps everyone become more cohesive, communicate better, and do a better job planning—which allows for enough time to build the code right.

Gary's right. The team's been burned too many times when they made ugly, hacky changes to the code, only to spend hours doing frustrating work that feels like it could have been avoided. Now they're afraid of making any changes at all. XP can help them <u>reach a new mindset</u> where they're not afraid of change.

Are any of the 12 agile principles especially appropriate here?

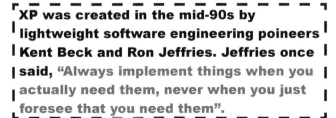

XP was created in the mid-90s by lightweight software engineering poineers Kent Beck and Ron Jeffries. Jeffries once said, "Always implement things when you actually need them, never when you just foresee that you need them".

you are here ▸ **181**

déjà vu this feels really familiar

Iterative development helps teams stay on top of changes

The second principle behind the Agile Manifesto is a good description of how XP teams think about change:

> Welcome changing requirements, even late in development. Agile processes harness change for the customer's competitive advantage.

But wait a minute... didn't we talk about this principle earlier in the book? Yes, we did—it's also a good way to explain iterative development and the idea of making decisions at the last responsible moment. So it shouldn't be a surprise that XP is an iterative and incremental methodology. XP uses **practices** that should feel very familiar to you now that you've learned about Scrum. XP teams use **stories**, just like Scrum teams do. They plan a backlog using a **quarterly cycle**, which is broken into iterations called the **weekly cycle**. In fact, the *only new planning idea here* is a simple practice called **slack** that XP teams use to add extra capacity to each iteration.

XP teams use stories to track their requirements

It's no surprise that XP teams use **stories** as one of their core practices, because they're a really effective way to keep track of what you're planning to build. They work exactly the same way that they do in Scrum.

Many XP teams use the "As a... I want to... so that" story format, and often write their stories on index cards or sticky notes.

They'll also typically include a rough estimate of how long the story will take. It's not uncommon to see an XP team use planning poker to come up with that estimate.

Here's an example of a user story that Ana wrote on an index card. The team used planning poker to come up with an estimate, which she wrote in the corner of the card.

> REMOVE A CLASS FROM A TRAINER'S SCHEDULE | 11 HOURS
>
> AS A TRAINER,
>
> I WANT TO USE THE MOBILE APP TO REMOVE A CLASS THAT I TEACH FROM MY SCHEDULE
>
> SO THAT MY STUDENTS KNOW THEY SHOULDN'T SHOW UP AT THAT TIME

XP teams plan their work a quarter at a time

The **quarterly cycle** practice makes a lot of sense, because doing long-term planning each quarter feels natural: we divide the year into seasons, and many businesses typically work in quarters. So once a quarter the XP team organizes meetings to do planning and reflection.

- ★ Meet and reflect on what happened in the past quarter
- ★ Talk about the big picture: what the company's focused on, and how the team fits into it
- ★ Plan the **themes** for the quarter to keep track of their long-term goals (where each theme is just an overall goal used to group stories together)
- ★ Plan the backlog for the quarter by meeting with users and stakeholders to pick the next quarter's worth of stories that they'll choose from

XP teams use themes to make sure they don't lose sight of the big picture. A theme is just like a sprint goal in Scrum: a sentence or two that describes what they want to accomplish.

xp extreme programming

XP teams use one-week iterations

The **weekly cycle** practice is a one-week iteration in which the team chooses stories and builds working software that's "Done" at the end the week.

Each cycle starts with a meeting where they demo the working software and plan out what they're going to accomplish by:

- ★ Reviewing the progress they've made so far, and doing a demo of exactly what they did last week
- ★ Working with the customer to pick the stories for this week
- ★ Decomposing the stories into tasks

XP teams sometimes assign the individual tasks when they plan the weekly cycle, but they'll often self-organize by creating a pile of tasks and have each team member pull his or her next task off of the pile.

The weekly cycle starts on the same day each week, usually a Tuesday or Wednesday (*Mondays are avoided*, so that the team doesn't feel pressured to work over the weekend), and the planning meeting is typically held at the same time each week. The customer is typically a part of the meeting to help the team select the stories, and to stay on top of the progress.

> **TASK** — 3 HOURS
> MODIFY THE DATABASE LIBRARY AND DATABASE TO ALLOW TRAINER CLASSES TO BE REMOVED AND LOG THE ACTION

> **TASK** — 6 HOURS
> ADD A "REMOVE CLASS" API CALL TO THE TRAINER SCHEDULE SERVICE

> **TASK** — 2 HOURS
> UPDATE THE MOBILE APP UI TO ADD A "REMOVE CLASS" BUTTON TO THE PAGE THAT DISPLAYS A CLASS, HAVE IT CALL THE NEW API CALL TO REMOVE THE CLASS

Ana's idea might improve performance, so that makes for a great "nice-to-have" feature to add as slack.

> I HAD AN IDEA FOR OPTIMIZING THE SCHEDULE SERVICE THAT COULD SERIOUSLY IMPROVE PERFORMANCE. I BET WE CAN **ADD IT AS A SLACK STORY** TO THE NEXT WEEKLY CYCLE.

Slack means giving the team some breathing room

Any time the team creates a plan, the team adds **slack**—another XP practice—by *including a small number of optional or minor items* that they can drop if they start to fall behind. For example, the team might include "nice to have" stories in the weekly cycle. Some teams like to block out "hack days" or even "geek weeks" during the quarter, where the team can work on their own work-related projects and follow up on good ideas that may have gotten swept under the rug. But don't go crazy with slack! Some teams only include one or two slack items, and it's very rare for slack to take up more than 20% of the weekly cycle.

you are here ▶ **183**

xp is focused *on programming*

Courage and respect keep fear out of the project

XP, like every agile method, depends on a team that has the right mindset. That's why XP comes with its own set of values. The first two values are **courage** and **respect**. Do those values sound familiar? They should—they're exactly the same values that you learned about earlier in Chapter 3, because Scrum teams also value courage and respect.

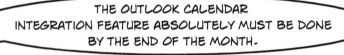

XP teams have the courage to take on challenges. Individual people on the team have the courage to stand up for their project.

> THE OUTLOOK CALENDAR INTEGRATION FEATURE ABSOLUTELY MUST BE DONE BY THE END OF THE MONTH.

> I UNDERSTAND HOW IMPORTANT THAT FEATURE IS, BUT WHAT YOU'RE ASKING JUST ISN'T POSSIBLE. LET'S FIGURE OUT WHAT WE CAN REALISTICALLY DELIVER.

Ryan doesn't like saying "no" to the boss, but he has the courage to do what's best for the project—which means not committing to a deadline he can't meet.

Respect

Teammates have mutual respect for each other, and every person on the team trusts everyone else to do their jobs.

Respect starts with listening to ideas and opinions you might not like and genuinely taking them into account.

> IT'LL BE A TOUGH SELL, BUT IF WE SHOW UPDATED APPOINTMENTS INSIDE THE APP'S UI, I THINK WE CAN MAKE IT WORK... FOR NOW.

It's easier for Ryan to have courage when everyone—especially Gary—respects his opinion. Respect goes both ways: Ryan thinks of Gary as not just the boss, but also an important member of the team, and respects his opinion and ideas too.

xp *extreme* **programming**

> LET ME GET THIS STRAIGHT. XP IS ITERATIVE, JUST LIKE SCRUM. AND IT HAS VALUES JUST LIKE SCRUM, INCLUDING **THE SAME EXACT VALUES OF RESPECT AND COURAGE**. THIS IS STARTING TO SOUND REALLY REDUNDANT. SO WHY USE XP AT ALL? WHY NOT JUST STICK WITH SCRUM?

XP and Scrum each focus on different aspects of software development.

XP, like Scrum, is an iterative and incremental methodology. But it ***doesn't*** have **the same strong focus on project management** that Scrum has—especially since it doesn't focus on empirical process control, which is a really powerful tool for teams to improve the way they manage their projects. It's also why Scrum teams feel very structured: each sprint starts and ends with their timeboxed meetings, and every day there's another timeboxed meeting at the same time.

The "P" in XP stands for **programming**, and everything in XP is optimized to help a programming team improve the way they work. XP is different from Scrum because it's focused on getting the team to work well together. XP has a lighter focus on project management, and more focus on improving the way the team builds code.

XP is focused on software development. There's nothing in Scrum that's specific to a software team—in fact, a lot of other industries have adopted Scrum to take advantage of its empirical process control.

you are here ▸ **185**

xp plus scrum is really powerful

> SO SCRUM HAS A DEEP FOCUS ON PROJECT MANAGEMENT, BUT DOESN'T REALLY GET INTO THE *DAY-TO-DAY DETAILS OF HOW A TEAM ACTUALLY BUILDS CODE.* THAT'S XP'S DOMAIN, SO IT'S LIGHTER ON THE PROJECT MANAGEMENT SIDE.

XP includes enough project management to get the job done.

What ties them together are common ideas and shared values like the ones in the Agile Manifesto. And since iteration in XP works like it does in Scrum, and XP shares the values of courage and respect with Scrum, **many agile teams use a *hybrid* of Scrum and XP** by combining the empirical process control of Scrum with XP's focus on team cohesion, communication, code quality, and programming.

What can people on a software team do to improve how well they communicate with each other?

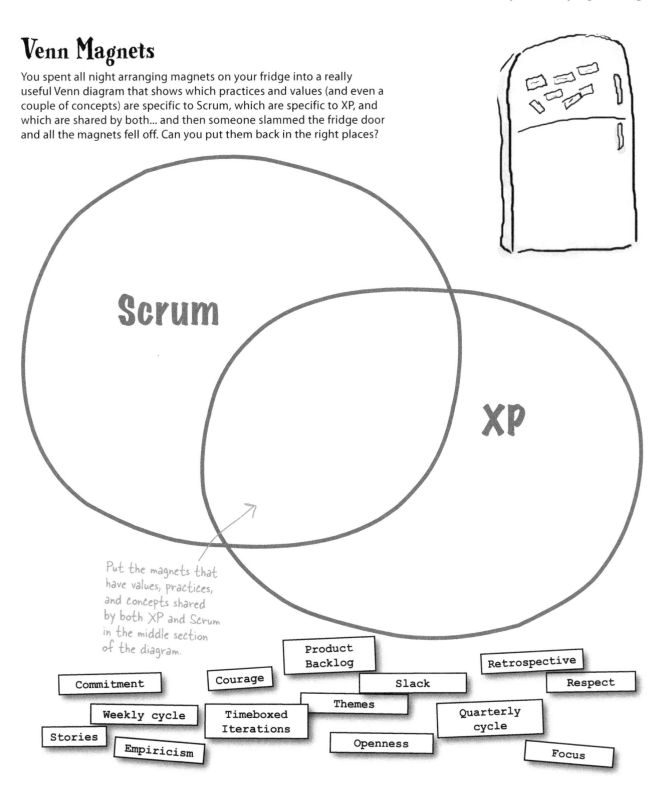

some overlap *many differences*

Venn Magnets Solution

XP teams have a different attitude towards planning than Scrum teams do, and it's reflected in the practices and values that they share and the ones that they don't.

It takes a lot of work to manage and refine a product backlog. That's why Scrum has a full-time Product Owner on the team.

XP teams don't have one single retrospective meeting that they hold during each iteration. Instead, they constantly talk about ways to improve how they work together as a team.

XP teams don't have a full-time Product Owner. Instead, the whole team meets with users and customers to do quarterly planning.

Scrum
- Empiricism
- Retrospective
- Product Backlog
- Focus
- Commitment
- Openness

Shared (both Scrum and XP)
- Respect
- Courage
- Stories
- Timeboxed Iterations
- Themes

XP and Scrum teams share the values of respect and courage. They also use stories in exactly the same way.

XP
- Quarterly cycle
- Weekly cycle
- Slack

Openness and commitment are really important, but only Scrum has them as part of the core value system.

Themes and timeboxed iterations are concepts, not necessarily practices or values.

> Slack is a really good way to get a sense of the difference between being on a Scrum team and being on an XP team. It's literally just throwing some extra stories or tasks into the weekly cycle: it lacks the structure, empiricism, and experimental approach of Scrum. But it's a "good enough" planning tool for a lot of teams.

xp extreme programming

there are no Dumb Questions

Q: How do stories get estimated?

A: Planning poker is very popular among XP teams, but they also use a lot of other methods for estimation. Early versions of XP included a practice that was called the **planning game** that guided the team through decomposing stories into tasks, assigning them to team members, and turning them into a plan for the iteration, which is still in use on a few XP teams here and there. But for most teams, estimation in XP is no different from estimation in Scrum. Techniques like planning poker are really useful, but in the end estimation is a skill: the results get better as the team gets more practice.

Q: How does a methodology "focus" on one thing or another?

A: Scrum is focused on project management and product development because the practices, values, and ideas of Scrum are specifically aimed at the problems of project management: determining what product will be built, and planning and executing the work. Scrum practices are primarily built to help the team get organized, to manage the expectations of the users and stakeholders, and to make sure everyone is communicating.

XP has a more limited approach to project management. It's still iterative and incremental, and the practices you've seen so far in this chapter—quarterly cycle, weekly cycle, slack, and stories—are an effective way to plan and manage those iterations. But they lack the structure and rigidity of Scrum: there aren't daily meetings, the meetings aren't timeboxed, and everything just feels "loose" compared to Scrum. A lot of teams find that the structure of Scrum works really well for them, so they'll opt for a **Scrum/XP hybrid** where they replace XP's quarterly cycle, weekly cycle, and slack with a **_complete_** implementation of Scrum. That means including all of the events, artifacts, and roles of Scrum.

Q: Wouldn't a "hybrid" of XP and Scrum break the rules of one of them?

A: Yes—but it's OK! If your team adopts a hybrid of XP and Scrum by replacing the planning practices of XP with the ones in Scrum, then obviously you're not performing every single practice in XP. But remember, the rules of a methodology are there to help you run your projects well. A lot of teams run into trouble when they modify an agile methodology because they don't really understand exactly why that methodology works. They often change or remove an element that may seem minor to them, but don't realize that it's one of the pillars that keeps the whole methodology up—like when teams try to replace the Product Owner in Scrum with a committee, which removes a critical piece of Scrum. Luckily, replacing XP's weekly cycle, quarterly cycle, and slack practices with a **_complete and unmodified_** implementation of Scrum impacts only the planning part of XP, and it doesn't remove any of the other pillars that make XP work, which is why so many teams have had success doing it.

SO WHEN TEAMS USE A **HYBRID OF SCRUM AND XP**, THEY COMBINE THE CODE-FOCUSED MINDSET AND PRACTICES FROM XP WITH THE COMMITMENT- AND VALUE-BASED MINDSET AND PRACTICES OF SCRUM TO GET THE BEST OF BOTH WORLDS.

no roles means everyone steps up

Teams build better code when they work together

A software team is more than just a group of people who happen to be working on the same project. When people work together, listen to each other, and help each other solve problems, they write a lot more code (sometimes 10 times as much!), and the code they build is much higher quality. XP teams know this, and the **whole team** practice helps to get them there. This practice is about everyone functioning as a team together. It means doing whatever it takes to make people feel like they belong, and helping each person support everyone else on the team.

> Everyone on an XP team feels like they're all in it together. They consider building a really supportive environment to be a core practice for the team.

A whole team is built on <u>trust</u>

When people on an XP team encounter obstacles, they all work together to overcome them. When they're facing an important decision that affects the direction of the project, that decision is made together. That's why trust is so important to XP teams. Everyone on the team learns to trust the rest of the team members to figure out which decisions can be made individually, and which decisions need to be brought to the rest of the team.

> UH... I COMPLETELY MISUNDERSTOOD HOW THIS FEATURE WAS SUPPOSED TO WORK. IT'S GOING TO TAKE ME AT LEAST A DAY TO FIX IT.

> Ryan knows that he can be open about this mistake, and the rest of the team will understand. But he also feels responsible, and will work hard to clean things up.

Trust means letting your teammates make mistakes

Everyone makes mistakes. When XP teams take the "whole team" practice seriously, people aren't afraid to make those mistakes, because each team member knows that the everyone else will understand that mistakes happen—and that the only way to move forward is to make those inevitable mistakes and learn from them together.

XP teams don't have fixed or prescribed roles

There are a lot of different jobs to do on any software project: building code, writing stories, talking to users, designing user interfaces, engineering the architecture, managing the project, and more. On an XP team, everyone does a little bit of everything—their roles change depending on the skills they bring to the table. That's one reason why XP teams **don't have fixed or prescribed roles**.

Roles can keep people from feeling a sense of belonging on the team. For example, it's not uncommon on a Scrum team for a Product Owner or Scrum Master to feel like they're not *really* part of the day-to-day work, as if their "special" role puts pressure on them to be more interested than committed. (Remember pigs versus chickens? Sometimes it's almost as if giving someone's role on the team a name can encourage some people to be more "pig" and others to be more "chicken" on the project.)

I'M GETTING BUSINESS CARDS FOR THE TEAM. WHAT'S EVERYONE'S TITLE? WHO'S THE ARCHITECT? WHO'S LEAD ENGINEER?

WE DON'T REALLY WORK THAT WAY. WHEN THERE'S A JOB TO DO, WE JUST MAKE SURE **THE RIGHT PERSON** STEPS UP AND DOES IT.

Ana does all sorts of jobs: she'll write code, do project management, work with users... whatever it takes to get the job done.

The "whole team" practice is all about making sure people are identified with the team, and not just locked into one specific role.

Teams work best when they sit together

Programming is a highly social activity. Yes, really! Sure, we're all familiar with the image of the lone coder who sits in the dark for hours on end, emerging from his hole after weeks with a complete, finished product. But that's not really how teams build software in the real world. Take another look at this agile principle:

> The most efficient and effective method of conveying information to and within a development team is face-to-face conversation.

When you're on a software team, you need information all the time: you need to know what you're building, how you and your team plan to build it, how the piece you're working on will fit into the rest of the software, and so on. And that means you're going to have a lot of face-to-face conversations.

I'VE GOT A QUESTION FOR ANA, BUT SHE SITS ALL THE WAY ON THE OTHER SIDE OF THE OFFICE. I'LL JUST SEND HER AN EMAIL INSTEAD.

What if Ryan has a really important question, but Ana doesn't check her email for an hour?

XP teams sit together because it's a lot easier for programmers to innovate when they can easily talk to each other, and don't have to spend a lot of time walking around in order to get the information they need.

So what happens if you and your team sit in entirely different parts of the office? This is really common: for example, when the person laying out the office space has the "coder in the dark hole" image of software development, the space gets laid out to give managers a lot of office space, and the programming team is sprinkled into whatever space is left over. This is a really ineffective environment for a software team.

That's why XP teams **sit together**. That's a simple practice where everyone on the team sits in the same part of the office so that it's easy and convenient for each person to find his or her teammates and have a face-to-face conversation.

xp *extreme* *programming*

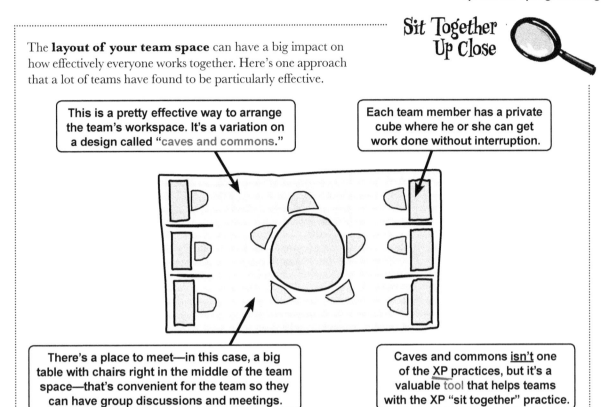

The **layout of your team space** can have a big impact on how effectively everyone works together. Here's one approach that a lot of teams have found to be particularly effective.

Sit Together Up Close

This is a pretty effective way to arrange the team's workspace. It's a variation on a design called "caves and commons."

Each team member has a private cube where he or she can get work done without interruption.

There's a place to meet—in this case, a big table with chairs right in the middle of the team space—that's convenient for the team so they can have group discussions and meetings.

Caves and commons <u>isn't</u> one of the XP practices, but it's a valuable tool that helps teams with the XP "sit together" practice.

BULLET POINTS

- **XP** is an agile methodology focused on team cohesion, communication, code quality, and programming.

- XP has **practices** that help teams improve the way they work and **values** to help them get into the right mindset.

- XP teams use **stories** to track their requirements. They work exactly the same way as they do in Scrum.

- The **quarterly cycle** practice helps teams plan the long-term work by talking about the big picture, choosing **themes** (or overall goals) for the quarter, and selecting stories for the quarter's backlog.

- The **weekly cycle** practice is a one-week iteration that starts with a planning meeting where the team gives a demo of the working software, works with the customer to pick the stories for the iteration, and decomposes them into tasks.

- XP teams add **slack** to each iteration by including optional "nice-to-have" stories that can be left out if the team falls behind so they can concentrate on delivering working software that's "Done".

- Some teams use a **Scrum/XP hybrid** by replacing XP's planning practices with a complete version of Scrum.

- The **whole team** practice is about giving everyone a sense of belonging on a team.

- XP teams **don't have fixed or prescribed roles**; each team member contributes whatever they can for the team.

- Everyone on the team **sits together** in the same space.

- **Caves and commons** is a common team space layout where each person has an area that gives some privacy, and a common area that's central to the space.

you are here ▶ **193**

code monkey *like you*

XP teams value communication

People on XP teams work together. They plan together, collaborate to figure out what they're going to build, and even code together. If you're on an XP team, you *truly believe* that when you're faced with a problem, you'll come up with the best solution if you communicate with your teammates. That's why **communication** is one of the XP values. One way that XP teams improve the way that they communicate is by having an **informative workspace**. This is an XP practice where the team sets up their working environment so that it's impossible not to constantly absorb information from those around them.

Communication

These are two more useful tools that help XP teams with the informative workspace practice.

XP teams take advantage of **osmotic communication**, which is what happens when you absorb information about the project because you're sitting near people who talk about it—almost like it's by osmosis.

XP teams make their team space informative by adding **information radiators**. Those are visual tools (like a big task board or burndown chart) posted on a wall that everyone can see from the work area. They "radiate" information because they're placed in a highly visible area of the workspace.

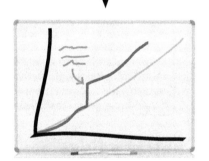

A lot of people code best when they wear headphones and tune out the rest of the world, and that's just fine. You don't need to absorb everything that goes on around you all the time.

> I TWEAKED THE SCHEDULING SERVICE BY CREATING AN OBJECT THAT CACHES THE DATA IN SOME OF THE TABLES.

Ryan overheard a useful snippet of Ana's conversation and it gave him a good idea. Then he saw the burndown chart for this week's cycle posted on the wall, and realized it made sense to push it to the next iteration. The team's informative workspace improved their project.

> I RAN INTO A SIMILAR PROBLEM. I BET I CAN **REUSE ANA'S OBJECT** IN MY CODE.

> HMM, LOOKS LIKE WE'RE RUNNING A LITTLE BEHIND. I'LL HAVE TO WAIT FOR THE NEXT WEEKLY CYCLE.

xp *extreme* **programming**

> *I'M NOT SURE ABOUT THIS. DON'T PROGRAMMERS JUST NEED UNINTERRUPTED TIME IN A DARK, QUIET ROOM TO START PUMPING OUT CODE?*

Team programming is social. It's not an isolated activity.

Programmers don't like being interrupted—and for good reason. When you're on a roll coding, you get into a kind of "zone," a state of high concentration where work seems to just flow. In fact, a lot of people have a name for this state: **flow**. It's actually pretty much the same thing as when an athlete is "in the zone" (where players feel like the baseball is the size of a watermelon, or the basketball hoop looks like it's 10-feet wide). This effect has actually been studied, and those studies found that it can take 15 to 45 minutes for a programmer to reach that state. An interruption, like a phone call or an annoying email, can completely break you out of flow. If you get two phone calls an hour, you can sit at your desk all day and get nothing done.

So wait a minute—doesn't that mean that to achieve maximum flow, the team should work in an environment of absolute silence? No—just the opposite! It's actually hard to concentrate in dead silence, because every time someone coughs or rustles some papers it feels like it's a freight train went by. If there's a little activity around you all the time, it's actually easier to tune it out. (And after all, athletes can get themselves in the zone even in front of screaming fans!)

Watch it!   **Don't fall into the "code monkey" trap. Programming is creative and intellectual work, not just rote typing.**

*If you haven't spent a lot of time writing code, you might think of programming as a "heads-down" activity: just put a programmer in a dark room in front of a computer for a few hours, and if there are no distractions he or she will just start spewing out lines of code. That's **not** how most professional software teams work. When people work together as a team, they can accomplish a lot more than if they work individually. (That's true of many kinds of teams, not just software teams!)*

you are here ▶   **195**

it's a marathon not a sprint

Teams work best with relaxed, rested minds

Software teams need to innovate all the time. Every day brings new problems to solve. Programming is a really unique job, because it's a combination of designing new products, implementing new ideas, understanding what people need, solving complex logical problems, and testing what you built. This kind of work requires a relaxed and rested mind. XP's **energized work** practice helps everyone on the team needs to stays sharp and focused every day. Here are some ideas for how to energize your work:

Leave yourself enough time to do the job

Crazy, unrealistic deadlines are the easiest way to destroy your team's productivity, as well as their morale and any joy they take from their work. That's one reason why XP teams use iterative development. When the team sees that they can't get all of the work "Done" for this week's cycle, they'll push some of it into the next iteration instead of trying to squeeze it all in by working late.

Get rid of interruptions

What would happen if everyone on the team turned off email notifications and silenced their office phones for two hours a day? Teams that try this find that it's a lot easier to get into flow, that state of deep concentration where you barely notice how much time has passed.

Just make sure everyone knows not to interrupt each other, because one tap on your shoulder can jar you right out of the zone.

Let yourself make mistakes

It's OK to make mistakes! Building software means constantly innovating: designing new features, coming up with new ideas, building code—and *failure is the foundation of innovation*. Every team goes down the wrong path every now and then; it's much more productive to decide as a team to think of it as a learning experience and an opportunity to learn important lessons about the code you're working on.

Remember this principle from Chapter 3? A good work-life balance is part of the agile mindset because it's the most productive way to run a team.

Work at a sustainable pace

Occasionally having a "crunch" period where you work long hours for a couple of days generally doesn't hurt, but no team can work like that forever. Teams that regularly do this find that they actually produce lower quality code, and end up writing less code and getting less done than they do under normal circumstances. A **sustainable pace** means working 40 hours a week, without long nights or weekends, because that's actually the best way to get the most productivity out of the team.

This is what the agile principle about sustainable development means.

Agile processes promote **sustainable** development. The sponsors, developers, and users should be able to maintain a constant pace indefinitely.

Who am I?

A bunch of XP practices and values are playing a party game, "Who am I?" They'll give you a clue, and you try to guess who they are based on what they say. Write down their names, and what kind of things they are (like whether they're events, roles, etc.).

And watch out—a couple of <u>tools</u> that are <u>not</u> XP practices or values might just show up and crash the party!

	Name	Kind of thing
I help XP teams get into a mindset where they know that they're best at solving problems when they share knowledge with each other.	_____	_____
I'm a great way to absorb information about the project from discussions happening around me.	_____	_____
I help you understand your user's needs, and I'm also used by a lot of Scrum teams.	_____	_____
I'm a team space that does a great job communicating information about the project.	_____	_____
I help the team work at a sustainable pace, because teams that work super-long hours actually build less code with worse quality.	_____	_____
I'm how XP teams do their long-term planning, by meeting with the users once a quarter to work on the backlog.	_____	_____
I help people get into a mindset where they treat each other well, and value each other's input and contributions.	_____	_____
I'm the reason people on XP teams will tell the truth about the project, even if it's uncomfortable.	_____	_____
I'm the way that XP teams do iterative development, and teams use me to deliver the next increment of "Done" working software.	_____	_____
I'm a large burndown chart or task board put up in the team space in a spot where everyone can't help but notice me.	_____	_____
I make sure that the team has a space where everyone is near their teammates.	_____	_____
I help give the team some breathing room in each iteration by adding optional stories or tasks.	_____	_____

➔ Answers on page 242

go home early *get more done* **who knew**

there are no Dumb Questions

Q: I don't believe this "energized" and "sustainable" stuff. Isn't it just an excuse that programmers use so they don't have to stay late?

A: Absolutely not! Modern workplaces didn't come up with a 40-hour week by accident. There have been many, many studies over the years run across many different industries that found that while teams can work long hours for a short burst, it doesn't take long for their productivity and quality to drop off a cliff. And if you've ever had to work three 7-day, 70-hour weeks in a row, you know exactly why—your brain gets tired, and is in no condition to do the kind of demanding intellectual work needed to build great software. That's why people on XP take work-life balance really seriously: they go home at a reasonable time every day, and have lives and families outside of the job.

Q: No, I still don't buy it. Isn't programming mainly just typing?

A: Programmers may spend the day in front of a keyboard, but building code is a lot more than just typing. A programmer can write anywhere from a dozen to a few hundred lines of code in a day. But if you hand that programmer a piece of paper with a few hundred lines of code on it, it might take 10 or 15 minutes to actually type them into a computer. The "work" of programming isn't the typing, it's figuring out what the code actually needs to do, and making it work correctly and efficiently.

Q: Doesn't osmotic communication interrupt people's work? Isn't it hard to work in a noisy environment?

A: Osmotic communication works best when people on the team are used to some noise. Our ears tend to perk up when someone is talking about something important and relevant—like how you can hear your name in a crowded room—so it's not hard to tune out conversations around you if you're used to them. It doesn't work so well in a "dead quiet" office environment where everyone feels compelled to whisper, or just not talk at all.

Q: I'm still not clear on how planning works in XP. When does the team meet? How do the stories get estimated?

A: The team estimates stories together during the quarterly planning meeting which happens at the beginning of the quarterly cycle, and they talk about those estimates during the weekly planning meeting at the start of the weekly cycle. They'll also discover stories along the way, so they meet up and estimate those stories together. As far as how stories are estimated, it's pretty common to see XP teams using planning poker, but they might also just talk about the story and come up with an estimate that makes sense.

Q: When does the team demo the software to the users?

A: At a meeting at the beginning or end of the weekly cycle, where the users see the software and discuss what the team will work on next. The relationship between the team and the users isn't as formal as it is in Scrum, which has a specific role—Product Owner—for a customer representative who can accept the software. XP doesn't have prescribed roles, but XP teams recognize that it's ideal to have real customers involved. Really effective XP teams feel that the "whole team" practice means treating users who help them understand what to build as a true part of the team.

Q: So wait—XP really doesn't have prescribed roles?

A: No, it doesn't. One of the basic ideas in XP is that if there's a job to do, someone will step up and do it. Everyone on the team brings something unique to the table, and each person's individual role on the project will change based on what's needed, and what their expertise is.

Q: I've heard programmers complain about being assigned "maintenance" tasks, like fixing bugs in old systems. Is that really creative or innovative?

A: Actually, maintenance work can be some of the most intellectually challenging work there is for a software team. Think about what "maintenance" actually means: fixing bugs, often in code you didn't even write yourself. That means taking a machine, one that may be really complex, figuring out how it works (often without much documentation and nobody to ask for help), tracking down how it's broken, and figuring out a way to fix it. Programmers often groan about having to do maintenance, and that gives it a reputation as being "grunt" work: it's intellectually demanding and, unlike coming up with a cool new feature, it's rarely rewarded or complimented by your boss or coworkers.

> **People on XP teams take work-life balance really seriously: they go home at a reasonable time every day so they can keep up a sustainable pace.**

xp extreme programming

SO FAR, XP'S BEEN ABOUT PLANNING THE PROJECT AND GIVING THE TEAM A RELAXED, CODE-FRIENDLY ENVIRONMENT. BUT WE HAVEN'T TALKED MUCH ABOUT **CODE QUALITY** OR **PROGRAMMING** YET, EVEN THOUGH YOU SAID EARLIER THAT THEY'RE A MAIN FOCUS OF XP. SO THE THINGS WE LEARNED SO FAR *MUST TIE IN WITH PROGRAMMING*, RIGHT?

Yes! XP teams give themselves space to build great code.

Everything we've talked about so far has been about removing things that kill the team's momentum. Iteration, slack, and stories help the team build the right software, and protect them from unnecessary schedule pressure. An energized workspace, a supportive team, and an informative team space give them the best possible environment to get work done. It's not an accident that XP focuses on these things—they're at the root of the vast majority of team problems, and eliminating them gives the team fertile ground for innovation.

So now the stage is set, and the team is ready. **It's time to dig into the code.**

BULLET POINTS

- **Communication**—one of XP's values—is what matters most in a software project.

- The **informative workspace** practice means anyone can walk into the team space and get a sense of how the project is going just by looking around.

- People absorb information via **osmotic communication** when they sit together and overhear useful discussions.

- The **energized work** practice is how the team stays relaxed, rested, and in the best mental shape for work.

- **Information radiators** are large visual tools like task boards or burndown charts that "radiate" information because they're posted in a place that's hard to miss.

- XP teams work at a **sustainable pace** so they don't burn out. This typically means working regular hours.

- An energized team has enough time to do the job, and **freedom to make mistakes**.

- Interruptions can break a developer's concentration and take him or her out of **flow**, a state of high concentration where he or she is "in the zone."

Question Clinic: The "which-is-NOT" question

YOU'LL SEE SOME QUESTIONS ON THE EXAM THAT LIST VALUES, PRACTICES, TOOLS, OR CONCEPTS AND ASK YOU TO DETERMINE WHICH ONE OF THEM IS NOT PART OF THE GROUP. USUALLY, YOU CAN FIGURE THEM OUT BY GOING THROUGH THE ANSWER CHOICES ONE BY ONE AND ELIMINATING THE ONE THAT DOESN'T BELONG.

XP and Scrum are both iterative. XP uses weekly cycles and Scrum uses sprints. So this isn't the right answer.

You definitely find stories on both Scrum and XP teams, so this one's not right either.

97. Which of the following is NOT shared by both XP and Scrum?

A. Timeboxed iterations
B. Stories
C. Respect and courage
D. Slack

The values of respect and courage are shared by both Scrum and XP teams. So the answers include values as well as practices and tools.

D's definitely the right answer: slack is NOT shared by both XP and Scrum. XP teams use slack by including extra stories in their weekly cycles that can be skipped if the other stories take longer than expected. Scrum teams have a much stronger focus on project management, and have much more detailed planning practices and tools.

> **Keep your eyes out for questions that ask for "all of the following except" (that's just another way of wording a which-is-NOT question).**

TAKE YOUR TIME AND THINK YOUR WAY THROUGH IT. ALL OF THEM WILL HAVE SOMETHING IN COMMON BUT ONE. AS LONG AS YOU REMEMBER THE GROUP YOU'RE FITTING THEM INTO, YOU WON'T HAVE ANY TROUBLE.

Take your time answering which-is-NOT questions.

HEAD 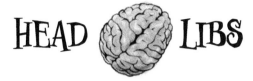 LIBS

Fill in the blanks to come up with your own "which-is-NOT" question!

Which of the following is NOT a _____?
(value, practice, tool, or concept)

A. _____
(value, practice, tool, or concept that is in the group)

B. _____
(the right answer)

C. _____
(value, practice, tool, or concept that is in the group)

D. _____
(value, practice, tool, or concept that is in the group)

Ladies and Gentlemen, We Now Return You To Chapter Five

putting the "p" in xp

```
File Edit Window Help XP
```

WARNING: The rest of this chapter talks about code

The P in XP stands for programming, and for good reason. While the XP practices you've seen so far apply to any team doing creative or intellectual work, the practices in the rest of this chapter are specifically focused on code.

If you're not a programmer, it's still worth reading the chapter! However, some of the material is a little more code-oriented than what you're used to seeing. But if you plan on working with a team that builds software, some familiarity with these ideas can be REALLY VALUABLE to you and your team. Understanding your teammates' perspectives better can help you reach a more agile mindset together.

If you don't have any background in programming, you should feel comfortable skipping over the sections of this chapter that contain snippets of code. Just make sure to read all the text, paying special attention to the words in boldface. If you do the exercises and test your knowledge with the crossword and exam questions at the end of the chapter, that's a very effective way to get the most important parts of XP into your brain.

And if you're using this book to study for the PMI-ACP® exam, don't worry - the exam does NOT require you to have programming knowledge.

xp *extreme* programming

> I ADMIT THAT I WAS SKEPTICAL ABOUT THIS SUSTAINABLE PACE STUFF, BUT NOW I'M CONVINCED. WE'RE NOT WORKING NIGHTS AND WEEKENDS ANY MORE, BUT I'M **GETTING A LOT MORE DONE!** CODING GOES A LOT FASTER WHEN MY **BRAIN'S NOT FRIED.**

LATER, BACK IN THE TEAM'S MEETING SPACE...

> UGH! THIS CHANGE IS REALLY GOING TO GIVE US A HEADACHE... AND IT WAS **AVOIDABLE.**

Ana: Quit your whining, Ryan.

Ryan: Hey, don't give me that attitude. This affects you, too.

Ana: OK, I'm listening. What's the problem?

Ryan: You won't like this. It's a change to personal trainer schedules on the mobile app.

Ana: Customers getting notifications about personal training sessions. What's the problem?

Ryan: The problem is they don't just want to get notifications. They want to schedule classes from the mobile app too.

Ana: Oh no. No, no, no. That is not going to work with the way we built this.

Ryan: Tell that to Coach. He's been promising that to the customers.

Gary: Did I hear someone mention me?

Ana: You promised customers that we'd let them schedule classes from the app?!

Gary: What's the problem, guys? How hard can it be to add that?

Ana: We're going to have to completely redesign the way data goes into the system.

Ryan: You know what's frustrating? ***If you'd just told us this was coming a few months ago***, we would have built a completely different backend for the last version.

Ana: Now we have to rip out the database entry code and replace it with a new service.

Gary: I know you guys can do it.

Ryan: Of course we can. But rewriting that much code will leave us with a ***giant mess***.

Ana: You know how they say rework creates bugs? This is a prime example.

Ryan: And that means a lot of ***totally avoidable late nights***. This stinks.

Back in Chapter 2 we learned that rework is a major source of bugs. Does that mean rework always causes bugs?

does rework always have to cause bugs

XP teams embrace change

Here's a basic fact about software projects: they change. A lot. Users ask for changes all the time, and typically have no idea how much work any one change will require. This wouldn't be too bad, but there's a problem: a lot of teams build code that's *difficult to modify*: changes require code modifications that are painful to make, and leave the code in very bad shape. This often leads to teams that push back against those changes. When teams *resist* change, the project suffers. The XP values and practices fix this problem at its source by helping teams to build code that's easier to modify. And when the code is easy to modify, programmers don't feel the need to resist change. That's why **XP has practices and values that are focused on programming**. Because these practices help teams build code that's easier to modify, XP helps them reach a mindset where they *embrace change* instead of resisting it.

People on software teams resist changes when they've had bad experiences with rework causing bugs... but rework doesn't have to do that. XP helps teams embrace change with practices and values that help them to build software that's easier to modify.

> Ever heard a programmer complain about spaghetti code? That's when the code's structure is complex and tangled (like a pile of spaghetti in a bowl). It's often the result of rework that results in many changes to the same part of the codebase. Programming doesn't have to be that way! XP teams have practices and values that make their code easier to modify, so the team can do rework without turning the code into a mess.

Frequent feedback keeps changes small

Talk to a group of programmers, and it often doesn't take long before someone starts to complain about how users always change their minds. "They ask for one thing, but when we build it exactly like they wanted they turn around and tell us they need something totally different. Wouldn't it be easier if we just built the right thing in the first place?"

But ask that same group of programmers how often they designed and built an API, only to find that some of the functions were awkward and difficult to work with. Wouldn't it be easier if we just built the right API in the first place? Obviously. But you don't really know if the interface you designed and built is easy to use until someone writes code that actually uses it.

An API ("application programing interface") is a set of functions that you build into your system so that another programmer can write code to control it.

When you put it that way, programmers recognize that it's really rare to build anything right the first time, so they try to get feedback early and continue to get it frequently. That's why XP teams value **feedback**.

And feedback comes in many forms:

Iteration

You've already seen a really good example of feedback: iteration. Instead of planning six months of work doing one big demo at the end of it, your team will do a small chunk of work and then get feedback from the users. That lets you **continually adjust the plan** as the users learn more about what they need.

Integrating code

When code files on your computer are out of date with the rest of your team, it can lead to frustrating problems. When you **integrate** your new code frequently with your teammates' code, it gives you early feedback. The more frequently you integrate, the earlier you catch conflicts, which makes them a lot easier to resolve.

Feedback

Teammate reviews

Open source teams have an old saying: "Given enough eyeballs, all bugs are shallow." Your team is no different. Getting **feedback from your teammates** helps you catch problems with your code—and it helps them understand what you built so they can work on it later.

That's called Linus's Law, named after the creator of Linux.

Unit tests

One really effective way to get feedback is to build **unit tests**, or automated tests that make sure the code that you built works. Unit tests are typically stored in files along with the rest of the code. When you make a change to your code and it breaks a test, that's some of the most valuable feedback you can get.

if this happened to you you'd fear change too

Bad experiences cause a <u>rational</u> fear of change

There's very little that's more frustrating to a programmer than being halted in your tracks by an annoying, frustrating problem. Most developers will recognize these very common—and very frustrating—problems that Ryan and Ana are running into.

> NOOOOOOOO! SOMEONE JUST COMMITTED THREE WEEKS OF CHANGES ALL AT ONCE, AND NOW ***I'M GETTING DOZENS OF CONFLICTS*** WHEN I TRY TO COMMIT MY CODE.

> IT'S GOING TO TAKE ME HOURS TO SORT OUT ALL THESE CONFLICTS. THIS IS A ***NIGHTMARE!***

The CircuitTrak code is stored in a version control system. That's a software tool that teams can use to work on a single set of files without overwriting each other's changes. It keeps track of every code change, letting you see who made the change and see what the code looked like before and after.

When you add your latest code changes to a version control system, it's called a commit.

Ana's been working on a change that affects a lot of different files. One of her teammates spent the last few weeks modifying many of those files, and just committed his changes. The good news is that when she tried to commit her code, the version control system detected the conflicts and rejected her update. The bad news is that now she has to painstakingly reconcile each of her changes with the ones her teammate made.

This is going to be so much work, Ana's considering just rewriting the whole thing from scratch rather than try to merge her changes.

> I STARTED FIXING A BUG, BUT TO GET IT TO WORK I HAVE TO FIX THESE OTHER TWO PARTS OF THE CODE...

> ...AND ONE OF THOSE CHANGES AFFECTS *THIS OTHER PIECE*, AND *THAT REQUIRES TWO MORE CHANGES OVER HERE...*

Ryan started making one little change to the code, but somehow it rippled out to many other areas of the software. By the time he gets all the way down the chain of fixes, he's practically forgotten what he was supposed to be doing in the first place. This is a really familiar feeling to a lot of programmers, and it's got a name: shotgun surgery.

> ...URGH, SO MANY CHANGES IT ***HURTS MY BRAIN!***

How a version control system works

Behind the Scenes

Ana, Ryan, and the rest of the team have been working on CircuitTrak for years, and the project now has thousands of files, including source code, build scripts, database scripts, graphics, and many other files. If they just used a shared folder on the network to hold the files, it would quickly become a mess:

10 a.m.: Ana copies the source folder to her computer so she can work on the code

11:30 a.m.: Ryan copies the source folder to his computer so he can work on the code, too

1 p.m.: Ana copies updated *TrainerContact.java* back to the shared folder

3 p.m.: Ryan copies updated *TrainerContact.java* back to the shared folder

Oops! When Ryan saved his changes, they overwrote Ana's changes. That's going to cause bugs later!

That's why the team uses a version control system, which provides a **repository** that contains not just the latest copies of each file, but also a complete history of changes. It even lets multiple people work on the same file at once:

10 a.m.: Ana checks out the source from the repository to a working folder on her computer

11:30 a.m.: Ryan has the source checked out, so he updates it with the latest changes

1 p.m.: Ana commits her *TrainerContact.java* changes back to the repository

3 p.m.: Ryan commits changes to a different part of *TrainerContact.java*

Ryan and Ana changed different lines of the file, so the version control system was able to merge their changes automatically.

Things are a little messy (but still manageable) when two people make **conflicting changes** to the same file.

10 a.m.: Ana and Ryan both have working folders updated with the latest source

1 p.m.: Ana commits changes to *TrainerContact.java* back to the repository

2:30 p.m.: Ryan tries to commit a conflicting change but it's rejected

> **When Ryan tried to commit his change, the system found that Ana had already checked in different changes to the same lines in the same file, so it rejected the change and updated his local file to show both sets of changes. Ryan had to resolve the conflicts before he could commit any more code.**

In this case, there were only a few conflicts, so Ryan was able to resolve them pretty easily and commit the updated code. But when there are a LOT of conflicts, the merge can be really difficult.

early feedback makes changes manageable

XP practices give you feedback about the code

A lot of agile practices are crafted to give the team feedback early and often—like the ones focused on iteration, which give the team feedback about the product they're planning to build and the work involved. Each iteration gives them more information, and they use that information to improve how they plan the next iteration. This is an example of a **feedback loop**: the team learns from each round of feedback, and makes adjustments and self-corrects, which changes what they learn about in the next round. These **next four XP practices** are especially good tools because they give the team really good feedback about how they design and build the code.

XP Practice
Pair programming
Two team members sit with each other in front of the same computer and work together to discuss, design, brainstorm, and write the code.

Team members give each other feedback about the code they're building. They switch pairs often, which helps everyone stay on top of how the whole codebase is changing.

XP Practice
10-minute build
The team maintains an automated build that compiles the code, runs the automated tests, and creates the deployable packages. They make sure that it runs in 10 minutes or less.

You learn a lot about your code when you try to build it and see what breaks. When the build runs quickly, everyone on the team is comfortable running it as often as they need to.

Under the Hood: Build Automation

Here's a quick overview of how an automated build works, just in case you haven't used one before.

Automated builds turn source code into packaged binaries

If you're not a programmer, you may not be 100% clear on the mechanics of how software is created. Exactly what do programmers type all day, and how does it turn into software that you can run? Here's what's going on:

- Software typically starts out as a set of **text files that contain code**. This is the source code of the project.
- Programming languages have compilers that **read the source code files and create a binary**, or the executable file the computer's operating system is able to run.
- The binary typically needs to be **packaged** into a single file that contains the binary and any additional files that are needed to run (like an executable installer, a mountable disk image, or a deployable archive file).
- Compiling the source code and packaging it up by hand can be **time-consuming and error-prone**, especially if there are multiple binaries and many files that need to be bundled into a single package.
- This is why teams **automate the compile and package steps** to create the binaries and other files. There are many tools and scripting languages that make it easy for teams to create automated builds.

xp *extreme* **programming**

> A system that gives a lot of feedback will often **fail fast**. You want your system to fail quickly so you can fix problems early, before other parts of the system are added which depend on them.

XP Practice
Continuous integration
Everyone on the team constantly integrates the code in their working folders back into the repository, so that nobody's working folder is more than a few hours out of date.

When each person has the latest code in a working folder, conflicts show up immediately, and they're a lot easier to fix when they're caught early.

A 10-minute build really helps with both continuous integration and test-driven development because it executes the unit tests, so you find out quickly if you add code that breaks an existing test.

XP Practice
Test-driven development
Before adding new code, the first thing a team member does is to write a unit test that fails, and only after that does he or she write or modify code to make it pass.

Unit tests set up a tight feedback loop: build the failing test, write code that makes it pass, learn more about what you're building, write another test, repeat.

⌐ **DICTIONARY DEFINITION** ¬
re-factor, verb
to change the structure of code without changing its behavior

*A particularly troublesome block of code that was much less frustrating to work with after Ryan took the time to **refactor** it.*

Developers typically run unit tests using a specialized program (often a plug-in for a build tool or a development environment). The unit test results are usually displayed with color codes: passing tests are green, failed tests are red. Teams that use test-driven development typically follow a cycle of adding failing tests that start off red, making them pass so they turn green, and then refactoring the code. Teams refer to this cycle as red/green/refactor, and consider it a valuable development tool.

Ready to take a closer look at each of these practices? Flip the page! ➜

you are here ▸ **209**

giant merge conflicts are really frustrating

XP teams use automated builds that run quickly

There's nothing more frustrating to a programmer than waiting. That's a good thing: a lot of innovation starts with a programmer saying, "I can't stand how long this takes." So it's especially frustrating when it takes a lot of time and effort to build the code—and very few things can kill a team's innovation as quickly as frustration. When something repeatedly takes a long time, the first thing that a good programmer thinks is, "How do I automate that?"

That's where the **10-minute build** practice comes in. The idea is straightforward: the team creates an automated build, usually using a tool or scripting language specifically made for automating builds—and they do it at the beginning of the project. The key here is that they make sure the whole build runs in under 10 minutes. That's pretty much the limit of patience most programmers have to wait for the build to finish—and it's long enough so that you can kick off a build, then go grab a cup of coffee and think. By keeping the build under 10 minutes, there's no hesitation in running it, which helps find build problems quickly.

When the build requires a lot of manual effort or takes longer than 10 minutes to run, it puts stress on the team and slows down the project.

The automated build reads the source files and packages up the binary.

When the build runs in 10 minutes or less, developers are comfortable running it frequently.

A 10-MINUTE BUILD IS JUST LONG ENOUGH FOR A PROGRAMMER TO **GRAB A CUP OF COFFEE** AND RELAX HIS OR HER BRAIN.

Continuous integration prevents nasty surprises

When you work on a team building code and committing it to a version control system, your day-to-day work follows a pattern. You do some work, then you update your working folder with the latest changes your teammates have pushed, then you push your own changes back to the version control system. Work, update, push... work, update, push... work, update... **merge conflict**! Uh-oh—one of your teammates made changes to the same line and committed it since the last time you updated. The version control system had no way of knowing whose change is right, so it modified the code files in your working folder with both sets of changes. Your job is to **resolve the conflict**: you'll look at them, figure out what the code is supposed to do, modify it so that it's correct, and commit the resolved change back to the repository:

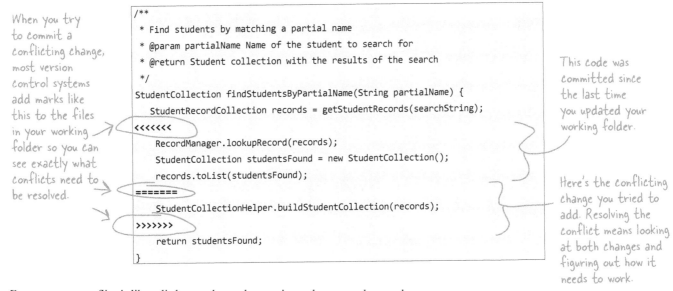

When you try to commit a conflicting change, most version control systems add marks like this to the files in your working folder so you can see exactly what conflicts need to be resolved.

This code was committed since the last time you updated your working folder.

Here's the conflicting change you tried to add. Resolving the conflict means looking at both changes and figuring out how it needs to work.

Every merge conflict is like a little puzzle, and sometimes those puzzles can be annoying to solve because you're not sure exactly what your teammate was trying to do.

Now flip back to page 206. Do you see why Ana ran into so much trouble?

Her teammate hadn't updated his working folder in weeks. Instead, he just kept making changes to an old version of the code—which got more out of date every day—and then committed all of those changes at once. Ana spent the last few hours working on a change to many of those same files. But instead of having one or two little puzzles to solve, now she has to contend with dozens of files marked up with conflicts. Few things are more frustrating to a programmer than resolving many merge conflicts at once.

That's why XP teams use **continuous integration**. It's a really simple practice: every person on the team integrates and tests their changes every few hours, so nobody's working folder is ever out of date. When everyone on the team does continuous integration, they're all working on a current version of the code. There will still be merge conflicts, but they're almost always small and manageable, and never a giant, frustrating monster of a change like the one that Ana has to deal with.

When everyone keeps their working folders up to date, the merge conflicts tend to be small and manageable.

The weekly cycle starts with writing tests

For many developers, XP brings a different way of working. One of the most obvious changes is that the team does **test-driven development** (or TDD). That's a practice where programmers write unit tests before they write the code that it tests. When you're in the habit of writing unit tests first, you think about what it means for your code to work correctly, which helps you write code that's *"Done"* done.

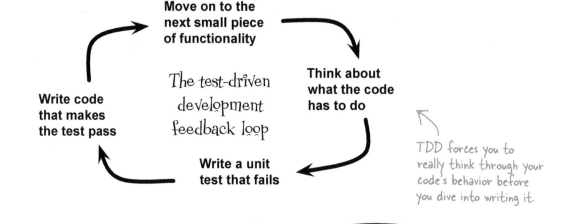

TDD forces you to really think through your code's behavior before you dive into writing it.

> OK, THIS DOESN'T MAKE SENSE. HOW DO YOU TEST CODE THAT YOU HAVEN'T WRITTEN YET? **WHY DOES IT MATTER** IF I WRITE THE TESTS FIRST OR WRITE THE CODE FIRST?

Unit tests change the way the team designs the code

All programmers have had the experience of writing code, only to wish later that they had written it a little differently—looking back, you might realize that a different argument would have worked better for a particular function, or that you could have used a different data structure, or made other choices. But now the code you wrote is called from five other places, and it will be more work to change it than it will be to just live with the poor decision.

In other words, some of the most annoying code problems happen when you make a bad design choice, then you add other code that depends on it. Do this enough and eventually you get that "shotgun surgery" feeling every time you touch that part of the code.

TDD makes it really obvious when you add extra unnecessary dependencies between different parts of your code, and those dependencies are exactly what cause the "shotgun surgery" feeling.

Unit testing helps prevent that problem. Design problems in your code often become apparent the first time that you write code that uses it. And that's exactly what you're doing when you write a unit test first: you **use the code that you're about to write**. And you do it in small increments, one bit at a time, smoothing out design problems as you encounter them.

xp extreme programming

Test-Driven Development Up Close

```
public class ScheduleFactory {
    public class TrainerManager {
public class UserInterfaceModel {
```

Code is always divided into discrete units

In some languages the units are classes; in others they're functions, modules, procedures... the specific unit varies from language to language, but every programming language works this way. For example, when you write Java code, most of your code goes into "chunks" called classes saved in *.java* files. Those are units of Java code.

Each unit gets its own unit tests

The name "unit testing" is pretty self-explanatory: you write tests for the units of code. For example, in Java unit testing is typically done on a class-by-class basis. Those tests are written in the same language as the rest of the code, and are stored in the same repository. The tests access whatever part of the unit is visible to the rest of the code—for Java classes, that means the public methods and fields—and use them to make sure the unit works.

```
public class ScheduleFactoryTest {
public class TrainerManagerTest {
public class UserInterfaceModelTest {
```

> The XP mindset helps you think differently about programming, design, and code because its practices give you **good habits** that keep your code clean, simple, and easy to maintain. TDD is one of those good habits.

This is just like how the Scrum mindset helps you think differently about planning.

Writing the unit tests first forces the developer to think about how the code is going to be used

Every unit of code is used by at least one other unit somewhere in the system—that's how code works. But when you're writing code, there's a paradox: in a lot of cases, you don't really know exactly how the unit you're working on will be used until you actually use it.

Test-driven development helps you catch problems in your code early, when they're much easier to fix. It's surprisingly easy to design a unit that's difficult to use later, and just as easy to "seal" in that poor design by writing additional units that depend on it. But if you write a small unit test every time you make a change to a unit, a lot of those design decisions become obvious.

*HMM, I DIDN'T REALIZE HOW WEIRD THIS CLASS WAS UNTIL I STARTED WRITING CODE IN A UNIT TEST THAT USES IT. I'M GLAD I CAN FIX IT NOW **BEFORE ANYTHING ELSE DEPENDS ON IT!***

you are here ▸ **213**

early feedback makes a better user interface

Agile teams get feedback from design and testing

Agile teams have great design and testing tools that help the teams get more feedback throughout the project. They can use **wireframes** to sketch out user interfaces before building them, **spike solutions** to figure out difficult technical problems, and **usability testing** to make sure they've made effective design choices. Some teams write a charter and list of test objectives as a very lightweight plan and then set out to break the feature or product that was just developed as a way of finding new combinations of actions that the developer might not have thought about. This kind of testing is called **exploratory testing** and it can be really effective at finding issues that users will run into. All of these tools are great at generating feedback, which is why XP teams integrate them into their weekly cycles—and rely on them to get feedback to help plan the next weekly cycles.

Wireframes help the team get early feedback about the user interface

Of all the things that software teams build, user interfaces seem to generate the most opinions from users and stakeholders, so they want to get feedback about the UI early and often. That's why teams use wireframes to sketch out user interfaces. There are a lot of different ways to create wireframes. Some are basic sketches of the system's navigation, while others are highly detailed representations of individual screens or pages. It's a lot easier to modify a wireframe than it is to modify code, so teams often review several iterations of each wireframe with the users.

Build spike solutions to get an idea of a feature's technical difficulty

It's not uncommon for a team to have trouble estimating a specific feature because they just don't know enough about what's involved in solving specific technical problems. That's where spike solutions come in handy. A spike solution is code written by a team member specifically meant to figure out a specific technical problem. The only purpose of the spike solution is to learn more about the problem, and the code is usually thrown out after it's done.

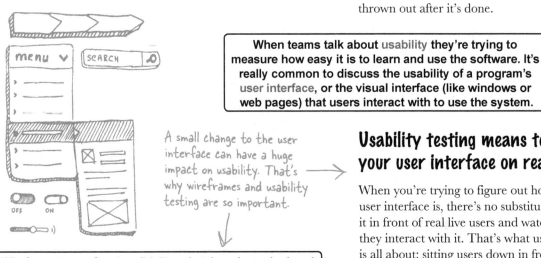

A small change to the user interface can have a huge impact on usability. That's why wireframes and usability testing are so important.

> When teams talk about usability they're trying to measure how easy it is to learn and use the software. It's really common to discuss the usability of a program's user interface, or the visual interface (like windows or web pages) that users interact with to use the system.

Usability testing means testing your user interface on real users

When you're trying to figure out how effective a user interface is, there's no substitute for getting it in front of real live users and watching how they interact with it. That's what usability testing is all about: sitting users down in front of an early version of the UI that the team has been building and having them use it to perform tasks they'll typically need to use it for. When XP teams do usability testing, it's often done near the end of the weekly cycle so that the information they learned from it can be used in the next one, setting up an extremely valuable feedback loop.

> Wireframes are often low fidelity: sketches drawn by hand or with a program that gives them a hand-drawn look. This encourages users to feel more comfortable suggesting changes than if they were more polished. Some users are hesitant to ask for changes if a UI looks really polished because it feels like they're asking the team to do a lot of extra work. Making wireframes look hand-drawn increases the amount of feedback that they generate.

xp *extreme* programming

Spike Solutions Up Close

A spike solution helps you solve a tough technical or design problem.

A spike solution is a simple program whose only purpose is to explore solutions to a problem. It's usually timeboxed to a few hours or even a few days, and after the spike is done the code is usually thrown away or set aside (so the team can use it later if they want). This gives the programmer a lot of freedom to focus single-mindedly on solving the problem and ignore the rest of the project. But even though the code is thrown away, the spike is still treated as real project work. The team will typically add a story for it to the weekly cycle.

Architectural spike

When XP teams talk about doing a spike solution, they're usually referring to an **architectural spike**. An architectural spike is used to prove that a specific technical approach works. Teams will often do an architectural spike when they have a few different options for designing a specific technical solution, or if they don't know if a certain approach will work.

Risk-based spike

Sometimes there's a problem that presents a project risk: the developers are pretty sure it will go well, but if it doesn't it could seriously derail the project. That's when the team will do a risk-based spike. It works just like an architectural spike, but with a different goal: it's done specifically to remove a risk from the project.

Spike solutions fail fast: if the programmer discovers that the approach won't work, the spike ends... and the team still considers it to be a <u>successful</u> spike.

Sharpen your pencil

Wireframes, usability testing, and spike solutions aren't specific to XP... but a lot of XP teams use them.

Here are three scenarios that Ryan and Ana are working on that have to do with getting feedback from their project. Write down the name of the tool being used in each scenario.

WE NEED A NEW WAY TO STORE TRAINER SCHEDULES THAT WILL REDUCE MEMORY. I'M BUILDING A PROOF-OF-CONCEPT SO WE CAN GET A SENSE OF HOW MUCH WORK IT WILL BE.

_____

I'M NOT HAPPY WITH HOW THIS CLASS GETS INITIALIZED – IT'S GOING TO BE HARD TO USE. ONCE ITS UNIT TESTS PASS, I'LL MODIFY IT.

_____

I FINISHED DESIGNING THE NEW USER INTERFACE. LET'S MAKE SURE THAT IT WORKS BY GETTING A BUNCH OF USERS IN THE ROOM AND OBSERVING THEM WHILE THEY USE IT.

_____

→ Answers on page 244.

two heads are *better than one*

Pair programming

XP teams use a pretty unique practice called **pair programming**, where two people sit at a single computer and write code together. This is a new experience for people who are used to thinking of programming as a solitary activity. But it can really be an effective tool for building high-quality code very quickly, because many people who do pair programming report that pairs get more work done together than they do when they work separately.

> Pair programming keeps everyone focused, helps the team catch bugs, makes it easy to brainstorm, and gets everyone on the team involved in every part of the codebase.

Ryan and Ana help keep each other focused.

When Ryan gets stuck, Ana can jump in and keep things moving, and vice versa.

They're constantly talking about the problems they're working on and brainstorming solutions.

You're a lot less likely to take shortcuts when someone else is working alongside you.

The whole team constantly rotates pairs, so everyone gets experience working with every part of the system.

You don't always know if you have a good handle on an idea until you explain it to someone else.

There are two sets of eyes on every change, and they catch a lot of those little mistakes that would cause headaches later if they were missed.

xp extreme programming

Sharpen your pencil

The XP practices are useful individually, but when you combine them they're especially effective. We've written down several of the XP practices, and drawn arrows between them. Each arrow has blank lines for you to write on. For each set of blank lines, write down one way that the practice the line is drawn FROM can interact with the practice the line is drawn TO in a way that reinforces and supports it.

We've started you out by filling in this blank to show how "sit together" impacts "informative workspace".

INFORMATIVE WORKSPACE ← _osmotic communication is more frequent when everyone sits close to each other_

SIT TOGETHER

SLACK

WEEKLY CYCLE

PAIR PROGRAMMING

ENERGIZED WORK

10-MINUTE BUILD

TEST-DRIVEN DEVELOPMENT

CONTINUOUS INTEGRATION

you are here ▶ **217**

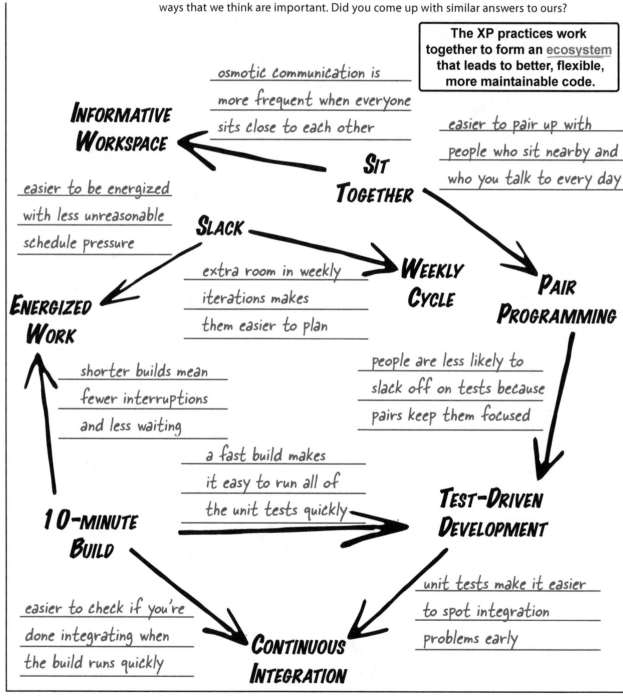

there are no Dumb Questions

Q: The "sit together" and "pair programming" practices require everyone to be in the same office. Does that mean global or distributed teams can't use XP?

A: Many global and distributed teams use XP. XP teams know that when they sit together, they have more face time, have fewer interruptions from phone calls, and can share an informative workspace. A distributed team where everyone works in different offices and communicates via email and phone can't take advantage of these things. But an important part of the XP mindset is that every practice is about making the team work better. If there are some practices that are simply impossible, they'll work with what they've got.

Q: But doesn't that mean they aren't doing "pure" XP?

A: Really effective teams know that there's **no such thing as "pure" XP**. XP teams are always looking for ways to improve. There's no "perfect" state that they're trying to get to; they're just trying to get better at what they do together. Mindlessly adhering to practices will de-energize the environment really quickly. And making people feel bad about not being "pure" enough is disrespectful. **Nagging people about XP "purity" is counterproductive.** It makes people feel like you're judging them and their work. That won't make anyone change—it'll just make them resent you, and resent XP.

Q: So does that mean it's OK to throw out practices I don't like?

A: No, that's not OK. The XP practices are carefully designed to work with each other, and when they're used together they help the team integrate XP values into their mindset. For example, teams really start to understand the XP value of communication when they sit together and have an informative workspace. When teams decide to throw out a practice, it's usually because their mindset is incompatible with one of the values, and that causes the practice to feel uncomfortable. When that happens, a really good thing to do is to **make a genuine effort to try the practice**. Often, that helps the team shift their mindset, which in turn helps everyone work better together and build better software.

Q: Doesn't continuous integration just mean setting up a build server?

A: No. A build server is a program that periodically retrieves the latest code from the version control system, runs the automated build, and alerts the team if there are any failures. It's a really good idea, and almost all agile teams use one. But a build server isn't the same thing as continuous integration. Continuous integration means that every person on the team actively (and continuously!) integrates the latest code their teammates wrote into his or her own working folder. The reason this is often given the same name as a build server is that the server is constantly "integrating" code from the version control system into its own repository, and will alert the team any time code is committed that won't compile or causes test failures. But that's no substitute for having each person keep his or her working folder up to date.

Q: I don't get it. If we have a build server that constantly integrates the code, isn't that less work for everyone?

A: It's true that having every member of the team continuously integrate the latest code from the version control into their working folder is more work than just setting up a build server. But if they just rely on email alerts from a build server to tell them when they're out of sync, it often ends badly. For example, you might discover that when you commit code that breaks the build, everyone gets really mad at you, so you commit your code a lot less frequently than you normally would. Or the team might just be so used to "broken build" emails from the build server that they start ignoring them and filing them into folders. On the other hand, if every person feels like they have a responsibility to stop what they're doing every few hours and integrate code from the version control system back into their own working folders, a broken build is rare—and when it happens, the team notices quickly and works together to fix it.

Q: So is doing continuous integration just a matter of making sure the team has enough discipline?

A: Not quite. When a team is really good at using practices like continuous integration, 10-minute builds, or test-driven development, from the outside it looks like they're really disciplined. But it's not really a matter of discipline at all. The team does those things **because they make sense to everyone**. Everyone on the team simply feels that the work will slow down if, say, they don't take the time to make the build faster, or build a unit test before writing code. They don't need to be nagged, yelled at, or reprimanded—in other words, disciplined—because it wouldn't occur to them *not* to do those things.

Q: I work with a QA team. Does test-driven development mean testers write my unit tests while I write the code?

A: No. You write your own unit tests first, then you write the code to make them pass. The reason that the unit tests should be written by the same person who writes code is that when you write the tests, you learn a lot about the problem that you're working on, which makes the code better.

a skeptical reaction to pair programming

> SOMETHING'S REALLY BUGGING ME. PAIR PROGRAMMING SEEMS LIKE A GIANT WASTE OF TIME. WHEN TWO PEOPLE WORK TOGETHER, DOESN'T IT CAUSE THEM TO **DELIVER CODE HALF AS FAST?**

Pair programming is actually a really efficient way to code.

Pairing up keeps you focused and eliminates a lot of distractions (like popping open a browser or checking your email). And there's always another set of eyeballs to catch bugs early instead of wasting time tracking them down later. But more importantly, it means you're **constantly collaborating with your teammates**. Programming is an intellectual activity: writing code means solving problems and puzzles all day long, one problem or puzzle after another. Talking through those puzzles and problems with one of your teammates is a really effective way to solve them.

That's why even people who are initially resistant to pair programming often find that they really like it after <u>genuinely</u> trying it out for a few weeks.

> Is it really fair for us to use the word "irrational"? We think so. Pair programming is a straightforward and—let's face it—unremarkable way to work, and many people do it every day. A very intense and negative emotional reaction to something so mundane is, by definition, not rational.

> OK, I GET YOUR POINTS. BUT... ARE YOU SURE? HONESTLY, *I JUST DON'T BUY IT.* PAIR PROGRAMMING SIMPLY DOESN'T FEEL RIGHT TO ME.

A practice "just doesn't feel right" when it clashes with your mindset.

Do you think of yourself as a better programmer than everyone around you? Is coding a solitary activity in your mind? If so, then you'll have an irrational dislike of pair programming. Do you think of yourself as a "rock star" surrounded by idiots who couldn't code their way out of a paper bag? Then you'll have an *extremely strong irrational hostility* to pair programming. The key word here is **irrational**: yes, you can think of reasons and rationalizations for disliking pair programming, but at the heart of it what you really have is a <u>*feeling* that *it's just not right*</u> for you or your team. And that's the definition of irrational: decisions driven by feelings, not reason.

But it turns out that a lot of really good programmers on real-world projects have discovered that not only can their "lesser" teammates (surprisingly!) keep up with them, but when they *genuinely* try pair programming—not just go through the motions, but *really try* to make it work—coding really does go a lot faster. Not only that, but their "slower" teammates start to pick up many of the skills and techniques they've learned, and the whole team improves together.

xp *extreme* **programming**

*SORRY, I STILL DON'T BUY IT. PAIR PROGRAMMING IS **BAD**, AND NOTHING YOU SAY IS GOING TO CHANGE MY MIND. DOESN'T THAT MEAN ALL OF XP IS **WRONG FOR ME AND MY TEAM**?*

Then you and your team don't value the same things that XP teams do.

XP teams value focus, respect, courage, and feedback. If you really value these things, pair programming makes a lot of sense. When you value focus, you appreciate how pair programming helps keep you and your teammates on track. When you value respect, you won't have an irrational response to the idea of pairing up with your teammates, because you have respect for them and their abilities. When you value courage, then you'll be willing to look past your own feelings of discomfort and try something that could potentially help the team. And when you value feedback, then having two eyes on every line of code that's written feels like a really great idea.

On the other hand, if the last few sentences seem cliché, oversimplistic, overly idealistic, or even stupid, **then you don't share the same values** as effective XP teams.

> When you try to adopt a practice that doesn't match your mindset or the culture of your team, it usually doesn't "take" and you just end up going through the motions.

SO WHAT? WHAT HAPPENS IF I DON'T SHARE XP'S VALUES?

Adopting new practices takes work, and shared values motivate everyone to do that work.

When teams try to adopt a methodology with values that don't match the team's culture, it usually doesn't end well. The team will try adding some of the practices, and a few of them may work out temporarily. But eventually it will just feel like you and your team are simply "going through the motions" of the practices. They'll feel like a burden without much benefit, and within a few weeks or months the team will go back to the way things were before.

But that doesn't mean there's no hope! It just means that you and your team should *talk about the values **before** you try the practices*. If you start working on the culture issues from the beginning, it makes XP (or any methodology!) a lot easier to adopt, and gives the whole effort a much better chance of sticking.

Speaking of improving the way the team functions, let's check in on Ryan and Ana ⟶

mo code mo problems

> OUR DEVELOPMENT IS REALLY HUMMING ALONG! I CAN'T BELIEVE HOW MUCH CODE WE'VE WRITTEN IN THE LAST FEW WEEKS.

> YEAH, THESE NEW PRACTICES MADE A HUGE DIFFERENCE. BUT... WELL, I'M WONDERING IF IT'S *TOO MUCH OF A GOOD THING*.

Ryan: Ha ha! Good one! ... um ... wait, you're not joking, are you?

Ana: No, I'm being serious here. We're adding a lot more code, but now we're building some pretty complex stuff.

Ryan: Yes!

Ana: That's not necessarily a good thing.

Ryan: Uh... what?

Ana: Like this centralized common automated build script that you created.

Ryan: How's that a problem? We had a bunch of nearly identical build scripts. That's a lot of duplicated code. I fixed it.

Ana: Yeah, you saved like 12 lines of code duplicated in eight different build scripts...

Ryan: OK.

Ana: ...by building this 700-line monstrosity that's impossible to debug.

Ryan: Um...OK?

Ana: And now any time I need to modify the build, I have to spend hours trying to debug through that enormous script. It's really painful.

Ryan: But it saves...well...OK. It saves like 12 duplicate lines in a couple of scripts. I get your point—duplicate code is usually bad, but in this case it would be a lot easier to maintain a few duplicate lines than keep working with the script I wrote.

Ana: And it's not just the builds. We built this super complex unit test framework.

Ryan: I see where you're going with this. I had to debug through it the other day because I had to update the test data for just one unit test. It took me two hours to do a really simple job that should have taken five minutes.

Ana: You know what? I think adding these new XP practices helped speed up our coding. But I'm starting to think that all of this complexity is starting to slow us down.

Ryan: So what do we do about it?

Complex code is really hard to maintain

As systems grow, they often become big and complicated, and **complex code tends to get more complex** as you work with it. And when code gets more complex, it gets harder to work with, which causes developers to take shortcuts that make the problem worse. That's exactly what happened when Ryan tried to make a change that customers needed:

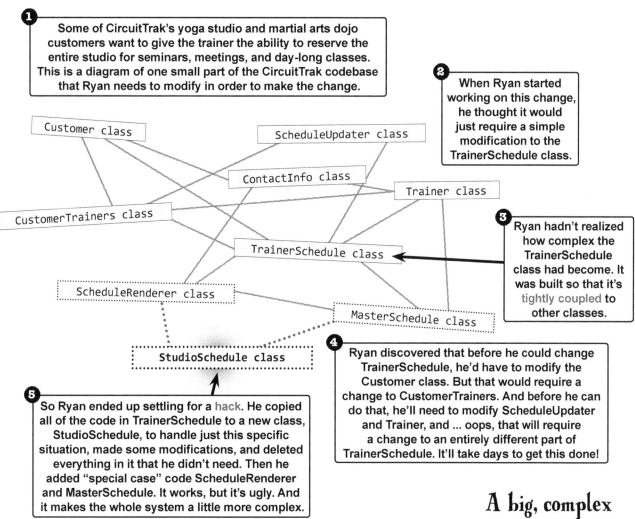

❶ Some of CircuitTrak's yoga studio and martial arts dojo customers want to give the trainer the ability to reserve the entire studio for seminars, meetings, and day-long classes. This is a diagram of one small part of the CircuitTrak codebase that Ryan needs to modify in order to make the change.

❷ When Ryan started working on this change, he thought it would just require a simple modification to the TrainerSchedule class.

❸ Ryan hadn't realized how complex the TrainerSchedule class had become. It was built so that it's tightly coupled to other classes.

❹ Ryan discovered that before he could change TrainerSchedule, he'd have to modify the Customer class. But that would require a change to CustomerTrainers. And before he can do that, he'll need to modify ScheduleUpdater and Trainer, and ... oops, that will require a change to an entirely different part of TrainerSchedule. It'll take days to get this done!

❺ So Ryan ended up settling for a hack. He copied all of the code in TrainerSchedule to a new class, StudioSchedule, to handle just this specific situation, made some modifications, and deleted everything in it that he didn't need. Then he added "special case" code ScheduleRenderer and MasterSchedule. It works, but it's ugly. And it makes the whole system a little more complex.

A hack (sometimes called a "kludge") is what programmers call a clumsy, quick-and-dirty solution that technically gets the job done but may cause problems down the road.

Ryan's solution is definitely a hack. The way he implemented it, if they ever need to change the way schedules work, they'll need to remember to change the new StudioSchedule class too—and that will probably cause a "shotgun surgery" cascade of changes.

A big, complex system gets more and more complex one little hack at a time.

refactoring often is a great habit

When teams value simplicity, they build better code

When you're solving a programming problem, there are an almost infinite number of ways that you can code the solution. Some of those ways are a lot more complex than others. They might have a lot more interconnections between units or add extra layers of logic. Units can grow far too big to understand all at once, or can be written in a way that's too convoluted to read and comprehend.

On the other hand, everything works better when your code is simple. It's easier to modify your code to add new behavior, or to modify it to change the way it works. When the code is simple, there are fewer bugs, and they're easier to track down when they happen.

So how do you know if a particular unit—like the `TrainerSchedule` Java class that Ryan was working on, for example—is getting too complex? There's no hard-and-fast rule that governs complexity. That's why instead of a rule, XP teams have a value. Specifically, people on XP teams value **simplicity**. Of the many ways to solve any particular coding problem, someone on an XP team will choose the simplest one that he or she can think of.

> Simplicity

Code gets complex when it does too many things.
One of the most common ways that your code can get complex is when one unit does too many things. Units of code tend to be organized by their behavior. When one unit does too many things, one of the most effective ways of reducing complexity is to **separate it into smaller units** that each do just one thing.

Refactor existing code to make it less complex.
There's no single "right" way to build a specific unit of code—there are many right answers, and it's rare to write code optimally the first time. That's why XP teams **refactor** their code as often as they need to. When they refactor (or modify the code to change its structure without altering its behavior), the code almost always ends up less complex than before.

Great habits are more effective than discipline.
If you try to nag your teammates (or yourself) into using practices like test-driven development or tools like refactoring, they typically won't stick. Instead, people on effective XP teams develop **great habits**. For example, they get into the habit of refactoring every time they see code that can be refactored, just like they get into the habit of writing unit tests first. This is part of the XP mindset.

XP team members are always on the lookout for units that are starting to get complex. They know it's worth taking the time to refactor as soon as they see anything at all that can be made more simple.

What kind of habits could help the CircuitTrak team avoid problems like the one Ryan ran into with the studio reservation modification?

224   Chapter 5

Simplicity is a fundamental agile principle

Let's take a closer look at one of the twelve principles behind the Agile Manifesto:

> Simplicity—the art of maximizing the amount of work not done—is essential.

Hmm... "maximizing the amount of work not done" sounds like a philosophical musing, or something the caterpillar said in *Alice in Wonderland*. What does that really mean?

When units are tightly coupled, it adds complexity to the project

When you're renovating a house, the most damaging thing that you can do is to take a sledgehammer and knock down a wall. That's one way that writing code is different from engineering physical objects: if you delete a bunch of code, it doesn't cause permanent damage to the project—you can easily recover it from the version control system.

If you really want to make your code worse, build some new code, modify a bunch of existing units so that they depend on it, and then modify some additional units so they're all tightly coupled to the ones you modified. That's practically a guarantee you'll spend many frustrating hours jumping from one unit to another trying to track down a problem.

WHAT WAS I THINKING WHEN I WROTE THIS CODE SIX MONTHS AGO? I THOUGHT I WAS MAKING IT REUSABLE, BUT IT'S JUST A MESS.

It's tempting to sacrifice simplicity for reusability

Developers love **reusable** code. When you're writing code, you often find that you need to solve the same problem in many different parts of the system. It's a really satisfying "a-ha" moment when you're working on a tough problem and you realize that you can call an existing method or use an object that already exists.

But there's a trap a lot of programmers fall into: optimizing code for reusability, while sacrificing simplicity. That's what Ana was talking about on page 222—Ryan created a very complex build script just to save a few lines of duplicated code, but the new script made it really hard to modify or fix problems in the build. Ryan wanted to avoid duplicate code, but ended up making the project harder to change.

> An effective way to maximize the amount of work not done is to only write code for a specific, concrete purpose that you know about right now. Avoid writing code just in case you might need it later.

pay down *your debt*

Every team accumulates technical debt

Little problems with your code add up over time. It happens to every team. All developers—even really good, highly skilled developers—write code that can stand to be improved. This is natural: when we're writing code to solve a problem, we often learn more about that problem as we work on it. It's very natural to write code that works, look at the results, *think about it for a while*, and then realize that there are ways that you can **improve and simplify it**.

But a lot of times developers don't always go back and improve their code—especially when we feel like we're under enormous pressure to get code out the door as soon as possible, even if it isn't as "done" as it could be. And the longer those "unfixed" design and code problems linger in the codebase, the more problems compound, which leads to complex code that's painful to work with. Teams refer to these lingering design and code problems as **technical debt**.

> I HAVEN'T TOUCHED THIS PART OF THE CODE IN TWO YEARS, AND IT'S A *GIANT PILE OF SPAGHETTI*. NOW I HAVE TO FIX A BUG IN IT. THIS IS GOING TO BE A *HUGE HEADACHE*.

USE SIMPLICITY AND REFACTORING TO PAY DOWN TECHNICAL DEBT

Technical debt happens to all teams. Why? Here's one reason: it's easy to write code that's a little too complex. Here are a few tips for achieving **greater simplicity** and **avoiding technical debt**:

★ **Care:** Simplicity is a **value**, which means that you have to genuinely care about it.

★ **Plan:** It takes work to simplify your code—in fact, a lot of people feel that it's harder to write simple code than it is to write complex code. That's why XP teams give themselves time to refactor when they plan their weekly cycles.

★ **Search:** It's not always easy to see that something is complex, especially when you're used to looking at it. Sometimes you need to work to find candidates for simplification.

★ **Act:** Found some complex code that could be simpler? Now it's time to refactor it!

And this brings us back to **slack**, which is more than just an agile way of padding your project schedule. It gives your team the time they need to pay down (or, better, prevent!) technical debt.

Watch it!

Don't be afraid to delete your code.

One of the most common traps that developers fall into is an **unwillingness to delete code** once they've written it. This can lead to **code bloat**: extra behavior, dead code, or other redundancies or inefficiencies that make code worse.

← You should feel comfortable deleting code, because you can always go back and recover it from the version control repository.

226 Chapter 5

xp *extreme* **programming**

XP teams "pay down" technical debt in each weekly cycle

Are you surprised to learn that programmers often don't get code right the first time they write it? It's true! Developers don't just "spit out" code and then move on to the next problem. Just like great artists, craftspeople, and artisans create sketches and preliminary designs before refining them into finished products, great programmers create initial versions of their code and then refactor that code, often many times.

That's why really effective XP teams make sure that they **add time to every weekly cycle** to "pay down" their technical debt and fix those lingering problems before they start to pile up. And the most effective way to do that is to refactor your code. XP teams have a name for this great habit: **refactor mercilessly**.

> I KNOW WE'VE GOT A DEADLINE, BUT IT'S **WORTH TAKING THE TIME** TO FIX THIS CODE **NOW**, BECAUSE THE NEXT CHUNK OF WORK WILL GO A LOT FASTER... AND IT'LL BE MUCH LESS FRUSTRATING!

Under the Hood: Refactoring

Let's look at a specific way that developers refactor their code

Refactoring means modifying the structure of your code without changing its behavior, and like most things that agile teams do, it's easy to get started (but it takes time and practice to master the subtleties). Here's an example of a common refactoring that Ana used to simplify her code—it's called **extract method**:

```
for ( StudioSchedule schedule : getStudioSchedules() ) {
    CustomerTrainers trainers = getTrainersForStudioSchedule( schedule );
    if ( trainers.primaryTrainerAvailable() ) {
        ScheduleUpdater scheduleUpdater = new ScheduleUpdater();
        scheduleUpdater.updateSchedule( schedule );
        scheduleUpdater.setTrainer( trainers.getPrimaryTrainer() );
        scheduleUpdater.commitChanges();
    } else if ( trainers.backupTrainerAvailable() ) {
        ScheduleUpdater scheduleUpdater = new ScheduleUpdater();
        scheduleUpdater.updateSchedule( schedule );
        scheduleUpdater.setTrainer( trainers.getBackupTrainer() );
        scheduleUpdater.commitChanges();
    }
}
```

These four lines of code update a class schedule so that the primary trainer is teaching it.

These four lines of code are almost identical. They do the same thing, except for the backup trainer.

Ana refactored the code by moving those four duplicate lines into a new method called createScheduleUpdaterAndSetTrainer()

Ana eliminated the duplicate lines of code, which made this code simpler. And now if she needs to do the same thing for other trainers, she can reuse the new method she created.

```
for ( StudioSchedule schedule : getStudioSchedules() ) {
    CustomerTrainers trainers = getTrainersForStudioSchedule( schedule );
    if ( trainers.primaryTrainerAvailable() ) {
        createScheduleUpdaterAndSetTrainer( trainers.getPrimaryTrainer() );
    } else if ( trainers.backupTrainerAvailable() ) {
        createScheduleUpdaterAndSetTrainer( trainers.getBackupTrainer() );
    }
}
```

design at the *last responsible moment*

Incremental design starts (and ends) with simple code

The practices we've talked about are really good for helping everyone on the team to develop habits to help them create small, decoupled units that work independently of each other. As they start to build up those habits, they can start practicing **incremental design**. This XP practice is exactly what it sounds like: the team creates the design for their project in small increments, building only the next bit of design needed for the current quarterly cycle, and concentrating mainly on what's needed in this weekly cycle. They build small, decoupled units, refactoring as they go to remove dependencies, separate units that get too big, and simplify the design of each unit.

When teams do incremental design they discover, uncover, and evolve the design bit by bit—just like when they do incremental development, they discover, uncover, and evolve the plan bit by bit.

When an XP team uses incremental design, the first set of units that they build typically evolve into a small, stable core. As the system grows, they add or modify a small number of units in each weekly cycle. They'll use test-driven development to make sure that each unit has minimal dependencies on the other units, which in turn makes the whole system easier to work with. In each iteration, the team adds only the design needed to build the next set of stories. When units interact in a simple way, it lets the whole system grow organically, bit by bit.

> INCREMENTAL DESIGN CAN'T POSSIBLY WORK. HOW CAN YOU BUILD A LARGE SYSTEM WITHOUT CREATING A BIG DESIGN ON PAPER FIRST?

All designs change. Incremental designs are built to change.

Generations of software engineers were taught in school that the design of a system needs to be complete before the team can start coding. This idea is built into the waterfall process: the project needs to complete the design phase before moving on to the development phase. Incremental design works when teams **make design decisions at the last responsible moment**, exactly the same way that teams using iterative development make planning decisions.

Incremental design <u>really does work</u> in the real world. One of the most successful examples is the Unix toolset (the set of Unix shell commands—`cat`, `ls`, `tar`, `gzip`, etc.). These tools weren't all developed at once. Instead, the Unix tools were based on a *philosophy of simplicity*: each tool does one specific, straightforward job, producing output that every other tool can use as input. This allowed thousands of different people to contribute to the entire toolset over many years. It grew incrementally, with individual tools being added one by one as the need for them arose.

XP teams take a very similar approach, starting with the same idea of embracing the value of simplicity. And just like with the Unix toolset, it's an effective way to work.

> SO BECAUSE *THE TEAM VALUES SIMPLICITY*, IT MAKES SENSE TO THEM TO BUILD ONLY THE UNITS THAT ARE NEEDED FOR THE NEXT SET OF STORIES, AND BECAUSE THE DESIGN IS ALREADY SIMPLE IT'S **EASY TO MODIFY.**

That's right. And software that's designed to be modified makes it easy for the team to embrace change.

The whole point of XP is to improve the way that the team writes code, and just as importantly, to improve and energize the working environment. When everyone on the team really "gets" incremental design, the whole system becomes much easier to work with. That makes the work **much more satisfying**: the most tedious parts of the software development job are reduced and often eliminated.

All of this leads to a very positive feedback loop: the weekly cycle and slack give the team enough time to do the work and constantly refactor the code, which lets them incrementally create a simple design, which helps everyone stay energized and approach problems with a fresh and clear mind, which lets them make progress quickly, which gives them success in the company. That success *lets the team work with the business more effectively*, and that gives them the ability to keep planning the project using weekly cycles and slack.

That feedback loop is what drives an XP team's ability to embrace change.

BULLET POINTS

- XP teams **embrace change** instead of resisting it.
- A **10-minute build** gives the team constant feedback about the build, and reduces frustration from waiting.
- The team uses **continuous integration** by making sure everyone's working folder is no more than a few hours out of date.
- **Test-driven development**, or building unit tests first and then building the code that makes the tests pass, helps teams keep units of code simple and reduce dependencies.
- People who do **pair programming**, with a pair of developers sitting at a single computer, produce better code more quickly than when they work separately.
- If it's not clear whether a technical approach will work, a **spike solution**, or a small, throwaway program to test it out, will help the team determine if it's a good approach.
- The XP practices **reinforce each other** to create an ecosystem effect.
- XP teams develop **good habits**, which leads to great software without forcing discipline on the team.
- When a methodology's practices **don't feel right**, it usually points to a clash between the values of the methodology and the mindset or culture of the team.
- Agile teams value **simplicity** because it leads to better code, and helps them to build less code.
- XP teams **don't sacrifice simplicity** for reusability.
- **Incremental design**, or building only the design that's needed for the current iteration, is an effective practice for helping to keep your system from getting complex.

exhaustion and boredom *slow down the project*

there are no Dumb Questions

Q: Does XP *really* make the job more satisfying?

A: Yes, really! Keeping the workplace energized means everyone watches for signs of exhaustion, boredom, and agitation. Those feelings are often indicators that team members are dealing with avoidable code problems, or are being forced to work late because of irresponsible planning.

Q: How do you know that Ryan's studio schedule change was a hack?

A: There were a few glaring warning signs. The first one was that he copied an entire class, left a bunch of it intact, and just deleted the bits he didn't need. That led to a lot of duplicate code. And then he added "special case" code to other parts of the system. That's code that looks for a particular state—in this case, scheduling a whole studio instead of a single yoga or martial arts class—and performs specific behavior just for that case. Developers try to avoid those things because they make the system more difficult to maintain. There's almost always a more elegant way to solve a problem like that.

Q: OK, now I'm confused about duplicate code. Ryan shouldn't have made a complex build script just to avoid a few duplicated lines, but he also shouldn't have made that hack with a class that had a lot of duplicate code. So is duplicate code good, or is it bad?

A: There's very little that's more aesthetically unpleasant to a programmer than a block of code that's duplicated in two (or worse, more!) places. It's almost always better to reuse the duplicate code by moving it into its own unit (like a class, function, module, etc.). But sometimes the situation isn't so straightforward. A few lines of duplicate code aren't necessarily particularly easy to reuse. Occasionally it takes a lot of work to extract them into their own unit. When we're coding, we sometimes go to such great lengths to avoid a small block of duplicate code that we end up adding complexity instead of removing it. That's the trap Ryan fell into with his build script.

Q: Hold on... "aesthetically unpleasant?" Since when do aesthetics have anything to do with code?

A: Code aesthetics matter a lot! If you're not a developer, it might seem weird to talk about code being "aesthetically pleasing" or not. But one sign of a great developer is a sense of aesthetics and even beauty in the code that he or she writes. Duplicate code is particularly aesthetically offensive to developers, because it's almost always a sign that something can be simplified.

Q: So how do I know if my code is too complex, or not simple enough? Is there a rule that I can apply?

A: No, there's no rule about how much complexity is too much. That's why *simplicity is a value, not a rule*. The more experience you have as a programmer on a team that values simplicity, the better you get at making your code simple. That said, there are definitely warning signs that your code might be too complex. For example, you know a block of code is probably too complex if you're afraid to touch it, or if there's a scary comment that says `Don't edit this!` at the top. Build scripts and unit tests are too complex if you find yourself avoiding a change because the change itself is easy, but modifying the build script or unit test will be really difficult or annoying.

Q: I still don't get the point about test-driven development and simplicity.

Does writing unit tests first really help keep code simple?

A: Yes. A lot of complexity happens when you build units of code that have many dependencies on other parts of the system. If you can avoid adding those dependencies, it makes your whole system a lot easier to maintain, and helps you avoid that "shotgun surgery" feeling when you're working on the code. Unit tests are really good at helping you avoid unnecessary dependencies, because a test for a single unit has to provide all of the input that the unit needs. If that unit has a lot of dependencies, it makes the test extremely annoying to write—and it becomes very obvious exactly which dependencies you really need. Often, that will also show you another part of the system that could be refactored. And it gives you incentive to do that refactoring immediately, because it will make the job at hand less annoying, tedious, or frustrating.

Q: And reducing annoyance and tedium... that's good for the team, right?

A: Yes! One of the best ways to make your team more productive is to make the work everyone is doing **less annoying, tedious, boring, or frustrating**. That's a really effective way to build an energized workplace. It's why people on XP teams really can't imagine working any other way.

> Exhaustion, boredom, and agitation can be early indicators of code problems that can be avoided.

Chapter 5

xp extreme programming

Here are some things we overheard Ana, Ryan, and Gary saying. Some of them are compatible with XP values, others are incompatbile. Identify the XP value that each of them is either compatible or incompatible with. Then draw a line from each speech bubble to either COMPATIBLE or INCOMPATIBLE, and another line to the appropriate Scrum value.

> THIS JAVA CLASS IS TOO BIG AND DOES TOO MANY THINGS. I'LL REFACTOR IT INTO TWO SEPARATE CLASSES.

COMPATIBLE / INCOMPATIBLE

Respect

> I ALWAYS RUN THE BUILD BEFORE I COMMIT MY CODE TO MAKE SURE EVERYTHING COMPILES AND ALL OF THE UNIT TESTS PASS.

COMPATIBLE / INCOMPATIBLE

Communication

> YOU'RE ASSIGNING THAT TO THE NEW GUY? HE'S PRETTY YOUNG, MAYBE WE SHOULD GIVE HIM SOME GRUNT WORK UNTIL HE GETS A LITTLE MORE EXPERIENCE.

COMPATIBLE / INCOMPATIBLE

Simplicity

> THERE'S A BUG IN MY CODE? ENTER A TICKET FOR IT, I'LL GET TO IT WHEN I HAVE TIME.

COMPATIBLE / INCOMPATIBLE

Feedback

→ Answers on page 243

XPcross

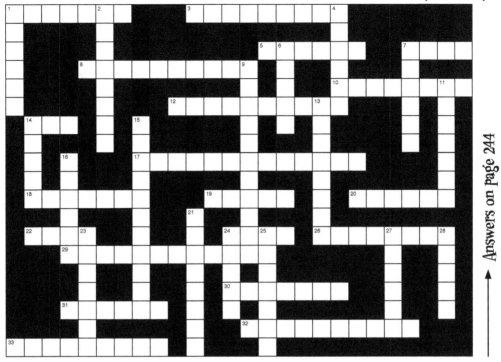

Answers on page 244

xp *extreme* **programming**

Across

1. Scrum has a strong focus on _____ management
3. The XP practices work together and reinforce each other to form this
5. XP teams create automated _____ that run in 10 minutes or less
7. A clumsy, quick-and-dirty solution
8. Everyone on the team continually _____ the code in their working folders back into the version control system
10. The kind of loop that teams use to repeatedly get useful information and make adjustments
12. What a version control system provides for the team to store their code
14. What XP teams do together to help them communicate well
17. When people have this value, they don't mind a little chatter in their office environment
18. What XP teams do with change
19. What teams add when they include optional or minor items
20. Another name for a clumsy, quick-and-dirty solution (rhymes with "stooge")
22. The kind of programming where two people sit at one computer
24. A programmer takes 15 to 45 minutes to reach this state of high concentration
26. What you do when you run across complex code that can be simplified
29. They add complexity to your code
30. Practice used in XP and Scrum to manage requirements
31. How often XP iterations happen
32. This value maximizes the amount of work not done
33. The _____ cycle is how XP teams do mid- to long term planning

Down

1. Nagging people to achieve this is not only annoying and ineffective, but actually counterproductive
2. Agile teams welcome _____ requirements, even late in development
4. It's a lot less stressful to work with code that's easy to _____
6. All code is broken down into these
7. They're better than discipline for making practices "stick"
9. The pace that XP teams strive for, and the kind of development that agile processes promote
11. If you agree to a deadline you know you won't meet because it's easier to apologize later, you lack this value
13. A burndown chart or task board posted where you can't help but absorb its data is an information _____
14. The kind of solution where you run an experiment by creating a small, throwaway program
15. The kind of design where teams make design decisions at the last responsible moment
16. When XP teams replace their planning practice with a complete and unmodified implementation of Scrum
21. When you try to commit a code change, but find that your teammate already committed a change to the same lines of code
23. XP and Scrum value that helps team members trust each other
24. TDD means writing unit tests _____
25. The kind of communication that happens when you absorb information from conversations all around you
27. A set of changes that you're pushing to a version control system
28. XP teams don't have fixed or prescribed _____

Exam Questions

> These practice exam questions will help you review the material in this chapter. You should still try answering them even if you're not using this book to prepare for the PMI-ACP certification. It's a great way to figure out what you do and don't know, which helps get the material into your brain more quickly.

1. Which of the following is NOT true about how XP teams plan their work?

    A. XP teams often self-organize by having team members pull their next tasks from a pile of index cards

    B. XP teams use week-long iterations

    C. XP teams focus on code, so they do very little planning

    D. XP is iterative and incremental

2. How do XP's values and practices help teams embrace change?

    A. By helping them build code that's easier to modify

    B. By placing strict limits on how users request changes

    C. By enforcing a change control process

    D. By limiting the amount of contact between the business users and the team

3. Amy is a developer on a team that builds mobile apps for commuters. They've adopted XP, but instead of using weekly cycles, quarterly cycles, and slack, they hold a Daily Scrum, do sprint planning, and hold retrospectives. Which of the following BEST describes Amy's team?

    A. They do not do adequate planning

    B. They are in the process of adopting XP

    C. They use a hybrid of Scrum and XP

    D. They are transitioning from XP to Scrum

4. Which of the following are NOT common to both XP and Scrum?

    A. Roles

    B. Iterations

    C. Respect

    D. Courage

Exam Questions

5. Which of the following is a valid way for XP teams to do estimation?

    A. Planning poker
    B. The planning game
    C. Traditional project estimation techniques
    D. All of the above

6. Evan is a project manager on an XP team. He noticed that over the last few weekly cycles, everyone had their headphones on and listened to music all day while coding. Evan is concerned that the lack of osmotic communication is making the workspace less informative. He called a team meeting to explain XP's informative workspace practice, and suggested that they adopt a rule against wearing headphones at work.

    Which BEST describes this situation?

    A. The team is not performing the informative workspace practice
    B. Evan has a responsibility to help the team adopt XP, and is demonstrating servant-leadership
    C. Evan needs to improve his understanding of the XP values
    D. The team is using a hybrid of Scrum and XP

7. Which of the following is true about test-driven development?

    A. Unit tests are written immediately after writing the code that they test
    B. Writing unit tests first can have a profound impact on the design of the code
    C. Test-driven development is used exclusively by XP teams
    D. Writing unit tests causes the whole project to take longer because the team spends more time writing code, but it's worth it for the extra quality

8. What is involved in continuous integration?

    A. Setting up a build server that constantly integrates new code into a working folder and alerts the team on build or test failures
    B. Using iteration to continuously produce working software
    C. Each person on the team keeps their working folders up to date with the latest code from the version control system
    D. Continuously reducing technical debt by improving the structure of the code without modifying its behavior, and integrating those changes back in

Exam Questions

9. Which of the following is NOT an example of an information radiator?

- A. The team sitting together so they can absorb information from conversations that happen around them
- B. Posting a burndown chart in a place where everyone can see it
- C. Keeping the team's task board on a wall in a common area
- D. Maintaining a list of stories the team has finished so far in the weekly cycle on a whiteboard that everyone can see

10. The following practices all establish feedback loops for XP teams except:

- A. Test-driven development
- B. Continuous integration
- C. 10-minute build
- D. Stories

11. Why do teams use wireframes that are low fidelity?

- A. Users give more feedback when a user interface mock-up looks less polished
- B. Agile teams rarely build software that contains detailed audio
- C. The team only builds and reviews one set of wireframes per weekly cycle
- D. They're only used for less complex user interfaces, and XP teams value simplicity

12. Which of the following promotes sustainable development?

- A. Thoroughly planning the next six months of work so that there are no surprises for the team
- B. Making sure everyone gets everything right the first time they build it, so no rework is required
- C. Making sure everyone leaves on time and nobody feels pressured to work weekends, so the team doesn't burn out
- D. Setting tight deadlines, so everyone is motivated to meet them

13. Which of the following is NOT a benefit of pair programming?

- A. Everyone on the team gets experience working with many different parts of the system
- B. There are two sets of eyes on every change
- C. Pairs help each other stay focused
- D. People take turns working, so fatigue is reduced

Exam Questions

14. Joanne is a developer on a team that constantly refactors, does continuous integration, writes unit tests first, and does many other XP practice. What BEST explains this team's culture?

    A. They have a strict manager who enforces the rules of XP
    B. They have good habits
    C. They are highly disciplined
    D. They are worried about getting fired if they don't work this way

15. What happens when a build takes longer than 10 minutes to run?

    A. It causes errors in the packaging process
    B. Team members run the build infrequently
    C. Merge conflicts occur that are difficult to resolve
    D. The unit tests fail

16. Joy is a developer working on a team building a mobile operating system. She tried to commit code for a feature that she's been working on, but the version control system won't let her complete the commit until she resolves many conflicts. Which practice will BEST prevent this problem in the future?

    A. Sustainable pace
    B. Continuous integration
    C. 10-minute build
    D. Test-driven development

17. Kiah is a developer on an XP project. Her team is doing quarterly planning. One of the features is extremely important, and failing to deliver it will have serious consequences for the project. Kiah is the expert on this part of the project, and she'll be the one doing the programming work. She feels that the design is relatively straightforward, and she's pretty sure that she knows exactly how to build it.

    What is the BEST action for Kiah and her team to take?

    A. Add a story for an architectural spike to an early weekly cycle
    B. Build a low-fidelity wireframe to get early feedback
    C. Add a story for a risk-based spike to an early weekly cycle
    D. Do extra usability testing

exam answers

Exam ~~Questions~~ Answers

> Here are the answers to this chapter's practice exam questions. How many did you get right? If you got one wrong, that's OK—it's worth taking the time to flip back and re-read the relevant part of the chapter so that you understand what's going on.

1. Answer: C

XP teams might not focus on project management to the extent that Scrum teams do, but XP is still an iterative and incremental methodology that values self-organizing teams. Those reasons are part of why it's an agile methodology.

This also helps keep rework from causing quite so many bugs.

2. Answer: A

It's a lot easier for teams to embrace change when they know those changes won't be a headache for them to make. XP helps with this by including practices and values that help teams build code that's easier to modify.

3. Answer: A

Teams that use a Scrum/XP hybrid have replaced the planning-related XP practices with a complete implementation of Scrum. Amy's team hasn't done that. They adopted some of the Scrum practices, but since they dropped the quarterly cycle without adding any sort of product backlog, they've pretty much stopped doing any sort of long-term planning.

They also don't have a Scrum Master or Product Owner. What other Scrum practices have they ignored? What do you think all of this indicates about the mindset among Amy's team members?

4. Answer: A

XP and Scrum both value respect and courage, and both use timboxed iterations for planning. But XP has no fixed roles, while Scrum teams must always have team members who fill the Product Owner and Scrum Master roles.

The planning game is a practice that was part of an early version of XP. It guided the team through creating an iteration plan by helping them decompose stories into tasks and assign them to team members. It's still in use by a few teams, but planning poker is a lot more popular.

5. Answer: D

XP teams use many different technique for estimating, and there is no specific rule that says the team must use any specific technique. So all of the techniques listed are valid. So is having the team simply meet and talk about how long they think the work will take.

Exam ~~Questions~~ Answers

6. Answer: C

It may seem weird to talk about feelings at work, but they're actually really important for getting a team to run smoothly. It's really difficult to innovate and do the difficult intellectual and creative work when you're distracted by negative feelings like resentment.

Evan has decided that the team is doing something wrong because they are not following his personal interpretation of the XP practices. When he called the team meeting and proposed a rule against wearing headphones, he was ignoring the fact that this is how the team prefers to work. This is very disrespectful, and shows that he doesn't trust them to find an effective way to work. Respect is a core XP value, and when people ignore it, that hurts the whole team by stirring up resentment and other negative feelings.

7. Answer: B

When you write unit tests first, it can have a profound impact on the design of the code. The reason is that when you're writing the tests, awkward constructions and unnecessary coupling between units can become much more apparent. Test-driven development is not exclusive to XP teams—many teams do it, even on waterfall projects. And even though it requires developers to write more code overall, most people who do test-driven development find that it actually saves time overall because it makes fixing bugs and making changes much, much faster.

The total time teams spend writing the extra code for the unit tests is more than made up for by the time saved making changes. This isn't a long-term effect—it's easily noticeable within days or even hours.

8. Answer: C

Continuous integration is a straightforward practice that can have an outsized effect on the project. The team continuously integrates the latest code from the version control system into their working folders every few hours. This prevents them from having to deal with time-consuming and annoying merge conflicts that span many files at the same time.

9. Answer: A

An information radiator is any sort of visual tool or display that conveys useful information about the project and is highly visible so that team members can't help but absorb the information on it as they walk by. The first answer describes osmotic communication.

Osmotic information and information radiators are both tools that can help with the informative workspace practice.

10. Answer: D

Stories are very useful, but they don't really establish a feedback loop the way some practices like test-driven development, continuous integration, or 10-minute builds do. The reason is that most of the time the story doesn't change very much once it's written, so there's no opportunity for feeding information back into it repeatedly. The other three practices establish feedback loops that occur many times over the course of a weekly cycle.

Exam Questions Answers

11. Answer: A

Wireframes are often low fidelity, which means they look like rough sketches or hand-drawn mock-ups. Users are often a lot more willing to give feedback about a sketch that looks like it was easy to draw than they are for a highly polished, accurate mock-up, because it feels intimidating to ask for changes to a design that looks like it required a lot of work. A low-fidelity wireframe can still capture all of the detail of a rich user interface, and is no more complex or simple than a mock-up that's highly polished.

> Low-fidelity wireframes are usually a lot less work than ones that are a lot more polished, which lets teams review several different versions with the users. They can help the team try out several iterations of the same UI in a single weekly cycle.

12. Answer: C

Sustainable development happens when the team works at a pace that they can comfortably manage, which almost always means working normal 40-hour weeks.

> A lot of teams have one or two people who make a point of staying late to show how "committed" they are (or to impress the boss). This often puts a lot of pressure on everyone else to stay late, too, which can easily create an unsustainable pace and burn the team out.

13. Answer: D

Pair programming is a very effective and efficient practice because two people working at the same computer keep each other focused, constantly collaborate, catch many problems, and get more done than if they were working alone. But both people are always working together at the same time—they don't take turns working.

14. Answer: B

XP teams use great practices every day because they have great habits. They don't do it out of a sense of discipline, and they certainly don't do it out of fear. Raw discipline and fear can cause temporary, short-term changes to the way teams work, but eventually teams revert back to their habits.

> The way to build great habits is to try out the practices, see great results, and use those to slowly change the way you think about your work. That's why adopting the XP practices helps get the team into the XP mindset.

KENT BECK, THE GUY WHO CREATED XP, ONCE SAID, "I'M NOT A GREAT PROGRAMMER. I'M JUST A *GOOD* PROGRAMMER WITH **GREAT HABITS**."

Exam ~~Questions~~ Answers

15. Answer: B

When an automated build takes a really long time to run, the team runs it a lot less frequently. That means the team gets less frequent feedback about the state of the build.

16. Answer: B

Continuous integration is a simple practice in which the team members keep their working folders up to date with the latest changes in the version control system. That prevents many merge conflicts, which can needlessly waste the team's time and cause a lot of frustration.

17. Answer: C

A risk-based spike is a spike solution that the team undertakes specifically to reduce project risk. In this case, Kiah already knows the technical approach that she will take, so there's no need for an architectural spike. But since the risk for this particular feature is very high, it makes sense to add a risk-based spike to a weekly cycle early in the project. That way the risk will be eliminated early on.

> And if it turns out there are unforeseen problems, it's a lot better to discover them early in the project than later.

exercise solutions

Who am I? solution

A bunch of XP practices and values are playing a party game, "Who am I?" They'll give you a clue, and you try to guess who they are based on what they say. Write down their names, and what kind of things they are (like whether they're events, roles, etc.).

And watch out—a couple of <u>tools</u> that are <u>not</u> XP practices or values might just show up and crash the party!

Clue	Name	Kind of thing
I help XP teams get into a mindset where they know that they're best at solving problems when they share knowledge with each other.	communication	value
I'm a great way to absorb information about the project from discussions happening around me.	osmotic communication	tool
I help you understand your user's needs, and I'm also used by a lot of Scrum teams.	stories	practice
I'm a team space that does a great job communicating information about the project.	informative workspace	practice
I help the team work at a sustainable pace, because teams that work super-long hours actually build less code with worse quality.	energized work	practice
I'm how XP teams do their long-term planning, by meeting with the users once a quarter to work on the backlog.	quarterly cycle	practice
I help people get into a mindset where they treat each other well, and value each other's input and contributions.	respect	value
I'm the reason people on XP teams will tell the truth about the project, even if it's uncomfortable.	courage	value
I'm the way that XP teams do iterative development, and teams use me to deliver the next increment of "Done" working software.	weekly cycle	practice
I'm a large burndown chart or task board put up in the team space in a spot where everyone can't help but notice me.	information radiator	tool
I make sure that the team has a space where everyone is near their teammates.	sit together	practice
I help give the team some breathing room in each iteration by adding optional stories or tasks.	slack	practice

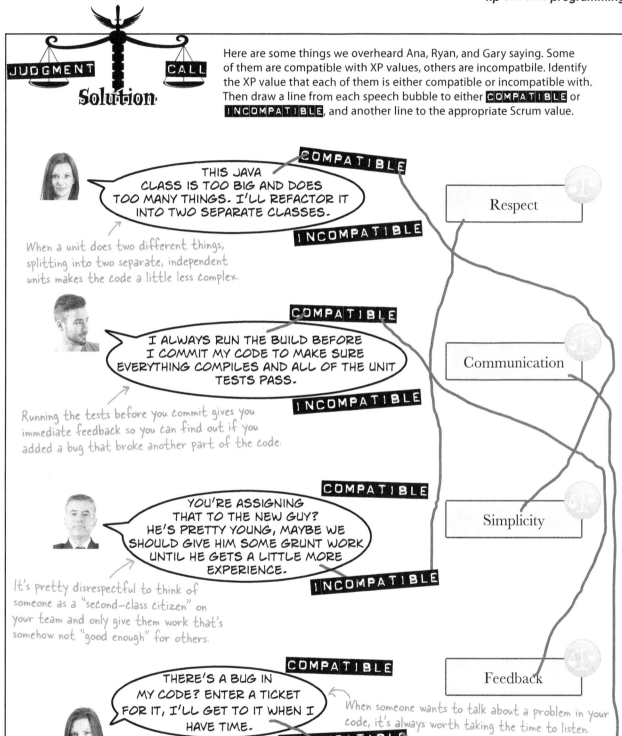

crossword *solution*

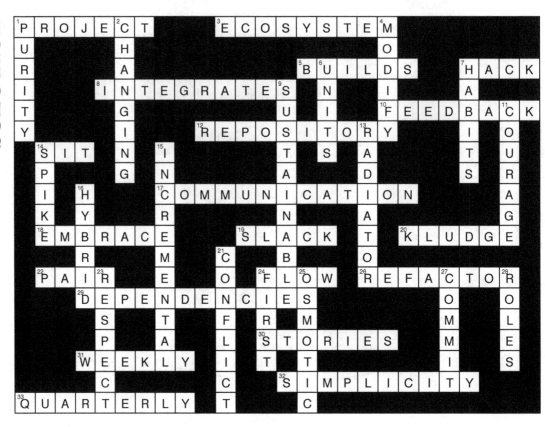

Sharpen your pencil

Here are three scenarios that Ryan and Ana are working on that have to do with getting feedback from their project. Write down the name of the tool being used in each scenario.

> WE NEED A NEW WAY TO STORE TRAINER SCHEDULES THAT WILL REDUCE MEMORY. I'M BUILDING A PROOF-OF-CONCEPT SO WE CAN GET A SENSE OF HOW MUCH WORK IT WILL BE.

architectural spike

> I'M NOT HAPPY WITH HOW THIS CLASS GETS INITIALIZED – IT'S GOING TO BE HARD TO USE. ONCE ITS UNIT TESTS PASS, I'LL MODIFY IT.

red/green/refactor

> I FINISHED DESIGNING THE NEW USER INTERFACE. LET'S MAKE SURE THAT IT WORKS BY GETTING A BUNCH OF USERS IN THE ROOM AND OBSERVING THEM WHILE THEY USE IT.

usability testing

6 lean/kanban

Eliminating Waste and Managing Flow

I'M NEVER TOO BUSY TO ELIMINATE WASTE!

Agile teams know that they can always improve they way they work. Team members with a **Lean mindset** are great at finding where they spend time on things that aren't helping them **deliver value**. Then they get rid of the **waste** that's slowing them down. Many teams with a Lean mindset use **Kanban** to set **work in progress limits** and create **pull systems** to make sure that people are not getting sidetracked by work that doesn't amount to much. Get ready to learn how seeing your software development process as a **whole system** can help you build better software!

new process same problems

Trouble with Audience Analyzer 2.5

Let's check back in on Kate, Ben, and Mike. The last release worked out really well because the team had a good idea of what their users needed from the beginning. When they started thinking about what features will go into Audience Analyzer 2.5, some of the problems they set out to correct using Agile practices came back.

> WE MADE ALL OF THESE AGILE CHANGES, BUT I'M STARTING TO THINK IT WASN'T WORTH IT. IT'S STARTING TO FEEL LIKE WE'RE RUNNING INTO **THE SAME OLD PROJECT PROBLEMS** THAT AGILE WAS SUPPOSED TO FIX! WE USED TO OBSESS ABOUT DEVELOPMENT HOURS AND HEAD COUNT. NOW WE'RE CONSTANTLY TALKING ABOUT HOW MANY STORY POINTS WE CAN FIT INTO EACH RELEASE. REMIND ME AGAIN... HOW IS THAT AN IMPROVEMENT?

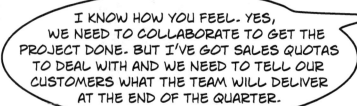

> I KNOW HOW YOU FEEL. YES, WE NEED TO COLLABORATE TO GET THE PROJECT DONE. BUT I'VE GOT SALES QUOTAS TO DEAL WITH AND WE NEED TO TELL OUR CUSTOMERS WHAT THE TEAM WILL DELIVER AT THE END OF THE QUARTER.

Everyone on Mike's development team was really happy when they first started with agile, but he's had to have a lot of talks with people lately about how stressful the project's become.

Ben loves seeing the team's progress at the end of each sprint, but as Product Owner he needs to know how much they can do in a quarter to make sure the product will hit sales projections, too.

lean kanban

> I AGREE. THIS DOES FEEL REALLY FAMILIAR, BUT NOT IN A GOOD WAY. I THINK THERE'S **SOMETHING REALLY WRONG** WITH THE WAY WE'RE WORKING.

Ben: Is it so hard for the team to tell me what features will be in the next major release? We can't just tell our customers to come to every demo and hope we'll get around to doing their requests.

Kate: Look, I get it. We all get it. You're asking for a really important feature, and everyone wants it yesterday.

Mike: The problem is that they're **all** really important features.

Kate: Right. And everyone wants all of them yesterday.

Ben: Hold on, guys. I'm not the bad guy here, but we've got a business to run. This used to be a well-oiled machine. The first few iterations were great. We all knew what we needed to do, I could always tell the senior managers exactly what would be in the next release. But lately, it seems like one, then two, then three features don't quite make it in. And worse, it seems like we keep finding bugs, and you guys never used to deliver low-quality work. What's the deal?

Kate: Honestly, Ben, I don't have a good answer. Something's just, well, off about the way we plan our work. I can see us slowing down, but I don't know what to do about it. I feel helpless here. We've got a backlog of work that we pull into each iteration, and that's good because it means we can deal with uncertainty. It's not that we're uncomfortable doing long-term planning, but... well...

Mike: I see where you're heading with this, Kate. We used to basically just be responsive: do this task now, then do this next task. Now everyone on the team has a good idea of what features we'll be doing in two months, three months, even six months. It's almost like that future work is so important that we feel like whatever we're working on today is just _blocking us from doing what we need to do tomorrow_. We feel like we're always late, and that's causing us to make mistakes and cut corners. It feels like there's all this work that hasn't been done yet, so we have to get everything done as fast as possible. It's really stressful, and it's affecting our quality.

Ben: I know that our projects don't have to be like this. I have no idea how to get there. Do you have any idea how we can improve things?

Kate sees the problem more clearly than anyone. The project manager often does. She can see how everyone on the team feels like things are getting slower and slower, and there's more and more pressure to get things done.

Have you ever been on a team that just doesn't seem to be running as smoothly as they used to? What do you think causes that? Is there any way for a team to improve the way they work?

you are here ▸ **247**

a mindset with a name

Lean is a mindset (not a methodology)

We've explored Scrum and XP, which each have a mindset component (values) and a method component (practices). Lean is different. It's not a methodology, and it doesn't include practices. **Lean is a mindset** based on principles that are all about making sure that the process your team is following is lined up to help you build products that are valuable to your customers. Where Scrum gives you a process to follow as a template (roles, planning meetings, sprints, sprint reviews, retrospectives), Lean asks you to take a look at the way you're working today, figure out where you're running into trouble, and apply Lean principles to correct those problems. Rather than telling you exactly what to do, Lean gives you the tools to see where the process you're using is actually getting in the way of meeting your goals.

Lean teams look at the way they're working today, and figure out where they are habitually running into problems. Then they make changes to improve the way they work.

Lean, Scrum, and XP are compatible

You don't have to have a Product Owner or a Scrum Master on your team to use Lean. You don't have to hold sprint planning meetings on the first day of your sprint, or retrospectives on the last day to do Lean. You don't have to do pair programming, or sit together, or ruthlessly refactor. But both XP and Scrum were developed with Lean in mind. So if you've started using XP or Scrum (or even if you haven't), you can use Lean to find ways to improve the way your team is working.

lean kanban

Lean principles help you see things differently

Lean is divided into **Lean principles** and **thinking tools** that help you use those principles when you build software. All of the Agile methodologies are influenced by Lean thinking, so many of the ideas you'll learn about in this chapter will be familiar to you or build on concepts you've learned in previous chapters. But Lean asks you to apply these concepts to the process you're using to develop software and to constantly improve the way you work. Teams that use Lean even have a name for applying these principles and thinking tools: they call it **Lean thinking**.

Eliminate waste

It takes a lot of work to build a product. But teams often do more work than they need to: they'll add unneeded features, waste time trying to multitask, and spend time just sitting around waiting. When a team has a Lean mindset, they try to find all of this waste and work hard to eliminate it from the project, often by removing anything that distracts the team from creating the software their users need.

Amplify learning

This principle is all about learning from what you do and using feedback to keep improving. As you make changes in the way your team works, observe what those changes do and then use your observations to decide what to change next.

Decide as late as possible

Make every important decision for your project when you have the most information about it—at the last responsible moment. Don't force yourself to decide things that don't need to be decided just yet.

If you need to refresh your memory on what "last responsible moment" means, take a minute and flip back to Chapter 3.

Deliver as fast as possible

Things that delay your project are costly. Keep track of how you work, where you're running into delays, and how those delays impact your team. Set up **pull systems**, **queues**, and **buffers** to even out the work because that will get your product done as quickly and efficiently as possible.

We'll learn more about these later in the chapter.

you are here ▸ **249**

not so different after all

More Lean principles

Remember the meeting of the minds at Snowbird we talked about back in Chapter 2? When they talked about the Agile principles of self-organization, simplicity, and continuous improvement, these last three Lean principles were an important part of that discussion.

> Just like Scrum applies transparency, inspection, and adaptation to your project to help your team get better Sprint by Sprint, Lean teams measure the effects of the changes they make and use those measurements to determine if those changes are working, or if they need to try a different approach. Lean teams use the principles and thinking tools to find the fastest and most efficient route from concept to actually putting software in their customers' hands.

Empower the team

There's no better expert in the world on how a team works than a member of that team. This principle is about helping every team member to have access to all of the information they need about the goals of the project and their progress, and about letting the team decide on the most effective way to work.

Build integrity in

Users best understand the purpose of the software that the team builds and can evaluate the quality of the work when the team builds software that meets the users' needs. If the software you're building is intuitive and does something valuable for your users, it's a lot more likely to be worth the effort everyone is putting into building it.

See the whole

Take time to understand how the team is working on the project—and take the right kind of measurements, so that everyone is exposed to all of the information they need to make good decisions about it. When everyone on the team can see what everyone else is doing (and not just their own contributions), the team can collaborate on the best way to get the work done.

A Lean mindset keeps your team focused on building the most valuable product possible in the time available.

lean kanban

Venn Magnets

You spent all night arranging magnets on your fridge into a really useful Venn diagram that shows which values and principles are specific to Scrum, which are specific to XP, which are specific to Lean, and which are shared by all... and then someone slammed the fridge door and all the magnets fell off. Can you put them back in the right places?

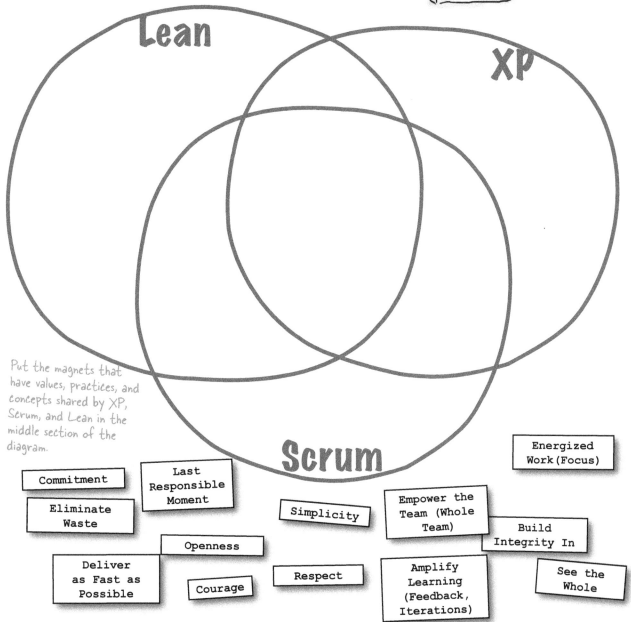

Put the magnets that have values, practices, and concepts shared by XP, Scrum, and Lean in the middle section of the diagram.

Commitment	Last Responsible Moment		Simplicity	Empower the Team (Whole Team)	Energized Work (Focus)
Eliminate Waste					Build Integrity In
	Openness			Amplify Learning (Feedback, Iterations)	
Deliver as Fast as Possible	Courage	Respect			See the Whole

tools to help you *understand how you work*

Venn Magnets Solution

Here are the magnets restored to the correct order. Lean and XP share a focus on empowering the team and all three of them focus on the last responsible moment and feedback loops. Lean and Scrum share an explicit focus on commitment.

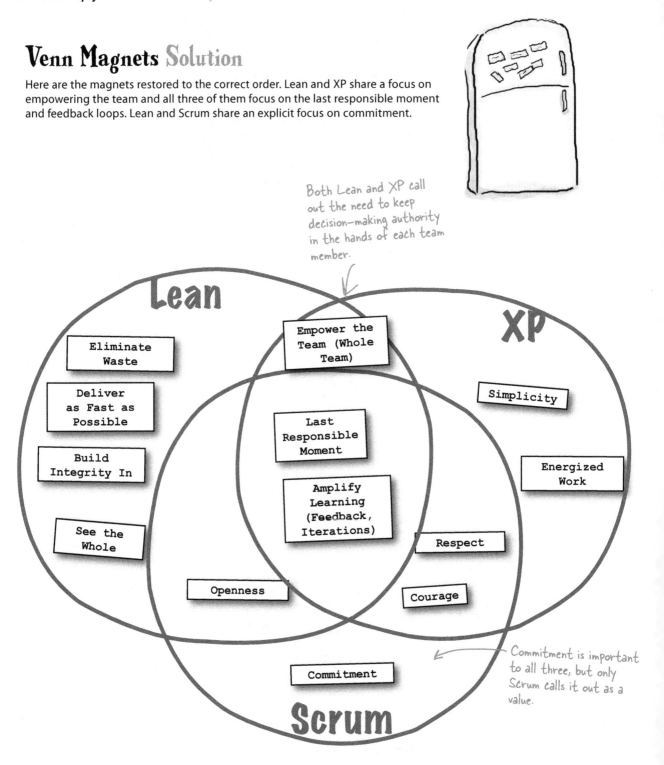

Both Lean and XP call out the need to keep decision-making authority in the hands of each team member.

Commitment is important to all three, but only Scrum calls it out as a value.

lean kanban

WHO DOES WHAT?

Since Lean was an important consideration when creating the Agile Manifesto. It's not surprising that many of the thinking tools are part of Scrum and XP as well. Here are some thinking tools you already know from previous chapters. Match the tool's name with its description.

The last responsible moment

Iterations and feedback

Refactoring

Self-determination, motivation, leadership expertise

Changing code to make it more readable and maintainable without changing its behavior.

Making decisions about which work will be done without requiring external approval.

Making decisions when you have the most information about them.

Delivering software in increments so that new features can be evaluated while still more features are being developed.

> SO THINKING TOOLS ARE SORT OF LIKE PRACTICES. THEY HELP LEAN TEAMS USE LEAN PRINCIPLES TO IMPROVE THEIR PROCESS.

That's right. They are the things Lean teams do to support the decisions they make.

The whole point of Lean is to see where the team is doing work that is getting in the way of building a valuable product. If teams make use of the **iterations and feedback** thinking tool, they'll be able to see the impact of the changes they're making and use that impact to **amplify learning**. If they use the **last responsible moment** thinking tool, they'll be able to **decide as late as possible** while building their product. That's how the thinking tools and the principles are linked.

eliminating waste with a stream of value

Some thinking tools you haven't seen before

Now that you're familiar with the Lean principles and some of the thinking tools that have been used in Scrum and XP, it's time to take a look at the thinking tools specific to Lean. Lean teams use these tools to get a handle of the root causes of their process problems, and then fix them.

Seeing Waste

Before you can eliminate waste, you need to see it—but that's easier said than done. Have you ever had a pile of clutter in your home that stubbornly stuck around for a while? After a few days, you don't see it anymore. That's exactly how waste in your process works, too, and it's why seeing waste is an important Lean thinking tool.

Find the work your team is doing that isn't helping you build a valuable product, and stop doing it. Is your team writing documents that nobody reads? Are you spending a ton of time manually doing work that could be automated? Putting a lot of effort into discussing and estimating features that probably won't make it into the finished product?

Value Stream Mapping

This tool will help a Lean team find the waste in the process they're using to build software. To build a value stream map, find the smallest "chunk" of the product that the customers are willing to prioritize on your backlog. Then think back through all of the steps the team took to build it, from when it was first discussed until it was delivered. **Draw a box** for every one of these steps, using **arrows** to connect the boxes. Next, **track the amount of time** it took you to perform each of the steps and the amount of time you waited in between each step. The time you spent waiting in between steps is waste. **Draw a line** that moves up to show that the project is working, and moves down to show that the project is waiting. Now you've got a visual representation of the work and the waste!

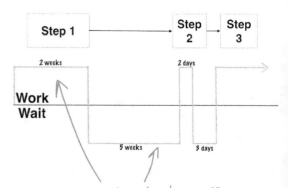

The line on the value stream map dips down to show waiting time, and rises back up to show working time. Now you can see exactly how much time this one "chunk" of work had to wait between spurts of progress.

lean kanban

> One way that Lean helps teams focus on delivering value as early as possible is by asking them to identify the smallest thing of value that they can deliver and then focus the team on delivering it as fast as possible. Lean teams often talk about Minimally Marketable Features (MMFs) as a goal for a release increment. Similarly, they'll try to identify the smallest product that they deliver that will be valuable to their customers—the name for that is a Minimally Viable Product (MVP). By focusing on delivering MMFs and MVPs, Lean teams make sure that they get a valuable products in the hands of their customers as soon as possible.

Queuing Theory

Queuing theory is the mathematical study of queues. Lean teams use queuing theory to make sure that people are not overloaded with work, so that they have time to do things the right way. The queue that needs to be optimized in software development could be a list of tasks, features, or to-do items for a team or for an individual developer. If you think of a backlog as an example of a queue, you'll recognize that tasks that go into the backlog first are usually the first ones to be completed... unless someone, like a product owner, explicitly changes the order. Lean tells us that making a team's queue of work public, and making it central to the decision-making process, helps teams to deliver more quickly. Queuing theory is often used by Lean teams to experiment with adding queues in a system to even out the flow of work through it.

Pull Systems

When a team organizes all of its work into a backlog and then has developers pick up a new task as they finish an old one, they're using a pull system. The backlog is a queue of work. Instead of having users, managers, or product owners push tasks, features, or requests down to a team, they'll add those requests to a queue that the team pulls from. Pull systems are about only working on one thing at a time and pulling work into the next phase of a process as a person frees up to work on it. That way people get to focus on doing the best work they can for each task they take up, and the product is built with attention to quality and efficiency.

Lean teams use value stream maps to find and eliminate waste.

you are here ▶ **255**

give yourself options

More Lean thinking tools

The rest of the tools help teams keep their options open when they're working, and constantly make sure that they are working on the most valuable thing.

Options Thinking

When a team is deciding which features will be included in an upcoming release, most people think of scoping as a commitment between the team and end users. Lean teams know that what the process is really doing is determining the options that the team might take in order to deliver value with each release. When we discussed Scrum, we talked about how a Scrum team doesn't spend a lot of time modeling and tracking dependencies between tasks. Instead, the team is free to add and remove tasks on the task board every day during the Daily Scrum. These tasks are options, not commitments: the tasks don't have due dates, and there's no such thing as a "late" task that causes the rest of the project to blow a deadline. By freeing up the team to think about their work plans as options, they can make changes when they need to and make sure that they can do what's best for the product instead of over-committing.

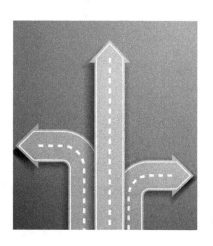

Cost of Delay

If a task is higher risk, it costs more to delay work on it than it would if it were low risk. Some features aren't useful at all if they're not completed within a certain time window. Understanding the cost of delaying each of the tasks in your team's queue can help you make better decisions about which tasks must be completed first.

This is one reason that Lean teams develop a **delivery cadence** for releasing new features. This means that instead of committing to deliver a certain set of features on a specific schedule, the team commits to delivering the most value that they can at regular intervals.

Delays happen. But if everyone understands the costs of those delays, they might give some things more priority. That's an important way that Lean thinking helps teams eliminate waste.

> I ASKED AMY A SIMPLE 10-SECOND QUESTION, AND I'VE BEEN STUCK WAITING FOR TWO HOURS FOR HER TO GET BACK TO ME.

Perceived Integrity/Conceptual Integrity

Lean teams are always looking to build integrity into their products from the very beginning. They divide their thinking about this Lean principle into perceived and conceptual integrity. Perceived integrity means thinking about how well a feature meets the needs of the user. Conceptual integrity means thinking about how well features work together to form a single unified product.

Set-Based Development

When teams practice set-based development, they spend time talking about their options, and change the way that they work in order to give themselves additional options in the future. The team will do extra work to pursue more than one option, trusting that the extra work will pay for itself by giving the team more information so that they can make a better decision later. This allows the team to gather more information about multiple options at once, and make the decision about which option to pursue at the last responsible moment.

> Lean teams establish a **delivery cadence** in which they release the latest set of completed work at regular intervals.

Measurements

Just like Scrum focused on transparency, inspection, and adaptation, Lean teams measure the way their system is working before making changes and then measure the impact of the changes they do make.

- ★ One measurement is **cycle time**, or the amount of time it takes for a feature or task to be completed from the time a developer begins to work on it until it's delivered.

- ★ Another measurement is **lead time**, or the amount of time it takes from when a feature is identified until it is delivered. Cycle time is often used to measure the effectiveness of changes to the development process, while lead time is used to measure changes to the requirements gathering and support processes as well.

- ★ Teams will also measure **flow efficiency**, or the percentage of the total time for a feature that the team spent actually working on it (as opposed to waiting).

Cycle time and lead time help you think about your process from two different perspectives.

When you're on a team working on a work item, you usually think in terms of cycle time, or the amount of time that elapsed from when you and your team started planning the work item until it's *"Done"* done.

But your customers don't really see cycle time. Let's say that a customer asks your team for a feature, but you don't actually get started on it for six weeks. Once you finally get started, it takes eight weeks to finish. The cycle time is eight weeks, but the lead time is fourteen weeks—and that's the measurement *that your customer really cares about*.

this seems theoretical *can it make a real difference?*

Cubicle Conversation

The people on the Audience Analyzer team know they've got a problem with their process. But they don't agree about what's causing it.

> I KNOW YOU WANT US TO BE REALLY ACCURATE ABOUT HOW LONG THINGS WILL TAKE TO BUILD. THE TEAM AND I TRY TO DELIVER EVERYTHING OUR BUSINESS USERS ASK FOR IN EACH ITERATION, BUT WE USUALLY FIND THAT SOME OF THE FEATURES AREN'T DONE WHEN THE RELEASE DATE ROLLS AROUND.

> MAYBE YOUR TEAM IS *JUST BAD AT ESTIMATING*. WE'RE RUNNING A BUSINESS HERE! YOU NEED TO COME UP WITH A BETTER WAY OF FIGURING OUT HOW MUCH WE CAN DO.

Ben: I'm fine with cutting down our scope. I'm happy to compromise. But the team really needs to get their estimates right.

Mike: OK...but it's not always that simple. Remember that privacy option feature in the last release? I think it's a pretty good example of why this isn't so easy. First you wanted it to be really intrusive, so that all users had to set their desired level of privacy. Then you wanted it to be set by our customers. Then, just about three days before we were going release it, you decided that we should default it to a strict privacy profile and allow both of them to change the profile at will.

Ben: Well, our initial market analysis told us to do it one way, but then our subsequent research showed that we needed to change our approach. Making those changes made our last release a huge success. I know they were late in the game, but we needed to react to what the market was telling us.

Mike: I'm glad we made those changes too. But you have to see that it's really hard to estimate how long it's going to take to build a feature weeks in advance when requirements are always changing.

Ben: But you're supposed to embrace change, right? Can't you get better at estimating the work and tracking dependencies so we can make more informed decisions when those changes come up?

lean kanban

WE TRY REALLY HARD TO ESTIMATE EVERYTHING WE NEED TO DO TO RELEASE FEATURES ON TIME. AND WE'RE ALWAYS WORKING REALLY HARD BUT MISSING OUR DATES. SOMETHING MUST BE GOING ON THAT WE'RE **NOT SEEING CLEARLY.** CAN LEAN HELP OUR TEAM TO MAKE OUR PROJECTS RUN BETTER?

Can you think of a way that Mike can use the Lean principles and Lean thinking to make a real-world improvement to the way the project is run?

understand your waste to eliminate it

Categorizing waste can help you see it better

Lean tells us that we need to eliminate waste. But what, exactly, does that waste look like? How do we spot it? One thing that can be really helpful is to think of **categories of waste** in software development projects. Many Lean teams use value stream maps to see how much time is wasted in the development of the features they deliver. Once the team sees that there's waste in their process, figuring out what type of waste it is can help them come up with improvements to eliminate it. Luckily, you don't need to come up with these categories on your own. Lean practitioners have identified **seven wastes of software development**:

★ **Partially done work**

If you start doing many different things at once and don't finish the tasks you start, you'll end up with a lot of work that's not ready to be demonstrated or released at the end of your iteration. Partially done work happens all the time on software projects, whether or not the team is thinking about the costs of multitasking. Sometimes it just feels *really productive* to start something new while you're waiting for information or approval on another task and that can lead to that first task being partially done when the timebox is complete.

★ **Extra processes**

Extra processes are those that don't actually help you deliver the software but add work to the team. Sometimes teams will put a lot of effort into documenting a feature that never gets delivered. Sometimes there are a never-ending series of status meetings that are meant to show the team that they've got management support, but which really just boil down to a lot of extra processes—like asking the team to create special status reports and gather information about each task in development.

★ **Extra features**

Team members often get really excited about a new piece of technology. Sometimes they insist on including extra features that they think are brilliant but which no one asked for. It's a common misconception that any kind of innovation on an existing idea is a benefit to a project. In truth, any new features that are added on by the team are taking time away from the features that the end users have asked for. That doesn't mean that teams can never be the source of good ideas, but those ideas need to be vetted and presented as options before the team wastes effort building them out, instead of what was asked for.

★ **Task switching**

It's really common for managers to lose track of the number of requests they've made to a team, and just assume that there's no cost to giving the team more work to do without adjusting anyone's expectations. Couple that with the fact that software developers often want to overcommit in order to impress their bosses and teammates, and you see how people end up switching between three, four, five or more tasks that are all supposed to get done at the same time. That's why task switching is an extremely useful concept in identifying software project waste. Any time a software developer needs to multitask between two competing priorities, time is wasted.

lean kanban

The seven wastes of software development help your team see the waste in your process, so that you can eliminate it.

★ **Waiting**

Sometimes teams need to wait for someone to review a specification and they can't get started until that's done. Sometimes they'll need to wait for infrastructure to set up physical hardware, or a database administrator to provision a database. There are many legitimate reasons that your team will need to wait during a software project. But all of that time is waste, and should be reduced wherever possible.

> Sometimes it's just not possible to eliminate certain waste—for example, if you're waiting for hardware to be delivered, and you couldn't pre-order it. That's why it's so important to identify as much waste as possible, so that you can eliminate what you can.

★ **Motion**

When teams sit in many different locations, just the effort it takes to get from one person's desk to another's can add a lot of waste to your project. Motion is all that time wasted in transit while working.

★ **Defects**

The later a defect is found in the process of building software, the more time that's wasted finding and fixing it. It's much better if defects are found as soon as they're injected, by the developer who put it there, rather than by any testing process later in the development process. The more a team focuses on quality-driven development practices and shared code ownership, the less time they waste on fixing defects.

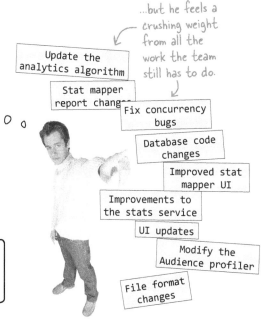

> I'M JUST STARTING THE DESIGN FOR THESE TWO FEATURES NOW. BUT EVERY HOUR I SPEND ON THEM IS AN HOUR THAT I'M *DELAYING THESE OTHER FEATURES* STILL WAITING TO BE DONE.

These are the features Mike is just starting now... → Feed in from data provider | Audience machine learning feature

...but he feels a crushing weight from all the work the team still has to do.

Update the analytics algorithm
Stat mapper report changes
Fix concurrency bugs
Database code changes
Improved stat mapper UI
Improvements to the stats service
UI updates
Modify the Audience profiler
File format changes

> Have you ever felt a huge amount of pressure to finish whatever you're working on right now because there's so much that still has to get done? Lean thinking helps you get past that. Taking the time to understand the types of waste on your project is the first step to fixing the problem.

you are here ▶ **261**

agile is *built on lean*

Q: It sounds like Lean teams never tell people when they're going to be done with any specific feature. Do they just decide what to do at the last moment and deliver as fast as they can? That would never fly where I work.

A: Teams that use a method that was developed with Lean thinking in mind—like Scrum or XP—focus on delivering small batches frequently. But rather than trying to figure out each and every task up front, they agree to build the most valuable thing they can build within the time frame. Lean teams know that just because customers say they need something at the beginning of the planning period, that doesn't stay true the whole way through the project. So Lean teams see it as positive that they can change their priorities when the business needs it.

A traditional team might create a Gantt chart that seems to predict exactly which team members will work on each task, which are on the critical path, and when each milestone will occur. While that gives everybody in the organization a lot of confidence that the team has spent time planning, the plan is almost always inaccurate before the work even gets started.

there are no Dumb Questions

A team that's using Lean thinking tools will focus on getting the process right. If the process is right, the highest value work rises to the top of the work queue at the start of every increment. That helps the team to focus on eliminating waste in the way that they approach and build each feature and deliver it as quickly as possible. A Lean team sees planning as a way to identify options, and they'll make the best choice from those options as new variables come into play.

Q: Oh, OK. So Lean teams develop in sprints, like a Scrum team?

A: Not necessarily. Lean teams establish a delivery cadence, which typically means setting a timebox for each delivery. It could be two weeks or two months—Lean teams deliver frequently and in small batches on predictable timetables. They plan their releases to coincide with those timeboxes. In Scrum, the sprints are a combination of planning and cadence. But many Kanban teams **decouple the cadence from the planning**: they'll plan new work items, move them through the workflow, and release whatever happens to be in releasable state when the cadence is done.

Some people refer to this as **iterationless development** because the timebox is not tightly coupled to the planning.

Q: So about those seven wastes. Do teams *really* build in extra processes and features? My team always feels like we don't have time for anything extra.

A: Yes, they really do! Sometimes the teams that are under the most pressure are the ones that create the most wasteful work. It's the projects that are under the tightest deadlines that often set up multiple status-gathering meetings a day, or create new practices for developers to log their time and scrutinize the number of hours spent on each activity. Those activities amount to waste that makes it harder to get the product delivered.

Options thinking helps you counter waste. The more complex a problem is, the less you know about it up front. That's why Lean teams commit to objectives, but not to specific plans. They agree to the overall objective, and to deliver the most valuable thing they can toward that objective within the timeframe. Thinking about the commitment in this way gives both the team and the organization the freedom to focus on delivery and that usually means higher quality products, faster.

*I'M GETTING IT. ALL OF THE AGILE METHODOLOGIES ARE INFLUENCED BY LEAN. FOR EXAMPLE, THE XP PRACTICE OF INCREMENTAL DESIGN IS ONE WAY THAT **XP TEAMS GIVE THEMSELVES OPTIONS** TO HANDLE CHANGES IN THE FUTURE.*

lean kanban

Here are some things we overheard Mike, Kate, and Ben saying. Some of them are describing waste in the project, and some are not. Identify the type of waste each of them is describing. Then draw a line from each speech bubble to either **WASTE** or **NOT WASTE**. If it's not waste, add another line to the appropriate type of waste in Lean.

WASTE

I HAVE TO DO SUPPORT FOR THE VERSION 1.8 WHILE I WORK ON THE CURRENT VERSION.

NOT WASTE

Task Switching

WASTE

WHILE I WAS WAITING FOR THE DEVOPS FOLKS TO DO THEIR WORK, I GOT STARTED ON ANOTHER FEATURE SO I'M NOT SITTING IDLE.

NOT WASTE

Extra Features

WASTE

I WANT TO MAKE SURE THAT PEOPLE ARE ALWAYS BUSY SO I ASK FOR MORE THAN I NEED.

NOT WASTE

Partially Done Work

WASTE

I MAKE SURE I FINISH A TASK BEFORE I PICK UP A NEW ONE.

NOT WASTE

➤ Answers on page 299

work floats down *the value stream*

Value stream maps help you see waste

A **value stream map** is a simple diagram that helps you see exactly how much of your project time is wasted waiting and not working. Sometimes it helps to draw out the process your team is using on a time line to show where waste is slowing you down. A value stream map can make it painfully clear how much of your team's time is spent on work that doesn't result in value for your customers. Once your team can clearly see the waste, they can work together to figure out ways to reduce the time waiting.

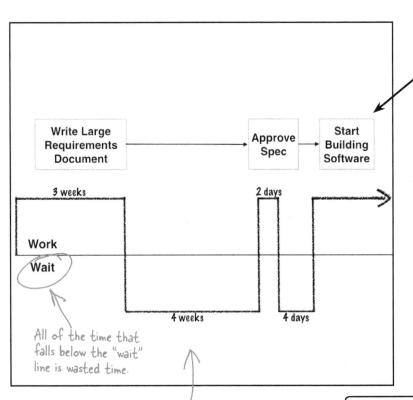

A value stream map shows all of the steps a specific feature went through from start to finish. Each of those steps is represented by one of these boxes at the top of the map. These steps represent what actually happened to that feature in the real world—which may be different than what the team had planned for it.

This feature took 7 weeks and 6 days from the time requirements gathering began until the team started developing it.

All of the time that falls below the "wait" line is wasted time.

It looks like the team waited for spec approval for a longer time than it took to write the specification in the first place!

But when you subtract the time spent waiting, only 3 weeks and 2 days of the total time spent on this feature were actually spent working on it.

The goal of value stream mapping is to help you understand the balance between work and waste. You can't eliminate all waiting time from every project—but seeing the waste by knowing exactly how long the team is waiting and when they do it is a valuable first step in eliminating that waste.

A lot of teams use flow efficiency to measure their value stream, expressed as a percentage:

100 * Time working ÷ Lead time %

Creating a value stream map of your most recent delivery is a great way to get your team thinking about how to improve their process.

lean kanban

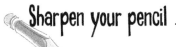

Sharpen your pencil

The Audience Analyzer team decided to create a value stream map for their most recent feature release, the Stat Mapper. Read the team's account of what happened and create a value stream map to show how much time was spent working versus waiting in that feature's delivery process.

Work

Wait

Step 1: Focus groups, research on the client requirements (3 weeks)
Step 2: Writing user stories, creating story map/personas (2 weeks)
Step 3: Wait for upper management spproval (3 weeks)
Step 4: Prioritize work in backlog (1 day)
Step 5: Wait for development to start (3 days)
Step 6: Develop and unit test the feature/write integration tests (5 days)
Step 7: Wait for integration test environment and automation (3 days)
Step 8: Integration Test (2 days)
Step 9: Fix bugs (1 day)
Step 10: Wait for integration test environment and automation (3 days)
Step 11: Integration test (2 days)
Step 12: Wait for demo environment installs (3 days)
Step 13: Deploy to demo environment (1 day)
Step 14: Demo/gather feedback (2 days)
Step 15: Wait for production release window (2 weeks)
Step 16: Release to production (1 day)

What is the lead time (the time from when the feature is identified until it is delivered) for this feature?_____

How much time was wasted building this feature? _____

What was the flow efficiency? _____

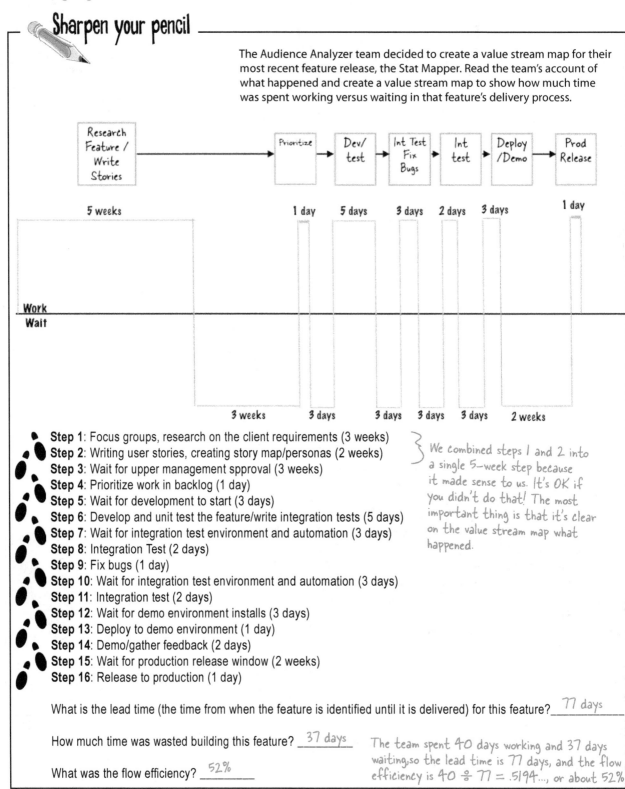

Trying to do too many things at once

After taking a look at the Stat Mapper's development time line, the team found a lot of time was getting wasted waiting for testing resources and environments. They investigated a little further, and realized that developers were starting up a new feature every time they finished the development and unit testing. That meant that each developer sometimes had four and five features at various places in the test and development pipeline. And since the design was highly coupled, features had to be released in batches that translated to test and deployment delays.

Once the team had the process mapped out as a value stream, it was easier for the whole team to come up with suggestions for how to spend more time working and less time waiting. When the team met as a group to take a look at the value stream map, they came up with a few easy-to-implement suggestions:

- Testers suggested automating their integration tests to cut down on time in test.
- Developers suggested simplifying the design of components so that they could be released independently.
- The operations team suggested automating the deployment pipeline so that it's easier to release frequently.

history of process improvement

Under the Hood: Toyota Production System

Lean thinking comes from the Toyota Production System (or TPS). From about 1948 to 1975, industrial engineers at Toyota, led by visionaries Taiichi Ohno and Kiichiro Toyoda, created a new way of thinking about manufacturing systems. They focused on being able to look at the entire flow of the manufacturing system, and eliminate places where work was not contributing to the end product. They found that by limiting the amount of work in each step and managing flow through the system, they could get much higher quality products in much less time than they had seen prior to the introduction of TPS.

Mary and Tom Poppendieck are software engineering experts, trainers, and consultants who adapted Lean thinking to software development, popularized in their book *Lean Software Development: An Agile Toolkit* (Addison-Wesley Professional, 2003). They started with these concepts from the TPS and applied them to the way teams work together to develop software. And as it turned out, the process for building cars and the one for building software have a lot more in common than you might think at first glance.

Three sources of waste must be removed

TPS is focused on the idea that there are three types of waste that slow down the workflow and must be eliminated:

* **Muda** (無駄), which means "futility; uselessness; idleness; superfluity; waste; wastage; wastefulness"
* **Mura** (斑), which means "unevenness; irregularity; lack of uniformity; non uniformity; inequality"
* **Muri** (無理), which means "unreasonableness; impossible; beyond one's power; too difficult; by force; perforce; forcibly; compulsorily; excessiveness; immoderation."

Software teams deal with all of these sources of waste as well. When teams start to look at the process they're using to develop, they find all kinds of things that people are doing to cope with futility, impossible goals, or unevenness. Finding those sources of waste and getting rid of them can help teams to develop smaller increments more frequently. That's exactly what TPS was trying to do, too.

Seven types of manufacturing waste

Another foundational development of TPS was the identification of the seven types of manufacturing waste. Shigeo Shingo documented them in his book, *A Study of the Toyota Production System* (Productivity Press, 1981):

Transportation

Inventory

Overproduction

Motion

Waiting

Defects

Extra Processing

The types of waste in software development are really similar to the types of waste in manufacturing. At first glance, it seems like the two processes would be really different. But they both require team members to constantly problem solve and think about the quality of their work.

> **Mary and Tom Poppendieck,** who adapted Lean thinking from manufacturing to the software development domain, developed the seven categories of waste that software teams run into from these seven types of manufacturing waste.

Foundational concepts of flow, continuous improvement, and quality

The Poppendiecks translated the seven wastes of manufacturing to develop the seven wastes of software development. They also applied the TPS goals of designing muda, mura, and muri out of the process that's used to develop a product. There are several other ideas from TPS that are translated directly into Lean for software development:

- **Jidoka**: *creating automated ways to stop production as soon as a problem is detected.* In TPS each step in the process has automated checks to make sure that problems are fixed right where they are found and not pushed to the next person on the manufacturing line.

- **Kanban**: *signal card.* Signal cards were used in the TPS manufacturing system to indicate that a step in the process was ready for more inventory.

- **Pull Systems**: *each step in the process indicates to the step before it that it needs more inventory when parts are depleted.* In this way, work was pulled through the system at the most efficient pace and never became uneven or backed-up.

- **Root cause analysis**: *figuring out the "deep" reason that something happened.* Taiichi Ohno talked about using the "5 Whys" (or repeating why five times) to figure out what caused a problem

- **Kaizen**: *continuous improvement.* Putting in place activities that improve all of the functions every day can only happen when the team pays attention to what's happening along the flow of work, and suggests ways to make things better.

Taiichi Ohno implemented TPS with a series of experiments. He worked team by team to figure out how to create the optimal flow through their system and create products as quickly and with as high quality as possible. He focused on empowering the team and giving them the freedom and responsibility to determine the best way to build products quickly and with as little waste as possible.

Lean thinking was foundational to all of the Agile methodologies, and it was explicitly discussed at Snowbird. That's why you'll find that many of these ideas appear in one form or another in both XP and Scrum.

> "There is no magic method. Rather, a total management system is needed that develops human ability to its fullest capacity to best enhance creativity and fruitfulness, to utilize facilities and machines well, and to eliminate all waste."
>
> – Taiichi Ohno
> (Toyota Production System, p. 9, CRC Press 1988)

so many *options*

Anatomy of an Option

Lean teams use **options thinking** to give them leeway to make decisions. They plan their projects just like any other team does—but the difference is the attitude each person on a Lean team has about those plans. They see all of the work they've planned to do *as options, not commitments*. The team commits to an objective, but not a plan. This allows them to focus on meeting the objective and change the plan if a better way of accomplishing the goal presents itself. By not committing to each step in the plan, the team leaves themselves free to change the plan when new information presents itself.

Here's how you might use options thinking on a project:

1 Define your objective and commit to that.

> **Goal: Release an Audience Analyzer enhancement that will store data from three new sources and serve that data to our Stat Mapper application.**

2 Define the tasks needed to accomplish that objective and consider the completion of those tasks **one option** for achieving the goal.

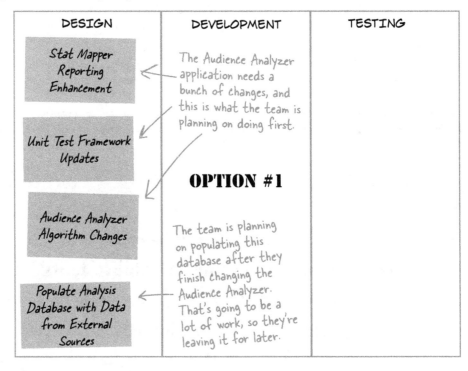

When the team did their planning, they had an idea of how they wanted to do the work. But they thought of this as just one option, and left themselves freedom to choose another path.

lean kanban

3 Start working!

...BUT WHEN MIKE STARTED WORKING WITH THE DBA TEAM, HE REALIZED THAT MOST OF THE DATABASE WORK WAS ALREADY DONE!

THE DBA LEAD LET ME KNOW THAT THEY ALREADY HAVE THE DATA WE NEED IN A DATABASE. WE JUST NEED TO MODIFY THE STAT MAPPER TO ACCESS THAT DATA.

4 Change the plan as needed.

DESIGN	DEVELOPMENT	TESTING
Stat Mapper Reporting Enhancement	Change Stat Mapper Database Code to Access New Data	It's a good thing the team gave themselves options for how they wanted to do this work! Based on what they learned from the DBA, they can replace the large database population task with this much smaller code change.
Unit Test Framework Updates	OPTION #2	
Audience Analyzer Algorithm Changes		

Options thinking lets Lean teams give themselves the freedom to decide as late as possible and reduce the cost of changes when they occur.

On a traditional team, a change in technical approach can be really disruptive. If you've committed to a detailed schedule, you might have set aside time and resources—say, development and DBA team members to work on a task that has programming and database work. You might have created milestones that you report status on to upper management for when the task was complete, or maybe you made this task a predecessor to work being done in the application, for example. That's a lot of up-front planning.

What if one of your team members discovers that a solution already exists? That could cause your whole schedule to be recalculated, and resources would need to be redistributed. That kind of thing happens all the time! By treating the team's original plan as an option, you don't have a number of tasks and resource assignments that are dependent on that task to re-assign and re-orient.

you are here ▸ **271**

it's all connected

Systems thinking helps Lean teams <u>see the whole</u>

Every team has a way of working. It might seem like you make it up as you're going along, but there are always rules that everybody on the team follows (even if you don't realize you're following them). That's what the Lean principle is about. It's called **see the whole**, and it starts with recognizing that each person's work is part of a bigger system.

When a team sees itself as a group of individuals with different functions, they tend to focus only on the improvements that can be made to their job. A programmer might focus on improvements that will make programming easier, a tester will focus on improvements to testing, a project manager might focus on improvements to scheduling or reporting on status. But when all of the team members see how their jobs contribute to a larger system, they can all start coming up with improvements that help the team reach its objectives together instead of optimizing one role above the others.

> SO IF I STEP BACK AND LOOK AT THE SYSTEM AS A WHOLE, IT SHOWS ME ALL OF THE LITTLE "HACKS" THAT INDIVIDUAL TEAM MEMBERS DID THAT MIGHT MAKE THEIR JOBS EASIER, BUT WHICH COULD BE SLOWING DOWN THE WHOLE TEAM.

When everyone sees the whole, the whole team gets better together.

Say for example, a project is running late, so a project manager optimizes his job function by asking for written status twice a day from every member of the team. That will probably make the project manager's job easier, because he always knows where the team is on every task. But it will also slow the team down and make it harder for them to get work done on time. Or, think of a programmer who believes that she can get so much more code written if she doesn't have to write unit tests. That programmer might produce a lot more code, but the cost of finding and fixing defects in it will probably outweigh her added productivity.

Lean teams work to **remove local optimizations** like those and optimize the system together. They look at the system together and remove any activities that are getting in the way of building software fast and with high quality. Lean teams know that limiting their work to the micro-problem of how to make each team member more productive actually gets in the way of operating as a team.

Behind the Scenes

When you create a value stream map, you're really mapping how long it takes for your system to take ideas and turn them into work products. Once you realize that a system exists, it's much easier to figure out how to make improvements that will really help you build higher quality software, faster. It's by recognizing that you are working within a system that you can start thinking about changes to the process your team is using, instead of just improvements to the job you're doing on your own.

Some "improvements" didn't work out

The team took a hard look at their value stream map and realized that they needed to focus on finishing each task before they start a new one. Next, they started to think of ways to improve their coordination with the Operations team. Instead of trying to see the whole system, they focused on just the work that they could control.

A failed experiment (and that's a good thing!)

Mike and Kate had a good idea! They could just get the code working, then dump it on another team. How do you think this worked out? Well, they tried this change for two weeks... until it started to blow up in their faces. All of the software they were building started to get backed up in the integration environment and customers started complaining that they weren't getting fixes fast enough. The team realized that their improvements needed to take the whole system into account (including the other teams they worked with) if they were going to see real improvements to their lead time.

← Not every experiment works out, and that's OK! With Lean thinking, teams are comfortable trying new approaches and ideas. A small failure is a stepping stone to a big success later on.

push me pull you

Lean teams use pull systems to make sure they're always working on the most valuable tasks

Traditional project teams push work through their systems. They attempt to plan all of the work up front and then control changes to that plan throughout the project's execution. By predicting exactly how much work will occur and who will need to do it, traditional projects attempt to maintain the right resource mix through meticulous planning.

Teams with a Lean mindset work in the opposite way: they use **pull systems**. In a pull system, each step in the process is triggered by the later step, pulling the output of that previous step only when it's completed. By having the later step in the process pull tasks from the former, Lean teams finish each feature as quickly as possible. Pulling work through the system establishes a constant flow of finished work through it and doesn't overload it with local optimizations.

In a pull system, the later step in the process pulls work from the step before it.

Moving to a pull system is as much about changing the way everyone thinks about the work as it is about changing the way they do it.

Set up a pull system by establishing WIP limits

Here's an example. Let's say a traditional test team is always overloaded trying to keep up with code being generated by a development team. If they moved to a pull system, work wouldn't begin on development items until the test team requested more work from development. So how would they do this?

One effective way is to limit **work in progress (WIP)**. A Lean team will establish a **WIP limit**, or a number of work items that can be worked on at any given time for each step in their system. If a team can only test four features at a time before their system begins to slow down, they can *establish a WIP limit in the previous stage* so that the development team is never building more than four distinct features at a time. By establishing a limit, they can define the shortest path through the process, and reduce the total lead time from when the feature is identified until it is released.

Kanban is a method for improving your process that implements pull systems and builds on Lean thinking. We'll talk about it a lot more in the rest of this chapter.

lean kanban

Pull Systems Up Close

Sometimes the development pipeline gets clogged with work

When you see your system as a whole, you start to figure out which steps happen in which order to get a product delivered. A pull system is a specific way that work flows through the system. When a team pushes work through a system they can find themselves running into logjams. That's the situation that Mike, Kate, and Ben are running into with their project.

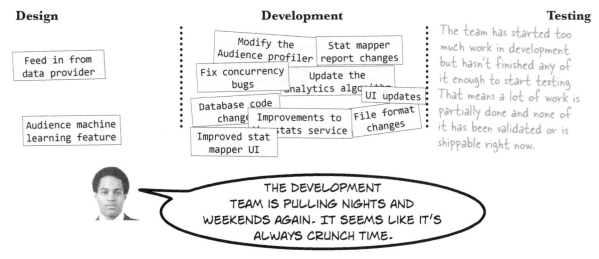

Adding a WIP limit sets up a pull system to smooth out the flow

Look what happened when everyone agreed to a WIP limit. Before, every feature that they finished designing got pushed straight into development. With the WIP limit, the developers only pull in the next feature when they're ready to work on it. This is how pull systems establish a smooth flow through the system—by making sure that the later step in the process **controls how much work the team can take on**.

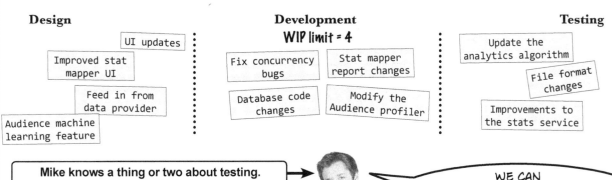

Mike knows a thing or two about testing. And that's really valuable right now—once development hit its WIP limit, he was able to test other features that were done. This is why Lean teams value generalizing specialists who are experts in one area but can help out in other stages of the process to move work along.

you are here ▶ **275**

understand the system

there are no
Dumb Questions

Q: Options thinking seems weird. Isn't it better to have a really accurate picture of what will happen on a project, and to be as predictable as possible?

A: Think about the last time you started a project. How much did you know about how it was actually going to play out? If it was like most projects, there were surprises from the time you started right up until you were done. Lean teams recognize this—rather than try to predict exactly what's going to happen and then measure against that prediction, they treat even the broad strokes of the project as optional. That doesn't mean they won't happen, but if a better option comes along, they'll take it.

Q: That sounds pretty theoretical. What does that mean in practice?

A: Options thinking in practice can mean starting down multiple paths at the same time to solve a tough technical problem that doesn't have a clear answer. It can mean being open to changing scope and strategy in order to meet an objective and deliver the most value. Lean teams keep their options open by focusing on the business outcome that's needed. They don't plan out every detail of the implementation of each task it will take to accomplish their goal and try to stick to that plan.

You've seen this idea before. Another name for options thinking is making decisions at the last responsible moment—just like you learned about in Chapter 3. And that's another example of how the Agile Manifesto signers had Lean thinking tools in mind when they wrote the agile principles.

Q: I have enough to worry about just developing my projects. Now I'm supposed to think about the whole system I'm working in, too?

A: A lot of people feel that way. But focusing just on the job you're doing can lead you to work in a way that makes it harder for your team to actually get work done. While it might seem to you like you're getting a lot more done when you make an improvement that's focused just on your role, it's more important that the whole process you're working in is producing high-value software than it is that you are able to do more in your functional silo.

While it might seem like each person focusing on doing their own jobs as well as they can will get a product out the door sooner, that's rarely true. The act of building software most often depends on a lot of people collaborating with each other to figure out what should be built, develop it, make sure it works, and make it available to the people who need it. Focusing too much on any of those steps can lead to problems in the software itself. Lean teams solve this problem by asking each person on the team to think about the whole system and not just the work they are personally responsible for. That's where the Lean thinking tools from earlier in the chapter can help—like seeing waste, using options thinking, taking measurements, and figuring out the cost of delay for each feature.

Q: I get some of the other thinking tools, but I'm still not clear on cost of delay. How do you figure that out?

A: For some features it can be pretty obvious. If you're building software to implement a new rule in a tax code, then the cost of missing tax day is probably pretty big. For other features, however, it can be a lot harder to understand. That's why Lean teams focus on trying to figure out not just the priority of the features they're working on, but each feature's cost of delay too. Making sure that the team talks about the cost of delay can make sure that teams are working on the highest-value feature at any given time.

One way to figure out cost of delay is to ask about it. When a product owner is describing the feature to the team in sprint planning, or when the team is committing to what will be in an upcoming release, have a discussion with the person who prioritized the features to make sure the team understands the cost of delaying that feature.

Sometimes asking questions about it is enough to cause the product owner to re-think the current priorities. Some teams use heuristics and practices like assigning a business value or cost of delay number to each of the features being planned in a sprint planning session so that it's easier for teams to understand the cost of not releasing it quickly.

The most important thing to remember here though is that you should consider the cost of delay when you're determining the priority order of work in an increment.

Q: What does that "queuing theory" stuff have to do with anything?

A: Once you see your software process as a system, it's not hard to start thinking of all of the work that goes through it as part of a queue. If you think of each feature as queued to go from the first step to the second, and then the third, and so on, you'll start to see how queuing theory applies to the work your team is doing. Have you ever noticed how some supermarket checkout lines move more quickly than others? The same principles that govern why some lines are faster or slower also explain why your team builds features quickly or slowly.

Lean teams think about their process as one big (mostly) straight line, and then set about trying to find the most direct route from identifying a feature to be delivered to releasing that feature with as little waste in between as possible.

276 Chapter 6

lean kanban

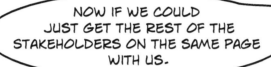

> NOW THAT WE'RE USING LEAN TO THINK ABOUT THE WAY WE DEVELOP, I CAN SEE THE TEAM GETTING A MUCH BETTER HANDLE ON WHAT WE CAN DO AND WHAT'S MOST IMPORTANT!

> NOW IF WE COULD JUST GET THE REST OF THE STAKEHOLDERS ON THE SAME PAGE WITH US.

Mike: We're setting aside a little time in each increment to make headway on the test automation, and we've started refactoring the components so we can release them independently. Everything is popping into place. In a few months we'll be cruising at twice the speed we are now.

Kate: Wait, what? A few months?! When we started tracking our velocity I thought we were giving everyone visibility into how much we can do. Now the whole company is obsessed with it. Whenever our velocity number goes down, I have to go to meetings with senior managers and explain why.

Mike: But... but that's not how velocity and story points work!

Kate: Tell that to the steering committee. They've asked us to come up with story point estimates for features earlier and earlier, and set schedules way out into the future based on them.

Mike: Ugh, that's so frustrating! The steering committee needs to understand that we're sharing our velocity with them so they can help us make the right choices, not so they can use it to push more work onto us. How is this any different from when we used to plan big releases with big specs?

How could you use the concept of a pull system to address the problem Kate and Mike are describing?

you are here ▸ **277**

Question Clinic: Least worst option

SOMETIMES A QUESTION ON THE EXAM WILL PRESENT YOU WITH *NO CLEAR RIGHT ANSWER*. IF ALL OF THE ANSWERS SEEM WRONG, **CHOOSE THE LEAST WORST OPTION** AND MOVE ON TO THE NEXT QUESTION.

> 109. You are a leader of an agile team. A stakeholder identifies an urgent change during a sprint. Which is an action you would take?
>
> A. Be a servant leader
> B. Add the change to the backlog
> C. Tell the team member to talk to the product owner
> D. Tell the team member to write up the change for the change management board

This answer isn't great, because it's not clear if the change would be added to the sprint backlog or the product backlog. It's the only one that looks like the change might get done, though.

The rest of the answers are worse, because they don't actually get the change implemented at all.

If none of the answers seem right, choose the one that seems the closest to right.

HEAD 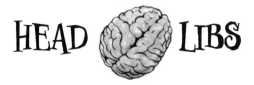 LIBS

Fill in the blanks to come up with your own least worst option question!

You are a _____ on a Scrum team.
(kind of agile practitioner)

Your team would like to use _____ as part of their regular sprint activities. How would you help them get started?
(agile practice)

A. _____
(wrong answer)

B. _____
(sort of wrong, possibly right answer)

C. _____
(ambiguous but correct answer)

D. _____
(ridiculously wrong answer)

illustrated instruction manual

Kanban uses a pull system to make your process better

Kanban is a *method for improving your process*. It's based on the Lean mindset, the same way that XP and Scrum are based on their specific values. But it's a little bit different from XP or Scrum, in that it doesn't prescribe roles or specific project management or development practices that tell you how to work on a team. Instead, Kanban is all about taking a look at the way you work today, figuring out how work is flowing through your system, and then experimenting with small changes and WIP limits to help your team establish a pull system and eliminate waste.

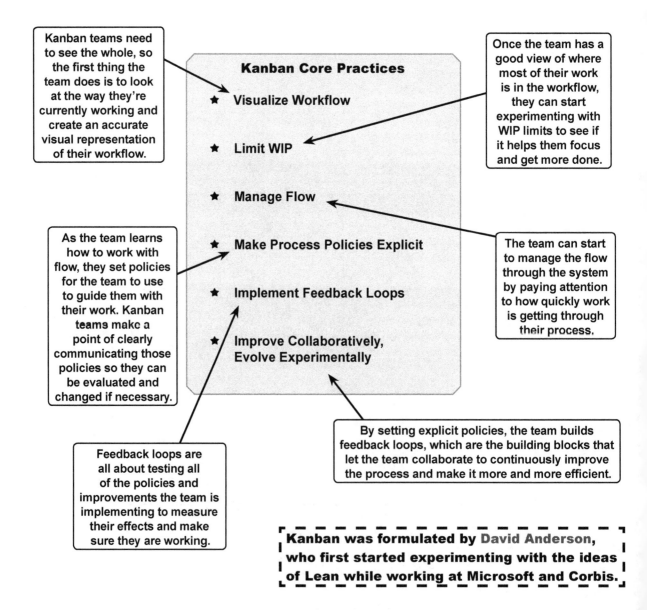

Chapter 6

Use Kanban boards to visualize the workflow

A **Kanban board** is a tool that Lean/Kanban teams use to visualize the workflow. It consists of a board—often a whiteboard—divided into columns, with cards in each column to represent **work items** flowing through the process.

A Kanban board looks a lot like a task board, but they're not the same thing. You've seen task boards in our discussions of Scrum and XP, so it's easy to look at a Kanban board and assume it's basically the same thing. It's not. The purpose of a task board is to make the state of current tasks clear to everybody on a team. Task boards help a team stay on top of the current status of their project. Kanban boards are a little different. They are created to help a team understand how work flows through their process. Because work items are kept at a feature level on a Kanban board, they aren't the best way to know exactly which task each team member is working on—but they're great for helping you see how much work is in progress in each state of your process.

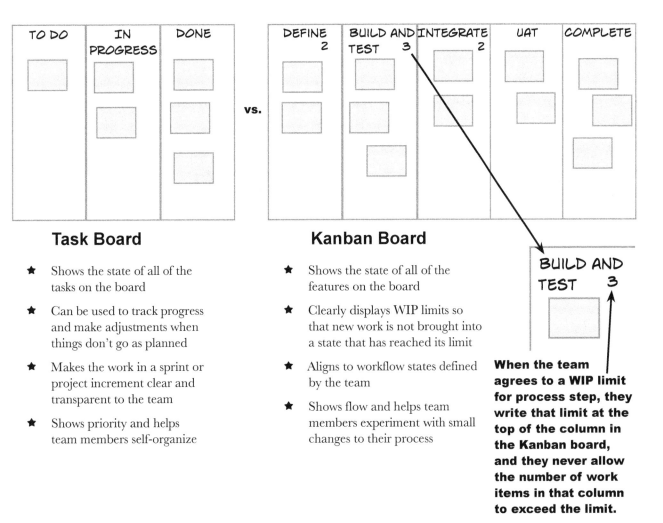

Task Board

★ Shows the state of all of the tasks on the board

★ Can be used to track progress and make adjustments when things don't go as planned

★ Makes the work in a sprint or project increment clear and transparent to the team

★ Shows priority and helps team members self-organize

Kanban Board

★ Shows the state of all of the features on the board

★ Clearly displays WIP limits so that new work is not brought into a state that has reached its limit

★ Aligns to workflow states defined by the team

★ Shows flow and helps team members experiment with small changes to their process

When the team agrees to a WIP limit for process step, they write that limit at the top of the column in the Kanban board, and they never allow the number of work items in that column to exceed the limit.

let's look at the board

How to use Kanban to improve your process

> A value stream map is a great way to create this picture! These are the same boxes that you'd see at the top of the map.

The core practices of Kanban are a series of steps that teams execute in order to see how their system is working, and then create a pull system that moves work efficiently through their workflow.

① Visualize Workflow: create a picture of the process you're using today.

Define → Design → Build → Test → Deploy

② Limit WIP: watch how work items flow through the system and experiment with limiting the number of work items in each step until work flows evenly.

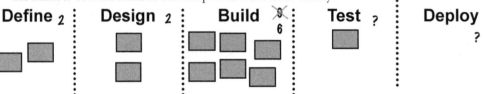

③ Manage Flow: measure the lead time and see which WIP limits give you the shortest time to delivering features to your clients. Try to keep the pace of delivery constant.

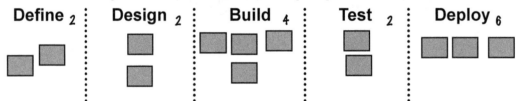

④ Make Process Policies Explicit: find out the unwritten rules that are guiding your team when they make decisions and then write them down.

⑤ Implement Feedback Loops: for each step in the process create a check to make sure the process is working. Measure lead and cycle time to make sure the process isn't slowing down.

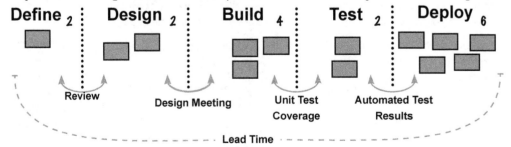

⑥ Improve Collaboratively: share all of the measurements you gather and encourage the team to come up with suggestions to keep on experimenting.

282   Chapter 6

Why Limit WIP?

Behind the Scenes

The fastest way to get something done is to start it, do the work, and finish it without letting anything get in your way. That seems pretty simple, right? But there are many reasons why teams don't do that. The most common reason is that they're focused on making sure that everybody on the team is busy. When a developer finishes coding and unit testing his or her work, for example, the next step is often to hand it off to a QA team to integration test it. What should the developer do while the feature is being tested? More often than not, he or she starts working on another feature. That sounds fine individually, but what if all of the developers on the team start new features halfway through an increment while they wait for testing? If they do that, there's a pretty good chance that there will be a number of half-finished features when the increment ends. But what if those same people **focused on having complete, shippable features** instead of starting new work? Then the team would get more done in each increment. If everyone on the team is motivated to keep themselves busy above all other priorities, they will keep themselves busy... but they won't finish each feature as quickly as they could.

The team that ends up with an iteration full of half-finished work is an example of what happens when the focus is entirely on resource utilization (keeping everyone busy) instead of finding the shortest path for features to take through their workflow. This happens a lot when teams have work requests coming in from many different sources, like several managers pushing features onto one team. If those managers don't know about each other, they can expect that their priorities are the most important ones the team faces. A sales person, for example, will usually have different feature requests than a support person. Unless they can clearly see how their requests stack up in relation to each other, they might each put pressure on the team to do more work than they planned.

Kanban's answer to this problem is to **visualize** the way the features are moving through the workflow, and to experiment with setting limits to the number of requests the team will work on at any given time. By showing how many features are currently in progress and limiting the number of requests the team can work on in each state, Kanban establishes a steady pull of work through the system and frequent delivery of completed features. Teams usually start by observing the **flow of work** on a Kanban board, and continue by experimenting with WIP limits in states where more work seems to be started than is getting finished. Usually a team will write a WIP limit in the column header on the task board and refuse to take work into that state when the limit is reached instead of overloading the system. People focus on helping to get the feature all the way to complete before starting work on a new one. Once a team is limiting the amount of work that can be in progress in a state at one time, they start to establish a predictable flow of work through their system. Teams find that their lead and cycle times go down, just by focusing everyone on finishing work, rather than on keeping themselves busy.

workflow time

The team creates a workflow

The first step in improving a workflow is to **visualize** it, and that's what's going on here. The Audience Analyzer team got together to talk through the set of steps the team follows every time they build a new feature. There are exceptions, of course: Sometimes the team never schedules a feature to be delivered. Sometimes the product manager asks for it to be built as an urgent priority because a user needs it right now, and everybody drops what they're doing to make it happen. Or bugs might get found in test too late, and they don't get fixed. Even though the process isn't always followed exactly in this order, the team tried to create a picture of phases a feature is supposed to go through when the team builds and delivers it.

After spending a while discussing those exceptions (and a few more) the team agreed that most often, they follow a workflow that looks something like this:

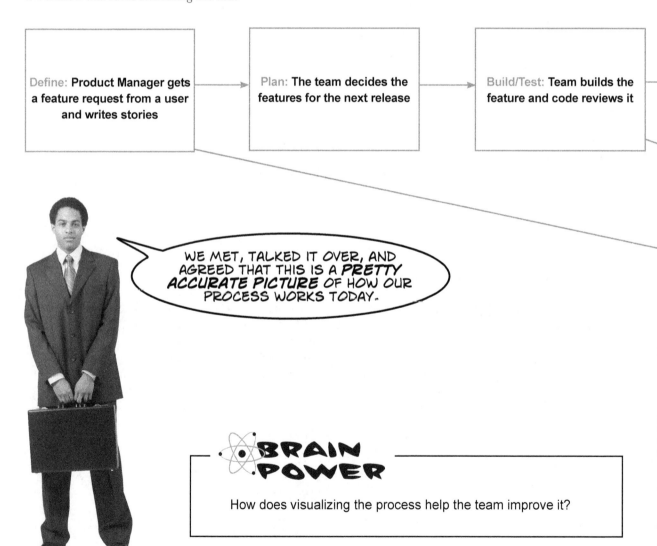

Define: Product Manager gets a feature request from a user and writes stories → **Plan:** The team decides the features for the next release → **Build/Test:** Team builds the feature and code reviews it

WE MET, TALKED IT OVER, AND AGREED THAT THIS IS A *PRETTY ACCURATE PICTURE* OF HOW OUR PROCESS WORKS TODAY.

> ### BRAIN POWER
>
> How does visualizing the process help the team improve it?

Next, they mapped out their process on a Kanban board so they could see which features were in which workflow state. They drew out columns on a task board that match the states they'd defined and then figured out which state each of their current features were in. For each feature, they created a sticky note and put it in the correct column on the board.

When they created their Kanban board, they **decided not to put a Plan column on the board**. The team always held a two-hour meeting at the beginning of each delivery increment to plan out the work they'd do, so features would never be in the "Plan" state for longer than two hours. It just didn't seem useful to track that meeting as a state.

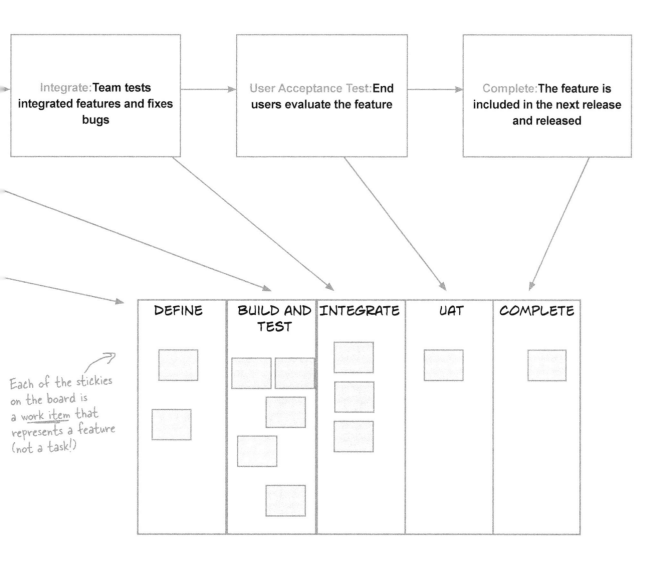

Each of the stickies on the board is a work item that represents a feature (not a task!)

method for improvement not a methodology for building software

> WAIT A MINUTE. ISN'T THAT "KANBAN BOARD" *JUST A TASK BOARD* WITH A FEW MORE COLUMNS?

Kanban is not a project management methodology. It's a method for process improvement that works by visualizing your team's actual, real-world process.

One of the biggest misconceptions about Kanban is that it's basically just Scrum without sprints. It's not. Kanban is **not intended to be a project management methodology**. You use a Kanban board differently than you use a task board: a task board is for managing a project, while a Kanban board is for understanding your process.

Remember how you used a value stream map to visualize the actual path that a real feature took through a project? A Kanban board does the same thing, except instead of just following a single feature, it follows all of the work. Once you and your team can see the whole workflow visualized on the board with work items moving through it, you make adjustments to it in order to make it as accurate a picture as possible.

The goal of Kanban is to help the team inspect its own way of working, so that they can make changes collaboratively in order to release small increments as frequently as possible. Kanban doesn't tell teams exactly how long their timeboxes should be, or what roles should be on their teams, or what meetings they should have on a frequent basis. It helps the team figure those things out for themselves.

Everyone on the team should take part in updating the Kanban board—more eyes means more chances to discover the extra states you didn't realize were there, so you visualize the workflow more accurately.

So how does the Kanban board fit in? Every team works a little differently, and ***those differences are reflected in the columns*** on the Kanban board. If your team always spends a few weeks building a proof of concept for each new feature and does a demo with a group of users before a project really gets started, then you'll add a column to your Kanban board that represents that state. And as you work, you'll discover new columns and add them, so that the whole team gets a clearer and clearer picture of how they work.

Teams that use Scrum, XP, or a hybrid will often use Kanban too. One common way that Scrum teams use Kanban is to visualize their workflow by creating a Kanban board alongside their task board. This helps them establish WIP limits and create a pull system ***within their Scrum implementation***. Kanban and Scrum can be combined to help teams focus on getting more done in their sprints and improving the quality of the work they do.

Cubicle Conversation

The team spent a few increments just watching how features moved through their process. As they observed, they learned a lot about how they were working.

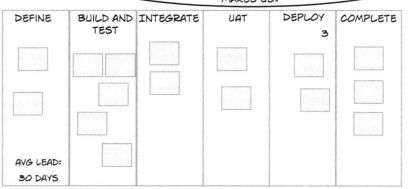

reading the signal cards

there are no Dumb Questions

Q: I've heard about "Lean/Kanban" before, and I see how Lean and Kanban work. But how do they fit together?

A: Lean is the mindset behind Kanban. Kanban relies on Lean, just like Scrum and XP rely on their values. Kanban builds on the thinking tools in Lean, and Lean teams use systems thinking to improve their process.

If you're working on a team that's been building software for a while, it's likely that you and the rest of your teammates are following an unwritten set of processes and policies to do your work. Because those rules and processes are not explicit, small misunderstandings can evolve over time that end up making the team less deliberate than they could be in the choices that they make during development.

Kanban helps teams to think about the way they're working, and to *really* examine the decisions that they they make when they're building a product. By asking the team to visualize the system together and then measure how work flows through it, the team can see the impact of the way that they work on the amount that they get done—and on the time it takes to do it.

By creating a visual map of their process, they can see the effect of any changes they might make together. Kanban's practices of visualizing the workflow, experimenting with WIP limits, and managing flow effectively get everyone on the team to start using the Lean thinking tools—like see the whole, cost of delay, pull systems, and queuing theory—collaboratively.

Q: Earlier you said that Kanban means "signal card." What does that mean? Are those the same as stickies?

A: Good question. In manufacturing, a Kanban is a physical token—like a card with a part number written on it—and it's the basis for a pull system in that environment. In the Toyota Production System, when the team is running low on a specific part, they put a kanban card (literally "signboard" in Japanese, 看板) with its part number into a bin. A supply team exchanges the kanban for more parts from a central supply. The team uses cards to pull parts as needed.

For a software team, Kanban uses work items on a board to implement a pull system. And the same ideas still apply, even though software teams are doing creative intellectual work and not assembly line manufacturing. Just like TPS uses Kanbans to cut down on wasteful processes and excess inventory in building cars, software teams using Kanban to improve their software development processes cut down on wasteful work using the same Lean principles.

Q: I'm starting to get it, but something doesn't make sense. Can you explain again how limiting the work you do gets more work done?

A: When people are focused on a lot of different goals and have their own ideas about how to work, they can end up working really hard toward different ends. When that happens, everybody seems busy, and they feel like they're doing the most they can to get a product released. But the truth is, until they are all focused on the same goal, they might actually be slowing the release of a product down.

By limiting the number of work items "in flight," Kanban asks teams to think about the end goal and eliminate any work they're doing that isn't related to it. Since Kanban teams measure the lead time of a feature before and after limiting WIP, they always know whether an experiment with WIP limits is slowing things down or delivering features faster. By focusing on delivering small increments as fast as possible, Kanban helps teams remove the waste in their process, and build software faster and with higher quality than they could before.

SO BY LIMITING WIP, TEAMS *ACTUALLY GET MORE DONE* BECAUSE THEY DO THE MOST IMPORTANT WORK FIRST AND THEY DON'T LET OTHER GOALS GET IN THE WAY.

lean kanban

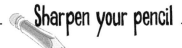

Sharpen your pencil

Take a look at the Audience Analyzer team's Kanban board and determine what to do next. Write the new WIP limit on the task board below.

DEFINE	BUILD AND TEST	INTEGRATE	UAT	DEPLOY 3	COMPLETE

Scenario: After visualizing their process, the team got together every day to map out how the features they were working on moved through each step for the next increment. Their team was used to working on a two-week cadence and releasing regularly, so they observed the features for two weeks. The team has a product manager, four developers, and one dedicated tester who integration tests all of the features. At the end of the two weeks, this is what their board looked like. Everyone on the team was very busy—in fact a couple of developers had put in a lot of weekend work to make progress on the features in build and test.

What should the board look like at the end of the two-week cadence?

....

Where should the Audience Analyzer team try adding a WIP limit?

....

What limit would you try first? Why?

....
....

you are here ▸ **289**

give it a try

Sharpen your pencil

Take a look at the Audience Analyzer team's Kanban board and determine what to do next.

There's no right answer to this exercise. Here are the answers that we came up with. The most important thing is that you think about how you might use WIP limits to manage flow.

DEFINE	BUILD AND TEST	INTEGRATE	UAT	COMPLETE
4	4	2		

Scenario: After visualizing their process, the team got together every day to map out how the features they were working on moved through each step for the next increment. Their team was used to working on a two-week cadence and releasing regularly, so they observed the features for two weeks. The team has a product manager, four developers, and one dedicated tester who integration tests all of the features. At the end of the two weeks, this is what their board looked like. Everyone on the team was very busy—in fact a couple of developers had put in a lot of weekend work to make progress on the features in build and test.

What should the board look like at the end of the two-week cadence?

Ideally, all of the features that were scoped would be in the Complete column.

Where should the Audience Analyzer team try adding a WIP limit?

Define, Build and Test, Integrate

What limit would you try first? Why?

4 for Define, and Build and Test because there are four developers. two in Integrate because there's one tester and a developer could help test after committing their code

290   Chapter 6

The team is delivering faster

It took a couple of tries, but after some experimentation (and a lot of lead time measurements) everyone could see real improvement. The first time they set a limit, it ended up slowing the team down a bit. So everyone got together and had a great discussion, and they decided to try making the number a little higher for the next increment. That seemed to do the trick. Once the team had that WIP limit in place, they found that they were starting and completing more features in each increment than ever before. Even better, they all started helping each other when they ran into problems. Pretty soon it felt like the work was more under control than it had ever been. Ben was especially happy, because the team was much more predictable than they had been in the past.

> OUR LEAD TIME IS GETTING BETTER AND **WE'RE GETTING MORE DONE!** OUR KANBAN BOARD SHOWS EVERYBODY WHERE ALL OF THE FEATURES ARE IN THE SYSTEM.

DEFINE 2	BUILD AND TEST 3	INTEGRATE 2	UAT	DEPLOY 3	COMPLETE

AVG LEAD: 15 DAYS

BRAIN POWER

Some teams find that counting the number of work items in each stage of the process can help them to get an understanding of the flow of work. Why do you think that might help the team understand their workflow better?

diagram *your flow*

Cumulative flow diagrams help you manage flow

Kanban teams use **cumulative flow diagrams** (or CFDs) to find out where they are systematically adding waste and interrupting their flow. They chart the counts of work items in each state over time, and use that to look for patterns that might be affecting the team's throughput (the rate that they can complete work items). CFDs give the team a visual way to keep track of how the system is working.

Once the team gets used to reading CFDs, they can get a good idea of the impact of process and policy changes a lot more quickly. Teams are always looking to have a stable development process with predictable throughput. Once a team has identified the right WIP limits to establish pull, they can start to look at how their working agreements and policies affect the way they work as well. If a team makes it a habit to constantly review their CFD, they all get a sense of how their suggestions and changes affect the amount of work the team can get done.

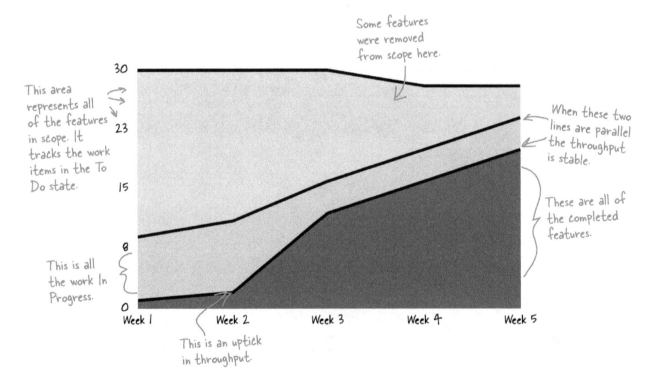

> This is just a broad overview of CFDs—we wanted to give you enough exposure to CFDs so that you can get a sense of how your process is working. Once you've been using Lean and Kanban for a while, you'll want to know more about common patterns in CFDs and how to interpret them. CFDs can be a really valuable tool, and they're actually pretty easy to build. We give you a step-by-step guide to building CFDs and using them to improve your workflow in our book, *Learning Agile*.

Kanban teams talk about their policies

One you've got the steps in your process written down and you've started looking at how features flow through it, you're ready to think about all of the rules team members are following when they do their day-to-day work. It's pretty common for people to come up with their own rules as they're working that guide their behavior and the decisions they make. Many of the misunderstandings and miscommunications that your team will run into while they're working come from these unwritten rules. But when you really start talking about the polices, and people on your team are having a truly open and collaborative discussion, your whole team can begin to work together to avoid many misunderstandings up front. Teams use these policy discussions to come up with **working agreements** together that make team interactions smoother and keep everyone focused on changing policies rather than fighting with each other when things go wrong.

Here's the working agreement the Audience Analyzer team came up with when they decided to make their policies explicit:

> **Audience Analyzer Team Working Agreement**
> - ★ All team members participate in estimation and use our agreed story point scale
> - ★ When a team member finishes a work item, he or she takes the highest ranked one from the backlog that they can work on next
> - ★ No one will pull a work item into a state that's reached its WIP limit
> - ★ All work items must satisfy our definition of "Done" to be considered complete
> - ★ No one will work on a work item that is not in our backlog

Initially the people who were testing the application didn't want to estimate. The team agreed that good discussions about the estimates came from discussions about both development and testing.

Team members had varied understandings of what to do when they finished a work item.

It's hard for team members to say "no" when someone calls and asks for a quick code change. Making this policy explicit helped them to know what to do.

Talking about the policies your team is following and writing them down helps you get everybody on the same page.

loop-*the-loop*

Feedback loops show you how it's working

Kanban teams are really focused on understanding the improvements they make to their process, so they explicitly create **feedback loops** to measure the impact of every change they make. They do this by taking measurements, then using the data from those measurements to make changes to the way they work. As they change their process, their measurements change, which they use to make more changes to their process, and over and over and over...

Teams use their feedback loops to establish a culture of continuous improvement, making sure that everyone helps to take measurements and suggest changes. When everyone is involved in measuring, changing, and repeating, the whole team starts to see each process change *as its own experiment*.

Kanban teams use lead time to create feadback loops

Kanban teams agree on how they'll measure all of the changes they make, and use the data they collect about their process to drive further decisions. And one of the most common ways that Kanban teams create feedback loops is to measure the lead time, make changes—by establishing WIP limits, but also by trying other things—and then see if those changes cause their lead time to go down. For example, let's say they want to try out a policy that team members should focus every Thursday on personal projects. The team can run an experiment by doing it for two releases, and then find the impact that policy has on their throughput by measuring their lead time before and after it goes into effect.

> **Sharpen your pencil**
>
> Which of these scenarios are examples of teams establishing feedback loops and which are changes the team is making to improve their process?
>
> 1. Kate knows the design documents sometimes add extra features and that can slow down the speed of development. She suggests that for all new features, they review the design team's architecture to make sure they're following the architect's vision for the product. They check to make sure this actually speeds things up.
>
>   ☐ Feedback Loop   ☐ Change
>
> 2. Mike realizes that the development team isn't writing enough unit tests to cover the functionality in the product. He sets a standard of 70% coverage for all new features.
>
>   ☐ Feedback Loop   ☐ Change
>
> 3. Ben uses his client meetings to talk through functionality with end users before writing a spec.
>
>   ☐ Feedback Loop   ☐ Change
>
> 4. Mike starts calculating the cycle time of all of the work items in each increment. He realizes that the team is getting faster across the board.
>
>   ☐ Feedback Loop   ☐ Change

lean *kanban*

Now the whole team is collaborating on finding better ways to work!

Now that the CFD and lead time are shared with the whole team, they're coming up with suggestions to make things work better every couple of weeks. Not all of them work, but that's OK, because the team learns together from every experiment they try. They've dramatically improved their lead time, and everyone feels engaged and in control.

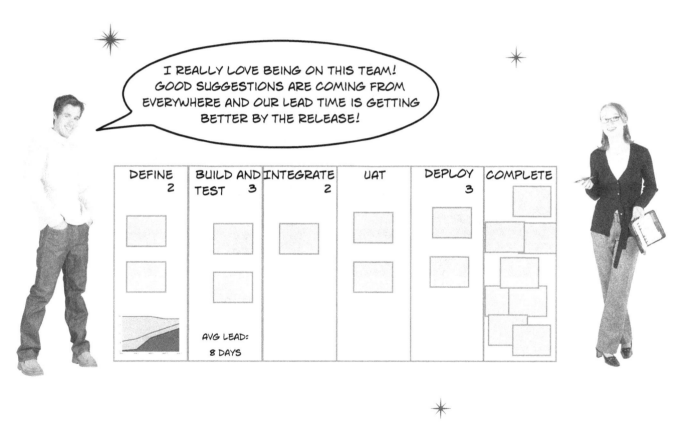

Now that the team is improving collaboratively, they have more control and they're getting more done!

check your knowledge

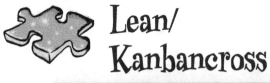

Here's a great opportunity to seal Lean and Kanban into your brain. See how many answers you can get without flipping back to the rest of the chapter.

Across
1. This type of waste is the Japanese word for unreasonableness
4. _____ thinking gives a Lean team choices until the last responsible moment
7. The first practice a team needs to master when using Kanban is to _____ their workflow
9. Lean teams try to quantify how time-critical a feature is using a metric called _____
10. Kanban means _____ card in Japanese
11. This type of waste is often identified through testing
12. Kanban teams set _____ limits on steps in their process in order to optimize for flow
16. Teams that pursue multiple options when developing features are using a Lean thinking tool called _____
17. Lean is derived from the _____ production system
18. Lean teams are always working to_____ waste
19. The Japanese word for continuous improvement is _____

Down
1. Unlike Scrum and XP, Lean is a _____, not a methodology
2. In Kanban, teams identify their _____ and make them explicit
3. Lean teams use _____ theory to analyze how work flows through their system
5. When the later step gets its work from the step before it in a process
6. Lean thinking asks people to "see the _____" when they analyze a process
7. Lean teams create a map of the _____ stream to find out how much time is spent waiting
8. A type of waste that you see when people try to do too many things at once
9. Lean teams don't use sprints, they develop on a delivery _____
13. When a product is intuitive to use and does what it is meant to do, it has _____ integrity
14. Kanban teams use metrics to establish _____ loops
15. Sometimes teams build _____ processes and features

lean kanban

Since Lean was an important consideration when creating the Agile Manifesto. It's not surprising that many of the thinking tools are part of Scrum and XP as well. Here are some thinking tools you already know from previous chapters. Match the tool's name with its description.

The last responsible moment — Changing code to make it more readable and maintainable without changing its behavior.

Iterations and feedback — Making decisions about which work will be done without requiring external approval.

Refactoring — Making decisions when you have the most information about them.

Self-determination, motivation, leadership expertise — Delivering software in increments so that new features can be evaluated while still more features are being developed.

Sharpen your pencil
Solution

Which of these scenarios are examples of teams establishing feedback loops and which are changes the team is making to improve their process?

1. Kate knows the design documents sometimes add extra features and that can slow down the speed of development. She suggests that for all new features, they review the design team's architecture to make sure they're following the architect's vision for the product. They check to make sure this actually speeds things up.

☒ Feedback Loop ☐ Change

2. Mike realizes that the development team isn't writing enough unit tests to cover the functionality in the product. He sets a standard of 70% coverage for all new features.

☒ Feedback Loop ☐ Change

3. Ben uses his client meetings to talk through functionality with end users before writing a spec.

☐ Feedback Loop ☒ Change

4. Mike starts calculating the cycle time of all of the work items in each increment. He realizes that the team is getting faster across the board.

☒ Feedback Loop ☐ Change

you are here ▸ **297**

exercise solutions

Lean/Kanbancross SOLUTION

Here's a great opportunity to seal Lean and Kanban into your brain. See how many answers you can get without flipping back to the rest of the chapter.

lean kanban

Here are some things we overheard Ana, Ryan, and Gary saying. Some of them are describing waste in the project, and some are not. Identify the type of waste each of them is describing. Then draw a line from each speech bubble to either **WASTE** or **NOT WASTE**. If it's not waste, add another line to the appropriate type of waste in Lean.

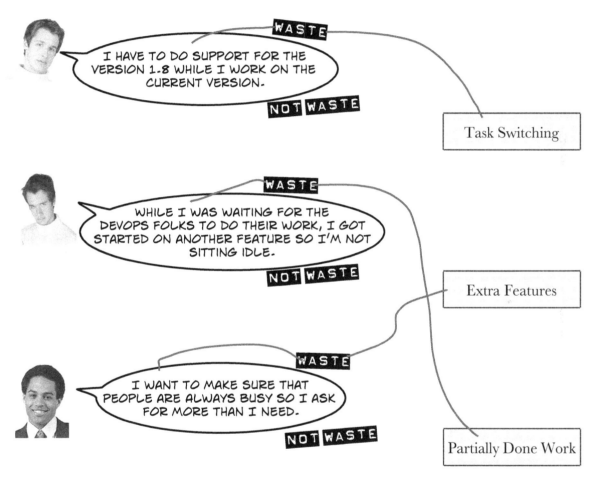

Exam Questions

> These practice exam questions will help you review the material in this chapter. You should still try answering them even if you're not using this book to prepare for the PMI-ACP certification. It's a great way to figure out what you do and don't know, which helps get the material into your brain more quickly.

1. Value stream maps are used for all of the following except:
    - A. Understanding a feature's lead time
    - B. Finding waste in a process
    - C. Discovering new features to build
    - D. Understanding a feature's cycle time

2. Sean is a developer on a team that's building financial software. His team has been asked to build a new trading system. He and his team had a meeting to come up with a picture of the workflow they're using. Then they put the process on a whiteboard with columns for each step in the process. After a few weeks of watching the work items the team was working on progress through the columns on the board, the team noticed that there were a couple steps in the process that seemed to get overloaded.

   What's the BEST thing for the team do next?
    - A. Work with the team to get better at doing the work in the steps where work is slowing down
    - B. Add more people to the steps that are slower
    - C. Focus on finishing the work on the board
    - D. Limit the amount of work in progress allowed in the steps that are overloaded

3. Lean is different than methodologies like Scrum and XP because it is a _____ with _____.
    - A. Mindset, thinking tools
    - B. Methodology, practices
    - C. Process improvement plan, measurements
    - D. School of thought, principles

4. A Lean team looked at all of the work items in their process and paid attention to how they were progressing between each stage in their process. They then focused on how to keep the line of work moving at an even rate. What thinking tool helps them see work as a line of features being removed from the system when they are done?
    - A. Seeing waste
    - B. Last responsible moment
    - C. Queuing theory
    - D. Measurements

lean kanban

Exam Questions

5. Which of the following BEST describes the Lean thinking tool "Cost of Delay"?

    A. Ranking features based on how soon your client needs them
    B. Assigning a dollar value to the time you spend building a product
    C. Understanding the time-criticality of each of the tasks in your team's queue so that you make better decisions about which tasks must be completed first
    D. Understanding how much money you are losing when you delay your project

6. What are the seven types of waste in software development?

    A. Partially Done Work, Extra Processes, Task Switching, Heroics, Over-commitment, Defects, and Extra Features
    B. Partially Done Work, Extra Processes, Extra Features, Task Switching, Communication, Waiting, and Defects
    C. Partially Done Work, Extra Processes, Task Switching, Waiting, Motion, Defects, and Extra Features
    D. Partially Done Work, Extra Processes, Task Switching, Detailed Planning, Motion, Defects, and Extra Features

7. Which of the following BEST describes Kanban's core practices:

    A. Visualize Workflow, Create Kanban Board, Limit WIP, Manage Flow, Make Process Policies Explicit, Implement Feedback Loops, Improve Collaboratively, Evolve Experimentally
    B. Plan Do Check Act
    C. Visualize Workflow, Observe Flow, Limit Work In Progress, Change Process, Measure Result
    D. Visualize Workflow, Limit WIP, Manage Flow, Make Process Policies Explicit, Implement Feedback Loops, Improve Collaboratively, Evolve Experimentally

8. Which of the following is NOT a Lean principle?

    A. Eliminate Waste
    B. Implement Feedback Loops
    C. Decide as Late as Possible
    D. See the Whole

Exam Questions

9. Which of the following BEST describes how a Kanban board is used?

    A. To observe how features flow through a process so that teams can determine how to limit WIP and identify the most even flow of work through the steps in a workflow
    B. To track WIP limits and current task status so that a team knows how much work they have left to do
    C. To track defects and issues and create the fastest path for resolving problems in a product
    D. To help a team self-organize and see where bottlenecks are in their workflow

10. What is the lead time and cycle time for this feature based on the value stream map below?

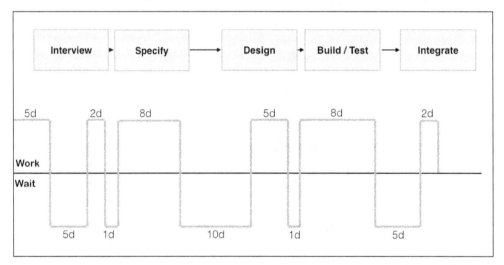

    A. Lead: 22 days, Cycle: 30 days
    B. Lead: 30 days, Cycle: 22 days
    C. Lead: 52 days, Cycle: 30 days
    D. Lead: 70 days, Cycle: 42 days

11. When a team has established WIP limits and started managing flow, what's the NEXT step in applying Kanban to their workflow?

    A. Implement feedback loops
    B. Make process policies explicit
    C. Improve collaboratively
    D. Deliver as fast as possible

Exam Questions

12. Members of your co-located team are having a hard time meeting their commitments. They keep committing to more and more work each increment, but they're not accomplishing their goals. Now there are many features in the first two columns of your Kanban board and very few in the later columns. Which of the following kinds of waste is NOT likely the cause of this team's problem with meeting commitments?

    A. Task Switching
    B. Defects
    C. Motion
    D. Partially Complete Work

13. Which of the following is NOT a principle that is shared between Lean and Scrum?

    A. Last responsible moment
    B. Iterations and feedback
    C. Self-determination and motivation
    D. Eliminate waste

14. When the later step in a process moves work from the step before it, this is referred to as...

    A. Queuing Theory
    B. Waste
    C. A pull system
    D. Inventory

15. Which of the following is NOT an example of options thinking?

    A. A team tries two parallel approaches to developing a high risk feature that they have little experience with
    B. A team identifies goals that can be delivered early in a project to validate the design approach
    C. A team sets a hard deadline and list of dependencies to deliver a feature they're unsure about
    D. A team spends time sharing knowledge together so that more people can work on all of the tasks that are planned for a sprint

exam answers

Exam ~~Questions~~ Answers

1. Answer: C

Teams use value stream maps to find waste in their process. Value stream maps can give us valuable information about lead time and cycle time, but rarely provide insight into future product needs.

2. Answer: D

This question is describing the initial steps a team takes when they are doing Kanban. The first step is to visualize the workflow, the second is to limit WIP.

3. Answer: A

Lean is a mindset with thinking tools to help you think about problems in a way that identifies and eliminates waste. XP and Scrum are methodologies with practices that help you deliver software while adhering to Agile principles.

Answer D is a pretty good answer, but answer A is definitely more accurate.

4. Answer: C

Lean teams use Queuing Theory to examine the order in which work is done in their system. Then they try to find waste that is causing the team to spend time waiting when they could be making progress and removing completed work from the line.

5. Answer: C

Cost of Delay is one of the criteria you can use to determine the order in which features should be developed. Understanding a feature's cost of delay means thinking about the risk involved in it as well as demand, and missed opportunities you're encountering because the team is working on the current feature.

6. Answer: C

Lean's seven wastes of software development (Partially Done Work, Extra Processes, Task Switching, Waiting, Motion, Defects, and Extra Features) are derived from the Toyota Production System's seven wastes of manufacturing (Transportation, Waiting, Motion, Defects, Over-Production, and Extra Processing). Both lists are a useful way to categorize the common types of behavior that slow down production when building products and features.

7. Answer: D

Only answer D includes all of the core practices of Kanban.

8. Answer: B

"Implement Feedback Loops" is a core practice of Kanban, but it's not a principle of Lean. The principles of Lean are: Eliminate Waste, Amplify Learning, Decide as Late as Possible, Deliver as Fast as Possible, Empower the Team, Build Integrity In, and See the Whole.

Exam Questions

9. Answer: A

Kanban boards are used to track how features progress through your workflow. When teams implement Kanban, they do it to understand how their process is working, not to track progress on a specific project. They are not used for tracking tasks, task boards are used for that.

10. Answer: C

> This is what people mean when they say that Kanban is a method for process improvement, and not a methodology for project management.

Lead time is the total time the feature is in the system. In this case adding up all of the numbers on the Value Stream Map gives 52 days. Cycle time is the total time spent working, that's 30 days.

11. Answer: B

The order of the practices in Kanban is: visualize workflow, limit WIP, manage flow, make process policies explicit, implement feedback loops, improve collaboratively, evolve experimentally.

12. Answer: C

The question mentioned that the team was co-located, so it's least likely that time in transit is causing the team's problem delivering. On the other hand, if they are constantly taking on new work without finishing the work they already have, it's likely they will run into problems with task switching, partially complete work, and defects.

13. Answer: D

The concepts of deciding as late as possible, releasing in short iterations in order to get frequent feedback, and self-organizing are common to both Scrum and Lean. Although Lean and Scrum are compatible with one another, Lean is focused on eliminating waste, while that's not something that Scrum emphasizes.

14. Answer: C

In a pull system, the later step in a process pulls work from the previous one. This keeps the workload even and is the least wasteful way to get work from the beginning to the end of a process.

15. Answer: C

By setting a hard deadline and determining which tasks will depend on which, the team is locking themselves into an approach and cutting down the options they have to achieve their goal.

PART 2: PMI-ACP® CERTIFICATION GUIDE

The first part of this book was all about agile. It turns out that learning about the principles and practices of agile and drilling down into Scrum, XP, Lean, and Kanban gets you 90% of the way to preparing for the PMI-ACP® certification! The second part of this book gets you that last 10%, so you're 100% prepared to take the PMI-ACP® exam... and the next step in your career as an agile practitioner.

7 preparing for the pmi-acp® exam

Check your knowledge

> BOY, EXAM DAY IS MY FAVORITE DAY OF THE YEAR! I ONLY WISH WE COULD HAVE SCHOOL ALL SUMMER, TOO.

Wow, you sure covered a lot of ground in the first part of this book!

You've delved into the values and principles of the agile manifesto and how they drive an agile mindset, explored how teams use Scrum to manage projects, discovered a higher level of engineering with XP, and seen how teams improve themselves using Lean/Kanban. Now it's time to **take a look back** and drill in some of the most important concepts that you learned. But there's **more to the PMI-ACP® exam** than just understanding agile tools, techniques, and concepts. To really ace the test, you'll need to explore how teams **use them in real-world situations**. So let's give your brain a *fresh look at agile concepts* with a ***complete*** **set of exercises, puzzles, and practice questions** (along with some new material) specifically constructed to help prepare you for the PMI-ACP® exam.

help your career *along with your projects*

The PMI-ACP® certification is valuable...

The PMI Agile Certified Practitioner (PMI-ACP)® credential is one of the hottest and fastest-growing certifications out there, and it's getting more and more valuable every day. But don't take our word for it! Go to your favorite job search website and do a search for jobs with the keyword "agile". You'll find that many of them prefer or require an agile certification—and employers recognize that job candidates who are PMI-ACP® certified are a perfect fit.

> The best way to prepare for the PMI-ACP® certification is to **understand agile**. Part 1 of this book was all about agile, but even though it didn't focus on the exam, it got you **90% of the way** to preparing for it! The rest of this book is about getting that last 10% into your brain.

...but you <u>really need</u> to know your stuff

The PMI-ACP® exam is all about understanding situations that teams experience in the real world. Agile teams use a lot of specific tools, techniques, and practices like user stories, value stream maps, information radiators, burn down charts—you've learned about them throughout this book. But cramming a bunch of tools into your head is not going to help you pass the PMI-ACP® exam, because it's based on **understanding situations** that agile teams run into.

> **The PMI-ACP® exam focuses more heavily on how teams react in specific situations than it does on tools, techniques, and practices that are used by agile teams—but you still need to know tools, techniques, and practices.**

Here's an example of a question that asks you about a situation. Can you figure out the answer?

---

63. You are an agile practitioner. A member of your team has asked you for clarification on one of the items that you added to the prioritized list of features, stories, and other items the team will build in future iterations. You don't know the answer to the question. What should you do next?

- A. Bring up the question at the next retrospective
- B. Advise the team to self-organize in order to find an answer themselves
- C. Meet with the stakeholder whose requirements are relevant to the item
- D. Update the appropriate information radiator

What do you think this question means when it asks about an "agile practitioner"? It mentions a "prioritized list"—what do you think that refers to?

308   Chapter 7

preparing for the pmi-acp

The PMI-ACP® exam is based on the content outline

The Project Management Institute put a LOT of effort into designing and maintaining the PMI-ACP® exam, and they work very hard to make sure that the material is correct and current, and that the exam has an appropriate level of difficulty. The main way that they've accomplished this is by creating the **PMI-ACP® Examination Content Outline**. The content outline tells you everything that's covered by the exam. Here's what you'll find in it:

★ The exam is divided into seven **domains** that represent separate aspects of agile projects that questions will focus on

★ Each domain contains a series of **tasks** that represent discrete actions that agile teams often take, or responses to specific situations that agile teams might find themselves in

★ The content outline contains a set of **tools and techniques** that might appear in exam questions

→ *But it's **not an exhaustive list**—there could be agile practices that appear on the exam but are not in that list! We've made sure to **cover those practices** in this book.*

★ It also includes a list of **knowledge and skills** that agile practitioners are expected to understand and apply to the situations that they encounter on the job

Questions on the exam are based on specific tasks in the content outline.

Most of these tools and techniques should be familiar, because they were covered in the first six chapters.

The content outline is an important preparation tool

If you understand all of the domains and tasks in the content outline, that will give you a huge advantage when you take the exam, especially when you combine it with all of the knowledge that you've already absorbed in the first six chapters of this book. We'll help you do this by giving you a series of carefully designed exercises, puzzles, and practice questions that combine the material in the content outline with the agile ideas, topics, tools, techniques, methodologies, and practices that you've already learned.

The PMI website, *http://www.pmi.org*, has two important PDFs that you need to read in order to effectively study for the PMI-ACP® exam. The PMI-ACP® Handbook tells you how to apply to take the exam, the specific exam requirements, how (and how much) to pay for the exam, what you need to do to maintain your certification, and other rules, policies, and procedures set by PMI that govern the exam. Make sure that you download and read this PDF.

But the most important information is in the PMI-ACP® Examination Content Outline, which tells you <u>what specific topics</u> are going to be on the exam. Understanding the material in the Examination Content Outline is an important key to doing well on the PMI-ACP® exam.

You can find both of these PDFs by opening *http://www.pmi.org* in your browser and entering "PMI-ACP examination content outline" or "PMI-ACP handbook" in the search box.

You can also find the content outline by using your favorite search engine to search for "PMI-ACP Examination Content Outline"

the "a" and "p" in *pmi-acp®*

"You are an agile practitioner..."

The PMI-ACP® exam is all about understanding real-world situations that happen to agile teams. You'll be asked about what someone on an agile team should do when specific things happen on a project. You'll be tested on your knowledge of different project situations, and how agile teams respond to those situations. One approach that many questions on the exam take is to ask you about how an **agile practitioner** would respond to a specific situation. Handling questions like this is an important key to passing the exam. Here's how to do that:

Do this!

❶ Understand what the question is asking

A really good starting point is to understand what kind of question this is. Is it a "which-comes-next" question, where you're being asked to figure out what happens next or how to handle a situation? Is it a "which-is-NOT" question, where you need to choose a response that isn't appropriate? Make sure you take the time to read the whole question.

> The better you understand the material in the first six chapters, the easier it will be to figure out what's happening in each situation. We recommend that you return to Chapters 2–6 and re-take all of the end-of-chapter exam questions before you do any of the exercises in this chapter.

❷ Determine what the team is doing

Understanding what the team is currently doing is an important key to figuring out the answer to the question. Is the team holding a retrospective? Are they in the middle of a daily stand-up meeting? Are they refactoring their code, doing continuous integration, or writing unit tests? Are they planning the next iteration, or giving a demo of the work they completed to the stakeholders? The answer to the question depends on which of these things are going on.

❸ Use clues in the question to figure out your role

Don't be surprised if terms like "scrum master" or "product owner" appear in lowercase on the exam. We'll sometimes use lowercase for these terms to get you used to seeing it.

A lot of questions lay out a specific situation and ask how you would respond. But your response will vary depending on what your role is on the project. You could be in the role of scrum master, product owner, team member, stakeholder, senior manager, or something else. So when you see a question that asks about an agile practitioner, always look in that question for any clue about what the practitioner's role is.

❹ Unless told otherwise, assume it's a Scrum team

The PMI-ACP® exam is not dependent on any specific agile methodology. Some questions might ask about a specific method or methodology, but often they won't. When this happens, it's really useful to assume you're working on a Scrum team. The question might not be asking about Scrum specifically, but following the rules of Scrum **will always help you figure out the correct answer.**

This is an example of a question that puts you in the role of an agile practitioner. It has enough information to tell you what your role is and what the team is doing, and that's the key to getting the correct answer.

> **63.** You are an agile practitioner. A member of your team has asked you for clarification on one of the items that you added to the prioritized list of features, stories, and other items the team will build in future iterations. You don't know the answer to the question. What should you do next?
>
> A. Bring up the question at the next retrospective
> B. Advise the team to self-organize in order to find an answer themselves
> **C.** Meet with the stakeholder whose requirements are relevant to the item
> D. Update the appropriate information radiator
>
> **Answer: C**
>
> This question is asking about what a product owner should do when a team member asks for clarification about a product backlog item. The question asked about a "prioritized list of features, stories, and other items that the team will build in future iterations"—that's a description of the product backlog. The clue to your role on the team is that you added items to the product backlog, and the product owner is responsible for updating the product backlog, and is the only member of the team who updates it. A team member asked you for more information about an item in the backlog, but you don't know the answer. So the best thing to do is to go back to the stakeholder who provided the requirements that drive this backlog item, and understand exactly what he or she needs so that you can communicate it to the team and help them understand it.

The key to answering this question is figuring out that you're the product owner, and knowing that when product owners need more information about a backlog item they talk directly to their stakeholders.

If you're asked about what an agile practitioner would do in a situation, use clues in the question to determine the practitioner's role—and if the question doesn't specify a specific method, it's safe to assume that the team is using Scrum.

understand the *exam content outline*

there are no Dumb Questions

Q: Why should the exam assume that you're using Scrum? Isn't that showing a specific bias toward Scrum?

A: Scrum is by far the most popular approach to agile—it's used by the majority of agile teams, according to recent surveys—which is why we focused so heavily on it in this book. However, *the exam doesn't necessarily assume that you're using Scrum*. The exam is testing you on specific situations that any agile team might run into. But if you imagine that the team is using Scrum, *it makes the question much easier to answer*.

Q: Do I need to spend a lot of time memorizing all of the tools and techniques from this book?

A: Not necessarily—but a good familiarity with them is definitely a really helpful staring point. The first six chapters covered most of the tools and techniques that you might see on the exam. That's why we recommend that you **do all of the exercises and puzzles** in the first six chapters before you start preparing for the exam.

But keep in mind that the PMI-ACP® exam is highly focused on situations. You will definitely see questions that involve tools, techniques, and practices, but they are almost always used as part of resolving a problem similar to one you would see on an agile team in real life.

Q: Did you just say that *"most"* of the tools and techniques in the content outline were covered in the first six chapters? Why not all of them?

A: There are a few things that appear in the "tools and techniques" section of the PMI-ACP® Examination Content Outline which are important and very useful, but not quite so common in the experience of a typical, day-to-day agile team—and in the first six chapters of this book, we concentrated on teaching real-world agile as much as possible. But don't worry, we'll definitely fill in the gaps and make sure that every tool and technique from the content outline that you haven't seen yet is covered in this chapter, so there won't be any surprises when you take the exam.

> Before you apply for the exam, make sure that you meet the eligibility requirements that are described in the Examination Handbook. You need to have at least 2,000 hours (or 12 months) of experience working on project teams in the last five years. And <u>in addition</u> to that, you need to have at least 1,500 hours (or 8 months) of experience working on a team using an agile methodology in the last three years. And <u>finally</u>, you need to have at least 21 contact hours of training in agile practices.

Do this!

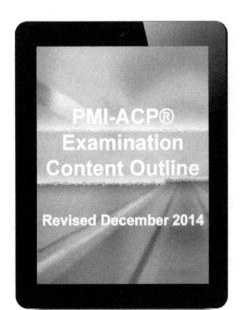

Download the PMI-ACP® Examination Content Outline from the PMI website right now. You'll use it as a study tool throughout the rest of this chapter. Make sure that you are using the PDF that was revised in December 2014 (that's the version the current exam is based on).

While you're there, download the PMI-ACP® Handbook.

You can find the handbook and content outline PDFs at http://www.pmi.org by entering "PMI-ACP" in the search box. You can also use your favorite search engine to search for "PMI-ACP 2014 examination content outline" or "PMI-ACP examination handbook" to find the PDFs.

preparing *for the* **pmi-acp**

A long-term relationship for your brain

Take a minute and think back over everything you've learned throughout this book. Does it seem just a little bit…well, overwhelming? Don't worry, that's absolutely normal. You've got all of this information that's floating around in your brain, and your brain is still trying to organize it.

Your brain is an amazing machine, and it's really good at organizing information. Luckily, when you feed it so much new data, there are ways that you can help make it "stick." That's what you're going to do in this chapter. Your brain wants its new information to be categorized, and we want to make sure everything you need to know for the exam sticks in your brain.

So for this study guide to be as effective as possible, **we need you to work with us**. We're going to concentrate on one specific area of the exam material at a time. But unlike the rest of the book, these areas don't necessarily line up with specific methodologies. Your job is to try to clear your mind of distractions, and concentrate only on the specific topic that we're presenting.

> Yes, we recognize that it can be hard to stick to a plan like this, especially when you've learned so much material already. But this is a highly effective way to get it all into your brain.

> I THINK I SEE WHAT YOU'RE SAYING! YOU'LL BREAK THIS CHAPTER INTO SEVERAL DIFFERENT "PIECES" THAT FEEL SIMILAR TO EACH OTHER, BUT REINFORCE DIFFERENT IDEAS THAT I'VE ALREADY LEARNED THROUGHOUT THE BOOK. AND ***IT'LL WORK BEST IF I DO MY PART*** BY CONCENTRATING ON ONE "PIECE" AT A TIME.

Yes! Cognitive psychologists call it *chunking*, and it's a really effective way of getting information into your long-term memory. When you have a collection of things that are strongly associated with one another, it gives your brain a sort of "guideline" for storing it. And the weaker associations with the other "chunks" give it a bigger framework for managing this large amount of information, so that it's all mutually reinforcing.

> Luckily, the PMI-ACP® exam content is already divided neatly into chunks that we can take advantage of—the domains that we talked about a few pages ago.

Let's get started!

domain 1 exercises

Domain 1: Agile Principles and Mindset — IN YOUR OWN WORDS

Writing things in your own words is one of the most effective ways to get a set of concepts and ideas into your brain. There's a description of each domain on page 4 of the *PMI-ACP® Examination Content Outline*.
In your own words, write down what you think Domain I ("Agile Principles and Mindset") is about:

....

The tasks in Domain 1 are listed on page 5 of the content outline. Write down what you think each task is about:

....

Task 1

....

Task 2

....

Task 3

....

Task 4

....

Task 5

....

Task 6

....

Task 7

....

Task 8

....

Task 9

preparing for the pmi-acp

Pool Puzzle

Your **job** is to take words and phrases from the pool and place them into the blank lines in the Agile Manifesto. You may **not** use the same word or phrase more than once, and you won't need to use all the words or phrases. Don't worry about the order of the values. See how much of this you can get without looking at the answer!

We are uncovering better ways of developing software by doing it and helping others do it. Through this work we have come to ____:

_____ and _____ over _____ and ____
_____ over _____
_____ over _____
_____ to ____ over _____ a ____

That is, while there is ____ in the items on the ____, we ____ the items on the _____.

Some of the blanks will contain a phrase that consists of several words from the pool.

Note: each word/phrase from the pool can only be used once! But some words appear more than once in the pool.

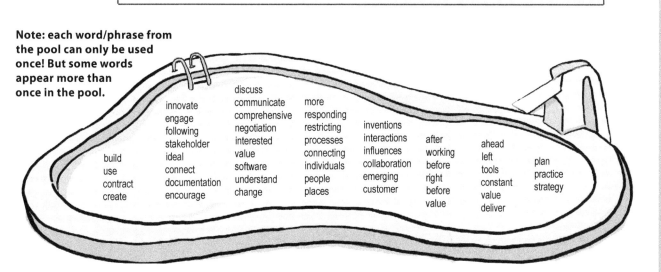

you are here ▶ 315

domain 1 exercise solutions

Domain 1: Exam Questions

IN YOUR OWN WORDS SOLUTION

Here's how we interpreted the tasks in our own words. It's OK if you used different words!

In your own words, write down what you think Domain I ("Agile Principles and Mindset") is about:

How you apply the values and principles of agile to your project, team, and organization

The tasks in Domain 1 are listed on page 5 of the content outline. Write down what you think each task is about:

Be an active advocate for agile ideas in your organization and with your customers
Task 1

Help everyone around you develop an agile mindset through your own words and actions
Task 2

Educate and influence people around your organization to help them become more agile
Task 3

Use information radiators to show your progress and build trust and transparency
Task 4

Make sure everyone feels comfortable making mistakes without being blamed feeling threatened
Task 5

Always keep learning and experimenting in order to find new and better ways to work
Task 6

Collaborate with your teammates so knowledge doesn't stay locked up with one person
Task 7

Help your team to self-organize and feel comfortable determining their approach to the work
Task 8

Use servant leadership to help everyone on the team to stay positive and keep improving
Task 9

preparing *for the* **pmi-acp**

Pool Puzzle Solution

Your **job** is to take words and phrases from the pool and place them into the blank lines in the Agile Manifesto. You may **not** use the same word or phrase more than once, and you won't need to use all the words or phrases. Don't worry about the order of the values. See how much of this you can get without looking at the answer!

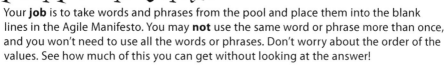

We are uncovering better ways of developing software by doing it and helping others do it. Through this work we have come to _value_:

Individuals and _interactions_ over _processes_ and _tools_

Working software over _comprehensive documentation_

Customer collaboration over _contract negotiation_

Responding to _change_ over _following_ a _plan_

That is, while there is _value_ in the items on the _right_, we _value_ the items on the _left more_.

It's OK if you put the values in a different order. They're all equally important.

Note: each word/phrase from the pool can only be used once!

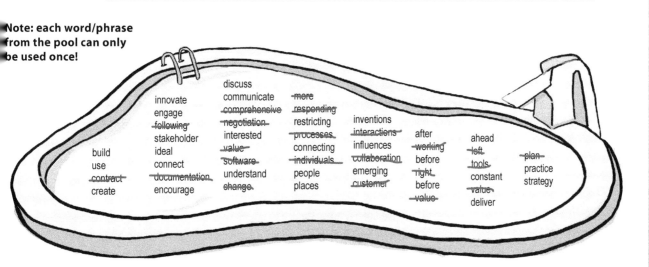

you are here ▶ **317**

domain 1 *exam questions*

Exam Questions

1. Agile teams highly value all of the following except:

    A. Customer collaboration

    B. Working software

    C. Responding to change

    D. Precise up-front planning

2. Joanne is a developer on a team that builds games. She put a lot of time into building a key feature of their upcoming release. When end users tested it, she discovered that it needed some fundamental changes and bug fixes. The users were still happy to play the game and gave it mid-range reviews during the test, but she knows the changes they've suggested would make it better. The game is due to be released in two weeks, but Joanne thinks that she'll be done with all of their requests in that time. What's the BEST thing to do next?

    A. Refuse to make the changes and release the features the way they are today

    B. Prioritize the work and let the team self-organize to get as many of the highest priority features in the first release as possible. Then release patches with the fixes and changes after the product is live.

    C. Do a root cause analysis on why the requirements were missed

    D. Delay the release by a few months so all of the features can be finished first

3. Ajay is an agile practitioner for a co-located software team. In his Daily Scrum meeting, the team often asks to review the current burn down chart and cumulative flow diagram. What's the BEST thing for him to do next?

    A. Create an information radiator where the team sits

    B. Ask the team to review the data prior to the meeting so it doesn't waste everybody's time every day

    C. Move the burn down and CFD review to the retrospective

    D. All of the above

4. Which is NOT a core focus of agile teams?

    A. Releasing early and often

    B. Simplicity

    C. Getting estimates right

    D. Self-organization

5. You are an agile practitioner on a Scrum team. Your team is being asked to create elaborate schedules and lists of milestones. Any schedule delay is treated as a cause for concern in weekly steering committee meetings mandated by your Portfolio Management group. What is the BEST thing for you to do?

    A. Make sure your team creates a plan and report any changes as requested

    B. Educate your upper management on agile principles and work with the Portfolio Management group on a different approach to status

    C. Refuse to cooperate with the Portfolio Management group

    D. Create a status report for the Portfolio Management group on your own and don't bother the team with it

Exam Questions

6. You are a Scrum Master on a five-person co-located software team. In the most recent retrospective, someone on your team suggested that the team might be overloaded with too much work in progress. What is the BEST thing for you to do?

    A. Tell the team to finish the work within the sprint
    B. Tell the customers to funnel all requests through you so the team doesn't get overloaded
    C. Experiment with setting WIP limits
    D. All of the above

7. You are a Scrum Master on a five-person co-located software team. Your team has a sprint planning meeting planned for four days from now. What is the BEST thing for you to do?

    A. Nothing, because the team is self-organizing so they can plan without you
    B. Make sure that every item in the backlog has complete documentation so that estimation is easier
    C. Talk to the end users about when they need all of the items in the backlog, and tell the team when they need to commit to each of them
    D. Work with the product owner to refine the backlog so that it's ready for the estimation session

8. Which of the following is NOT an agile principle?

    A. Satisfy the customer by releasing early and often
    B. Under-commit and over-deliver
    C. Focus on technical excellence
    D. Work at a sustainable pace

9. What is the best indicator of success on an agile project?

    A. Status reports that show no critical issues
    B. A well-developed plan
    C. Working software delivered to customers
    D. Happy teams

10. How do agile teams create the best architectures and designs?

    A. Prototyping
    B. Self-organizing
    C. Documenting
    D. Planning

domain 1 exam answers

Exam ~~Questions~~ Answers

1. Answer: D

While agile teams do value the act of planning, they focus on responding to changing conditions rather than sticking to up-front plans. That's why the more precise the plan is at the beginning of the project, the less flexible it is in the face of change.

2. Answer: B

If the users are happy to play the game in its current state, so delaying the release is a bad choice. The features can be added in future frequent releases. It wouldn't make sense to refuse changes that the end users request or to spend time trying to figure out why the requirements were missed instead of making the changes.

3. Answer: A

An agile practitioner should focus on creating information radiators so that teams have access to all of the data about their work and can make decisions about how to keep their work on track on their own.

4. Answer: C

Agile teams focus on delivering software frequently, self-organizing, and simplicity in design and approach. Those are all principles that drive an agile mindset; getting estimates right is not.

5. Answer: B

You can't expect that everyone will understand agile immediately. If you're working on a Scrum team in an organization that hasn't fully embraced agile, the best thing you can do for your team and your company is to help to educate people around you and influence the processes you work with to align with the principles and mindset your team is using.

6. Answer: C

It's important for a Scrum Master to be open to experimenting with new practices to make the team's work more efficient. Just telling the team to finish the work doesn't seem like it would solve the problem and telling the customers to talk to you instead of the team would make it hard for the team to self-organize and collaborate with the customer.

Exam Questions Answers

7. Answer: D

As a servant leader, you are not responsible for documenting all of the backlog items or committing the team to dates. The best thing you can do is help the product owner refine the backlog and get it ready for the team to estimate.

8. Answer: B

Under-committing and over-delivering is not an agile principle. Agile teams try to commit to what they can deliver, giving their customers an accurate picture of their project, and satisfying their customers by delivering frequently.

9. Answer: C

The best indicator of success on an agile team is working software delivered to customers.

10. Answer: B

Teams do their best work when they're able to self-organize. That's how they create the best architectures, designs, and products.

> I'VE NOTICED THAT THERE ARE A LOT OF WAYS TO PHRASE A "WHICH-IS-BEST QUESTION: "WHAT IS THE BEST WAY", "WHICH IS THE BEST OPTION", "WHAT IS THE BEST THING TO DO NEXT". THEY'RE ALL ASKING THE YOU TO CHOOSE THE BEST OPTION FROM THE FOUR CHOICES THAT ARE OFFERED.

get some practice

Domain 2: Value-Driven Delivery — IN YOUR OWN WORDS

In your own words, write down what you think Domain II ("Value-Driven Delivery") is about:

....

The tasks in Domain 2 are listed on pages 6 and 7 of the content outline. Write down what you think each task is about:

....
Task 1

....
Task 2

....
Task 3

....
Task 4

....
Task 5

....
Task 6

....
Task 7

....
Task 8

....
Task 9

....
Task 10

→ Answers on page 330

preparing for the pmi-acp

.... Task 11 ....
.... Task 12 ....
.... Task 13 ....
.... Task 14 ....

WHO DOES WHAT?

Let's reinforce some of the ideas behind Value-Driven Delivery. Match each of the items on the left with the descriptions of what they do or how they impact the project on the right.

Operational work   Repairing bugs and defects, and fixing
                                        other problems in the software

Maintenance   Activities related to the day-to-day
                                        functioning of the organization

Technical debt   A deliverable with just enough features to
                                        meet a specific, basic need

Backlog grooming   Activities related to hardware, networking,
                                        physical equipment, and facilities

Infrastructure work   Work that needs to be done to make code
                                        more maintainable in the long term

Minimal Viable Product (MVP)   Adding, removing, and reprioritizing items
                                        in the list of features to be developed

→ Answers on page 331

Watch it!

You might see the word "grooming" on the test in place of "product backlog refinement." We avoided it in the book because in some cultures, that word has a very negative association. But keep an eye out for it!

domain 2 exercises

You might see any of the tools and techniques that you learned about in this book on the exam—but the questions won't necessarily refer to them by name. Instead, the question or answer might describe a tool or technique using words. In this exercise, we'll describe a tool or technique. Your job is to choose the right one from the bottom of the page and write it in the blank.

Putting backlog items in order based on what the stakeholder most needs ....

Getting the customers' hands on the product to find problems they have operating it ....

The smallest possible piece of functionality that is still coherent and deliverable ....

A tool for visualizing the progress of individual work items through an iteration ....

Everyone keeping working folders up to date with the source code repository ....

A complete deliverable that is as small as possible but still meets stakeholder needs ....

An agreed-upon condition which, when satisfied, means a feature is complete ....

Activities to verify whether or not a particular approach will work ....

Constantly checking that requirements are correct and that deliverables meet them ....

customer-valued prioritization

minimal viable product (MVP)

Oh no! Someone was careless and spilled a bottle of ink on the answers. Can you figure out the solution without seeing all of the words?

minimum marketable feature (MMF)

definiti...

exploratory testing

frequent verification and validation

usability testing

continuous integration

task board

Answers on page 329

preparing for the pmi-acp

Agile teams use customer value to prioritize requirements

The very first principle of the Agile Manifesto does a great job of describing the attitude that agile teams have toward their customers and stakeholders:

> Our highest priority is to satisfy the customer through early and continuous delivery of valuable software.

This is why agile teams—and especially teams using Scrum—pay so much attention to the product backlog, and how the items in it are ranked. That's why the PMI-ACP® exam might cover **customer-valued prioritization** tools and techniques, including these:

MoSCoW method

This is a simple technique where requirements or backlog items are divided into "Must have," "Should have," "Could have," and "Won't have"—the first letter of each option form the word MoSCoW to make it easier to remember.

Relative prioritization/ranking

When teams use relative prioritization or ranking, they take work items or requirements, assign a numeric value to each one to represent customer value, and sort them using that value.

Kano analysis

The Kano model was developed in the 1980s by Noriaki Kano, a Japanese professor who studies quality and engineering management. His model of customer satisfaction can be used to track how innovations that previously delighted customers become basic needs over time that will disappoint them if not present in a product.

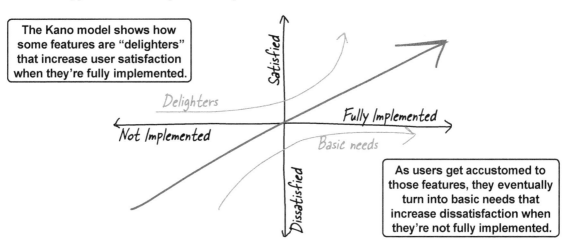

The Kano model shows how some features are "delighters" that increase user satisfaction when they're fully implemented.

As users get accustomed to those features, they eventually turn into basic needs that increase dissatisfaction when they're not fully implemented.

you are here ▸ 325

a few tools and techniques you might see on the exam

Value calculations help you figure out which projects to do

There are a few types of calculations that will appear on the PMI-ACP® exam as definitions. You won't need to calculate these, but you should know what each term means. All of these numbers can help a team determine which projects have the most value. If you're trying to decide between two projects, these can help you determine which one is the best.

Return on Investment (ROI)

This number is the money you expect to make from a project that you are building. Ranch Hand Games expect to sell a million units of *CGW5* within the first month of release. Of course, the longer it takes to develop, the higher the cost to get that return.

This number is just a straight total amount that the company expects to make on the investment it makes.

Net present value (NPV)

This is the actual value at a given time of the project minus all of the costs associated with it. This includes the time it takes to build it and labor as well as materials. People calculate this number to see if it's worth doing a project.

Money you'll get in three years isn't worth as much to you as money you're getting today. NPV takes the "time value" of money into consideration, so you can pick the project with the best value in today's dollars.

> In the real world, agile teams typically only do value calculations when they're required by the company. They're much more likely to use the relative sizing techniques from Chapter 4 like story points or T-shirt sizing. They'll sometimes use a technique called affinity estimating to come up with the estimates. That's where the team members divide a whiteboard into groups (like XS, S, M, L, XL T-shirt sizes or Fibonacci sequence story points) and take turns sticking each item to estimate into a category.

Earned Value Management (EVM) for Agile

If you studied for the PMP® exam, you learned about earned value calculations, which measure how well the project is performing by putting an actual cost in money or hours on value the product is delivering, and figuring out how much of that value has been delivered so far. Agile projects can use this too.

Internal rate of return (IRR)

This is the amount of money the project will return to the company that is funding it. It's how much money a project is making the company. It's usually expressed as a percentage of the funding that has been allocated to it.

The higher the rate of return, the better for the project.

preparing *for the* **pmi-acp**

> HOLD ON! WHY DIDN'T YOU COVER THESE TOOLS AND TECHNIQUES IN THE FIRST SIX CHAPTERS?

We kept chapters coherent to get them into your brain.

Most of the tools, techniques, and practices that we cover in this chapter but nowhere else in the book are on the PMI-ACP® exam mainly because they're traditional project management techniques—they're relevant to the exam because they are also used by some agile teams, but they're not really a core part of Scrum, XP, or Lean/Kanban. Including them would have been a distraction... and **distractions reduce the coherence of a topic** that you're trying to learn, and that can keep information from getting into your brain as effectively. (This is actually an application of chunking, which we talked about in the beginning of this chapter!)

The PMI-ACP® exam is much more heavily focused on understanding specific situations than it is on specific tools and techniques. The tools in this chapter are not as common on agile teams, so they're more likely to appear on the exam as incorrect answers than they are to be the direct subject of a question.

test your knowledge

WHAT'S MY PURPOSE

Rick, the product manager from Ranch Hand Games in Chapters 3 and 4, is using value calculations. Match each scenario to the cost numbers that Rick is using in each one.

1. The minute the demo was done, Ranch Hand Games released it as a playable demo for $1.00 on all of the major gaming consoles. The company is making about $1,000 a week while they're working on the game.

A. Return on Investment

2. Alex, the product owner, was having trouble figuring out whether or not certain features should be in the product backlog, so Rick helped him use MoSCoW to choose the ones that were really important.

B. Internal rate of return

3. Even though the team is developing on the latest console hardware, they know that new generations of the consoles will be released within the next three years. That means that all of the games developed on the current hardware will sell for about half of their current price once the upgrade happens.

C. Net present value

4. Rick wants to figure out how much the project is worth so far. So he adds up the value of all of the materials and licenses the team has used, and subtracts the labor and any depreciation that needs to be accounted for. The number he ends up with gives the value of the overall project right now. Then he subtracts that from the projected sales for the game.

D. Relative prioritization/ranking

5. Alex divided gamers who might buy CGW5 into categories and used Kano analysis to determine how much each of them might like different features.

6. Before the team decided to do the original demo, they compared how much the project was going to cost to how much they thought they'd make from it once it was released.

→ Answers on page 331

preparing for the pmi-acp

Tools Solution

You might see any of the tools and techniques that you learned about in this book on the exam—but the questions won't necessarily refer to them by name. Instead, the question or answer might describe a tool or technique using words. In this exercise, we'll describe a tool or technique. Your job is to choose the right one from the bottom of the page and write it in the blank.

Putting backlog items in order based on what the stakeholder most needs	customer-valued prioritization
Getting the customers' hands on the product to find problems they have operating it	usability testing
The smallest possible piece of functionality that is still coherent and deliverable	minimal marketable feature (MMF)
A tool for visualizing the progress of individual work items through an iteration	task board
Everyone keeping working folders up to date with the source code repository	continuous integration
A complete deliverable that is as small as possible but still meets stakeholder needs	minimal viable product (MVP)
An agreed-upon condition which, when satisfied, means a feature is complete	definition of done
Activities to verify whether or not a particular approach will work	exploratory testing
Constantly checking that requirements are correct and that deliverables meet them	frequent verification and validation

Oh no! Someone was careless and spilled a bottle of ink on the answers. Can you figure out the solution without seeing all of the words?

customer-valued prioritization
minimal viable product (MVP)
~~minimal ma~~rketable feature (MMF)
definiti~~on of done~~
e~~xplor~~atory testing
frequent verification and validation
usability testing
~~continuou~~s integration
task b~~oard~~

Domain 2: Exam Questions

IN YOUR OWN WORDS SOLUTION

Here's how we interpreted the tasks in our own words. It's OK if you used different words!

Using iterative and incremental development to deliver as much value as possible to stakeholders

The tasks in Domain 2 are listed on pages 6 and 7 of the content outline. Write down what you think each task is about:

Break the work down into minimal units and build the ones that deliver the most value
Task 1

Figure out what "Done" done means for each work item at the last responsible moment
Task 2

Use a methodology with practices and values that match the culture of the team and organization
Task 3

Break your product down into MMFs and MVPs and deliver the most valuable ones first
Task 4

Adjust the length of your iterations so that you can get frequent feedback from stakeholders
Task 5

Review the results of each iteration with stakeholders to make sure you're delivering value
Task 6

Make sure your stakeholders are helping you prioritize the work so you deliver value quickly
Task 7

Build maintainable software and constantly fix technical debt to keep long-term costs down
Task 8

Operational and infrastructure factors can impact your project, so take them into account
Task 9

Meet with your stakeholders frequently and make corrections to the work and the plan
Task 10

preparing for the pmi-acp

Take the time to identify project risks, and add items to reduce risk to the backlog
Task 11

Your stakeholders' needs and environment change all the time, so keep refining the backlog
Task 12

Make sure the team understands non-functional requirements like security and operations needs
Task 13

Continually inspect and test all of your artifacts (including the plan), use the results to adapt
Task 14

WHO DOES WHAT? Solution

Let's reinforce some of the ideas behind value-driven delivery. Match each of the items on the left with the descriptions of what they do or how they impact the project on the right.

Operational work — Repairing bugs and defects, and fixing other problems in the software

Maintenance — Activities related to the day-to-day functioning of the organization

Technical debt — A deliverable with just enough features to meet a specific, basic need

Backlog grooming — Activities related to hardware, networking, physical equipment, and facilities

Infrastructure work — Work that needs to be done to make code more maintainable in the long term

Minimal Viable Product (MVP) — Adding, removing, and reprioritizing items in the list of features to be developed

WHAT'S MY PURPOSE Solution: 1–B, 2–D, 3–C, 4–A, 5–D, 6–A

domain 2 *exam questions*

Exam Questions

1. For an agile team, the most important attribute of a product is...

    A. Technical excellence
    B. Quality
    C. Frequency of delivery
    D. The value it brings the customer

2. Which of the following BEST describes the goal of an agile product release?

    A. To release the smallest increment that provides value to a customer as soon as possible
    B. To release the largest increment that a team can produce in a time frame
    C. To include as many customer requests as possible
    D. To find the smallest market possible for a product

3. Which Scrum ceremony provides customer feedback on working software?

    A. Planning
    B. Backlog refinement
    C. Sprint retrospective
    D. Sprint review

4. Some agile teams use a practice called _____ to collaboratively prioritize work based on value to a customer.

    A. Fist-of-Five voting
    B. Planning poker
    C. Backlog refinement
    D. Joint design sessions

5. You and your team are holding a backlog refinement meeting. In the course of the discussion, it occurs to a few of your teammates that one of the features in the backlog could have multiple technical approaches. Your team is concerned that if they don't take the correct approach, it might result in serious performance problems. What's the BEST thing for your team to do next?

    A. Research and document the right approach before starting work on it
    B. Move the risky feature to the end of the backlog so the team has more time to think about the solution
    C. Move the risky feature to the start of the backlog so the team focuses on it first
    D. Write down the risk in a risk register and report it to upper management

Exam Questions

6. Paul is a developer on an agile software team. During a planning session, the Product Owner tells everyone that customers have asked for performance improvements in their upcoming release. Performance problems have caused a few cancellations recently and the situation is quickly becoming a priority for many customers. What should the team do next?

    A. Prioritize the performance improvement toward the top of the sprint backlog so the team focuses on it
    B. Create a persona for the user who requested the feature
    C. Add the feature request to the product backlog for later consideration
    D. Create a non-functional requirements document and include the performance requirements in it

7. In XP, developers use the _____ practice in order to review code changes as early as possible.

    A. Sit together
    B. Pair programming
    C. See the whole
    D. Regression testing

8. You are an agile practitioner on a team that is using Scrum. Halfway through a sprint, you find out that the main feature you're working on is no longer needed by clients. Which is the BEST thing to do next?

    A. Finish the sprint and take the new priorities into account at the next backlog refinement session
    B. Re-prioritize the sprint backlog and have the team start working on the next highest priority as soon as possible
    C. Try to understand why the change happened so you can avoid having a change like that again
    D. A and C

9. Your team is getting ready to start a new sprint. The product owner refers to a requirements document for a large feature, and begins to break it down into user stories that can be planned in small increments. What is this practice called?

    A. Work breakdown structure
    B. Big requirements up front
    C. Just-in-time requirements refinement
    D. Waterfall approach

10. A tool for keeping agile teams focused on building small increments that solve a user need is:

    A. Kano analysis
    B. User stories
    C. Short subjects
    D. Emergent design

domain 2 exam answers

Exam ~~Questions~~ Answers

1. Answer: D

The reason for developing a product is the value it brings to a customer. Value is what makes the product viable, and drives all of the decisions an agile team makes during development.

2. Answer: A

Agile teams try to break the product down into increments that provide value to the customer but can be released as soon as possible. These increments are sometimes called minimally marketable features (or MMFs).

3. Answer: D

In the sprint review, the team demonstrates working software to the customer and gets feedback.

4. Answer: C

Backlog refinement (also called product backlog review, or PBR) is an opportunity for the Product Owner to prioritize work collaboratively with the rest of the team.

5. Answer: C

Doing the highest-risk work first is the best way to approach this problem. That way, if the project is going to fail and you can't find the solution, it will fail fast and you'll have the information the team learned by working on the feature to help you figure out what to do next.

6. Answer: A

Non-functional requirements, like performance and quality concerns, should be prioritized along with features in the team's backlog.

7. Answer: B

Pair programming is a core XP practice that helps developers find defects before they're more permanently embedded as technical debt in the codebase.

Exam Questions Answers

8. Answer: B

There's no point in finishing out the sprint if the work the team is doing won't be useful. The best thing to do is to tell the team about the priority change right away, so that they can help determine the best way to deal with it. The sooner they can get working on the next high priority feature, the better.

9. Answer: C

Decomposing the work into stories just before you plan an increment is called just-in-time requirements refinement. By breaking down the work right before the work begins, you're sure to take any changes that might have happened to the requirements into account before you build.

10. Answer: B

Teams use user stories to stay focused on building small, valuable pieces of software that solve specific user needs.

Kano analysis and other tools are more likely to appear as incorrect answers than they are to appear as correct ones.

domains 3 and 4 *in your own words*

Domain 3: Stakeholder Engagement

IN YOUR OWN WORDS

In your own words, write down what you think Domain III ("Stakeholder Engagement") is about:

....

The tasks in Domain 3 are listed on page 8 of the content outline. Write down what you think each task is about:

....
Task 1

....
Task 2

....
Task 3

....
Task 4

....
Task 5

....
Task 6

....
Task 7

....
Task 8

....
Task 9

Domain 4: Team Performance

preparing for the pmi-acp

IN YOUR OWN WORDS

In your own words, write down what you think Domain IV ("Team Performance") is about:

....

The tasks in Domain 4 are listed on page 9 of the content outline. Write down what you think each task is about:

....
Task 1

....
Task 2

....
Task 3

....
Task 4

....
Task 5

....
Task 6

....
Task 7

....
Task 8

....
Task 9

domains 3 and 4 *in your own words solutions*

Domain 3: Exam Questions

IN YOUR OWN WORDS SOLUTION

Here's how we interpreted the tasks in our own words. **It's OK** if you used different words!

In your own words, write down what you think Domain III ("Stakeholder Engagement") is about:

Building trust with your project's stakeholders by engaging and collaborating with them

The tasks in Domain 3 are listed on page 8 of the content outline. Write down what you think each task is about:

The team identifies stakeholders and routinely meets with them to review the project
Task 1

Share all project information early and often with stakeholders so they stay engaged
Task 2

Help important stakeholders form a working agreement with each other to collaborate
Task 3

Stay on top of changes to your organization in order to identify new stakeholders
Task 4

Help everyone make better decisions more quickly through collaboration and conflict resolution
Task 5

Build trust with stakeholders by working with them to set a high-level goal for each increment
Task 6

Make sure everyone agrees on a definition of "done", and what trade-offs are acceptable
Task 7

Make your project transparent by clearly communicating status, progress, roadblocks, and issues
Task 8

Give forecasts so stakeholders can plan, help them understand how certain the forecasts are
Task 9

Domain 4: Exam Questions

preparing for the pmi-acp

IN YOUR OWN WORDS SOLUTION

Here's how we interpreted the tasks in our own words. It's OK if you used different words! :

Helping the team collaborate, trust each other, and create an energized work environment

The tasks in Domain 4 are listed on page 9 of the content outline. Write down what you think each task is about:

The team should work together to set ground rules that bring everyone together
Task 1

The team is committed to building up technical and interpersonal skills needed for the project
Task 2

Team members strive to be "generalizing specialists" who can contribute to all aspects of the project
Task 3

The team is self-organizing, and feels empowered to make important project decisions
Task 4

Teammates find ways to keep each other motivated and keep from demotivating each other
Task 5

The team should be co-located in the same office if possible, and use collaboration tools
Task 6

Distractions should be kept to a minimum to make sure the team achieves "flow"
Task 7

Everyone "gets" the project vision and understands how each piece of work contributes to it
Task 8

Measure project velocity and use it to figure out how much the team can do in each iteration
Task 9

you are here ▸ 339

Exam Questions

1. Scrum teams demo working software at the end of each sprint in an event called _____

    A. Sprint demo
    B. Sprint retrospective
    C. Sprint review
    D. Product demo

2. You are an agile practitioner on a newly created Scrum team. As part of the preparation for the team's first sprint, you meet with some of the stakeholders for the product you're going to build to understand their objectives. In that meeting, the group creates a list of features categorized as "Must have," "Could have," "Should have," and "Won't have."

    What is the BEST way to describe the method of prioritization they're using?

    A. Relative prioritization
    B. Stack ranking
    C. Kano analysis
    D. MoSCoW

3. What are the three questions each team member answers in a Daily Scrum meeting?

    A. What did I commit to doing today? What will I commit to doing by tomorrow? What mistakes have I made?
    B. What am I working on today? What will I work on tomorrow? What problems have I run into?
    C. What did I do today that brings us closer to our sprint goal? What will I do tomorrow that brings us closer to our sprint goal? What obstacles are in the team's way?
    D. None of the above

4. Who makes decisions on behalf of business stakeholders on a Scrum team?

    A. Scrum master
    B. Product owner
    C. Agile practitioner
    D. Team member

5. Julie is working on a team that's using Kanban to improve their process. Every day they put index cards on a board to show how many features are in each state of their process. Next, they add up the number of features in each column on the board and create an area chart that shows those totals over time. What tool are they using?

    A. Cumulative flow diagram
    B. Task board
    C. Burn down chart
    D. Burnup chart

Exam Questions

6. Agile teams commit to business _____ and sprint _____. They know that plans can change, and welcome those changes no matter when they happen. By focusing on what the team needs to achieve, they keep their options open.

    A. Leadership, deadlines
    B. Objectives, goals
    C. Forecasts, plans
    D. Demands, retrospectives

7. Which of the following is NOT a tool for providing stakeholder transparency in an agile team?

    A. Information radiators
    B. Feature demos
    C. Task boards
    D. Net present value

8. An agile team updated a daily burn down chart on the wall where they sit. What can stakeholders understand from looking at this chart?

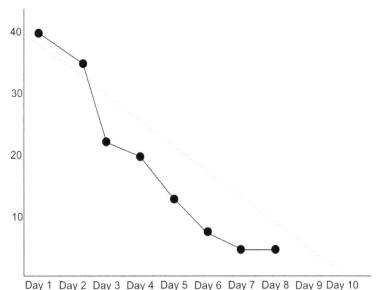

    A. The project is running late
    B. The team is on track to meet its sprint goal
    C. The team ran into trouble on day 3
    D. Only project managers need the information in this chart

Exam Questions

1. You are an agile practitioner on a newly created Scrum team. As part of the team's first planning session they decide to come up with a series of policies that the team will follow while working on their upcoming project. These policies include the following: "The Daily Scrum meeting is timeboxed to 15 minutes and will start on time every day. A team member will never mark a feature complete until it satisfies the team's definition of 'done'. The team will use pre-defined coding standards and check code in for nightly builds."

   What is the BEST way to describe the list of policies they've defined?

   A. Definition of ready
   B. Administrative guidelines
   C. Team charter
   D. Working agreements

2. You are an agile practitioner on a team building software for an advertising company. About halfway through your first increment, you notice that many features are showing as "In Progress" on your task board, but very few are in the "Done" column. One closer inspection, you see that developers are starting work on new features whenever they are waiting on code review or testing. What is the BEST thing for you to do next?

   A. Ask the team to help code review and testing to finish existing tasks before starting a new one
   B. Assume that the work will get done by the end of the sprint because the team has made progress on so many features
   C. Bring more testers onto the team to deal with the glut of features
   D. Tell the stakeholders that the team will not have anything ready to demo at the end of the sprint

3. Kim is an agile team member. Her team has four people on it and is four days into a two-week sprint. She just completed work on "Story 1," the highest priority story in a five-story priority-ranked sprint backlog. This image shows the current state of her team's task board. What should she do next?

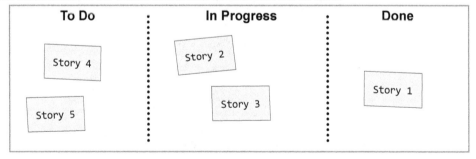

   A. Move story 4 or story 5 to the "In Progress" column and start working on it
   B. Find a way to help work on story 2 or story 3 if possible
   C. Add a feature from the product backlog to the "In Progress" column and start working on it
   D. Wait for the Scrum Master to assign a new story

Exam Questions

4. A software company is transforming its organization from using traditional software development practices to using agile methodologies. Which of the following is NOT a factor when deciding how to form agile teams?

    A. Teams should be co-located where possible

    B. Teams should be small

    C. Teams should have a documented change management process for dealing with late changes

    D. All of the above

5. You are an agile practitioner on a five-person team that is currently six days into a two-week sprint. An external stakeholder calls a team member to request an urgent change. What is the BEST thing for the team member to do?

    A. Drop what she is doing and make the change that the stakeholder requested

    B. Ask the stakeholder to work with the product owner to prioritize the change

    C. Advise the stakeholder to wait until the next sprint planning session and then bring it up

    D. Remind the stakeholder to prioritize the change in the product backlog

6. You are an agile practitioner on a team that builds financial software. During a sprint planning session with the team, a tester and a developer get into an argument about how big a story is. The developer says that the story is a small code change, so it should happen immediately. The tester says that it impacts many crucial areas of the software, and many tests will need to be run to make sure that it works. What should you do next?

    A. Side with the tester and recommend that the team allot more time for the feature

    B. Side with the developer and cut tests to make sure that the feature is delivered sooner

    C. Suggest that the team use planning poker to discuss their assumptions and come up with an approach and size for the feature that everyone agrees to

    D. Ask the product owner to prioritize the tests based on how important the functionality is to the end users

7. You are an agile practitioner on a team that builds software. One of your teammates is expected to work on two projects at once. The person's functional manager has asked the team to expect him to be 50% allocated on the agile team and 50% allocated to a functional support team. What is the BEST thing for you to do next?

    A. Help your teammate find a way to avoid over-committing to stories in sprint planning

    B. Tell the functional manager that agile teams require team members focus on their tasks, and the team member should not be allocated to two teams at once

    C. Make sure the team doesn't over-commit because resources are not fully allocated

    D. Make sure that the person only spends four hours per day working on the other team

---

> We gave you questions for domains 3 and 4 back to back. See if you can do them all in a row, then look at the answers. This will help you get ready for the big final exam in Chapter 9.

domain 3 exam answers

Exam ~~Questions~~ Answers

1. Answer: C

The sprint review is the main opportunity for the team to demonstrate working software at the end of each sprint. All of the project stakeholders attend the demo and provide feedback on the software. That feedback is incorporated back into the product backlog, which the team uses to plan future sprints.

2. Answer: D

These stakeholders are using the MoSCoW method for prioritization. It helps the team understand the business perspective on the features in the backlog.

3. Answer: C

The Daily Scrum questions focus on what the team is doing to get closer to achieving this sprint goal. Just telling the team what each individual is working on can cause team members to focus too much on their own perspective, and to lose sight of the goal the team is working toward.

4. Answer: B

The Product Owner acts as a proxy for business stakeholders in a Scrum team. The Product Owner communicates business priorities, and makes decisions for the team that will help them meet each sprint goal.

5. Answer: A

The question describes the process that a team would take to create a cumulative flow diagram (CFD). They're also using a Kanban board as a means of figuring out the numbers to map on the CFD, but "Kanban board" was not one of the choices available for this question.

6. Answer: B

Agile teams commit to business objectives and sprint goals. They try to decide on the exact route they'll take to achieving those objectives as late as possible.

Exam Questions Answers

7. Answer: D

Net Present Value (NPV) might help your team decide whether or not to do a project, but it's not a practical tool for keeping stakeholders informed about what's going on with your project.

8. Answer: B

A burn down chart is an effective way to keep everyone on the team and all of your stakeholders in the loop about daily progress toward a goal. This burn down shows that the team will most likely complete the work they committed to by the end of the sprint because the line that shows the amount of work left is below the line.

domain 4 exam answers

Exam ~~Questions~~ Answers

1. Answer: D

Agile teams work to define their working agreements when they start working together. That way, all team members know what to expect when they're working with the group.

2. Answer: A

Teams work most efficiently when they focus on getting each task completely finished (*"Done"* done) before moving on to the next one. That's why agile teams value members who focus on being "generalizing specialists" that do whatever they can to move their work through all stages of development. By focusing on collaborating and making the work flow, the whole team gets more done and creates a higher-quality product together. In this case, it means that everyone on the team has the skills to help with code review and testing.

3. Answer: B

The team works on the stories in order of priority, so we know that stories 2 and 3 in the "In Progress" column are more valuable than stories 4 and 5 in the "To Do" column. Teams should focus on completing the highest priority work as soon as possible and on finishing work before starting new work. If Kim can help her teammates get story 2 or 3 done faster, she should do that instead of starting work on another story. Since agile teams are self-organizing, she shouldn't have to wait for the Scrum Master to tell her which story to work on next.

4. Answer: C

Teams should be small and co-located so that they can easily collaborate with one another and find new and better ways of working. Agile teams also value responding to change, especially changes in priority, as urgently as possible. However, they typically do not focus on creating well-documented change management processes, as they are often focused on slowing down the rate of change in a project.

5. Answer: B

The product owner maintains the backlog and the priority order of work for the team. If an external stakeholder needs a change, the team member should ask that stakeholder to work with the product owner to figure out where the change fits into the team's backlog.

Exam Questions Answers

6. Answer: C

Planning poker is built for situations like this, because it helps teams to come together on an approach and agree on the size of the effort. In this case, the tester could explain why the code change affects so many tests, and the developer and tester could jointly come up with approaches for dealing with the problem.

There are a lot of different approaches that this particular team can take. Developers can write automated unit tests (serving as generalizing specialists and helping to shoulder some of the quality work). They can also pay special attention to pair programming and unit testing, if the area they are modifying is critical to the product. Testers could write tests in parallel with development, and take part in code reviews so that they know what to expect from the feature when it's delivered. These are all ideas that might come up during a planning poker session. Choosing a specific approach will help them come up with a relative size for the story, and get a more accurate sense of what they need to do in order to deliver it.

7. Answer: B

Agile teams expect 100% of a team member's focus and time, and recognize that compromising on that can lead to significant problems. When one person is shared between teams, it leads to a large amount of task-switching that causes a lot of waste. As an agile practitioner, you should work to influence the person's functional manager, and try to convince him or her to allocate that person 100% to the agile team.

Domain 5: Adaptive Planning — IN YOUR OWN WORDS

In your own words, write down what you think Domain V ("Adaptive Planning") is about:

....

The tasks in Domain 5 are listed on page 10 of the content outline. Write down what you think each task is about:

....
Task 1

....
Task 2

....
Task 3

....
Task 4

....
Task 5

....
Task 6

....
Task 7

....
Task 8

....
Task 9

....
Task 10

➡ Answers on page 372

preparing for the pmi-acp

Adapt your leadership style as the team evolves

You may see a few questions on the PMI-ACP® exam about **adaptive leadership**, a useful theoretical concept that can help leaders improve how they lead their teams. Applying adaptive leadership to a specific team starts the **stages of team formation**.

Every team goes through these stages during a project

Forming: People are still trying to figure out their roles in the group; they tend to work independently, but are trying to get along.

Storming: As the team learns more about the project, members form opinions about how the work should be done. This can lead to temper flare-ups in the beginning, when people disagree about how to approach the project.

Norming: As team members become better acquainted, they begin to adjust their own work habits to help each other and the team as a whole. Here's where the individuals on the team start learning to trust one another.

Performing: Once everyone understands the problem and what the others are capable of doing, they start acting as a cohesive unit and being efficient. Now the team is working like a well-oiled machine.

Adjourning: When the work is close to completion, the team starts dealing with the fact that the project is going to be closing soon. (This is sometimes called "mourning" by teams.)

> Researcher Bruce Tuckman came up with these five stages in 1965 as a model for team decision making. Although this is the normal progression, it's possible that the team can get stuck in any one of the stages.

Situational leadership

People have a tough time creating team bonds initially, but a good leader can use his or her adaptive leadership skills to help the team to progress through the stages quickly. Paul Hershey and Kenneth Blanchard came up with the **situational leadership theory** in the 1970s to help guide leaders through this. That theory includes four different leadership styles. Adaptive leadership means *matching different leadership styles* to stages of team formation:

* **Directing:** Initially, the team needs a lot of direction to help get used to the specific tasks they need to accomplish. They don't really need a lot of emotional support yet. This is matched with the **forming** stage.

* **Coaching:** A good coach knows how to give a lot of direction, but also provide the emotional support the team needs to get through the temper flare-ups and disagreements. This is matched with the **storming** stage.

* **Supporting:** As everyone on the team gets used to each other and their work, the leader doesn't need to direct as much, but still needs to provide a high level of support. This is matched with the **norming** stage.

* **Delegating:** Now the team is running smoothly, and the leader doesn't need to provide much direction or much support, only handle specific situations as they come up. This is matched with the **performing** stage.

adaptive leadership *exercise*

The PMI-ACP® exam focuses on specific scenarios, so we can use scenarios to explore adaptive leadership. Each of the following scenarios demonstrates one of the stages of team development. Write down which stage each scenario describes. Then write down the leadership style, and fill in either "High" or "Low" for the level of direction and support that leadership style provides.

1. Joe and Tom are both programmers on the Global Contracting project. They disagree on the overall architecture for the software they're building, and frequently get into shouting matches over it. Joe thinks Tom's design is too short-sighted and can't be reused. Tom thinks Joe's design is too complicated and probably won't work. They're at a point right now where they're barely talking to each other.

   Stage of development: _____

   Leadership style: _____ Level of direction: _____ Level of support: _____

2. Joan and Bob are great at handling the constant scope changes on the Business Intelligence project. Whenever the stakeholders request changes, they shepherd them through the change control process and make sure the team doesn't get bothered with them unless it's absolutely necessary. That leaves Darrel and Roger to focus on building the main product. Everybody is focusing on their area and doing a great job. It seems like it's all just clicking for the group.

   Stage of development: _____

   Leadership style: _____ Level of direction: _____ Level of support: _____

3. Derek just got to the team, and he's really reserved. Folks on the team aren't quite sure what to make of him. Everybody's polite, but it seems like some people are a little threatened by him.

   Stage of development: _____

   Leadership style: _____ Level of direction: _____ Level of support: _____

4. Danny just realized that Janet is really good at developing web services. He's starting to think of ways to make sure that she gets all of the web service development work and Doug gets all of the client software work. Doug seems really happy about this too—he seems to really enjoy building Windows applications.

   Stage of development: _____

   Leadership style: _____ Level of direction: _____ Level of support: _____

⟶ Answers on page 354

A few last tools and techniques

There are just a few more tools and techniques that you might see on the PMI-ACP® exam. Luckily, they're very straightforward, and should fit in easily with what you've already learned.

Risk-adjusted backlog, pre-mortem, and risk burn down charts

When a team maintains a risk-adjusted backlog, they include risk items in the backlog and prioritize them along with the other backlog items. This means:

★ When the team encounters a risk, it gets added to the backlog—risk backlog items are prioritized by value and effort, just like any other backlog item

★ One way teams identify risks is to hold a **pre-mortem**, where they imagine that the project has failed catastrophically and brainstorm the reasons that caused the failure

★ When the team plans an iteration, they include risk items along with other items in the iteration backlog

★ They create an estimate for each risk backlog, using exactly the same estimation techniques that they already use to generate estimates for other product backlog items

★ When the team performs backlog refinement (which they sometimes call "grooming" or "PBR"), they update, review, re-estimate, and re-prioritize the risk backlog items along with the rest of the items in the backlog

★ You might see risks referred to as **threats and potential issues** on the exam

★ When the team maintains a risk-adjusted backlog that includes relative size estimates for risk items, they can use those estimates to create a **risk burn down chart** for each iteration (so, for example, a Scrum team might create a risk burn down graph that shows the sum of the story points for all risk items left in the sprint backlog)

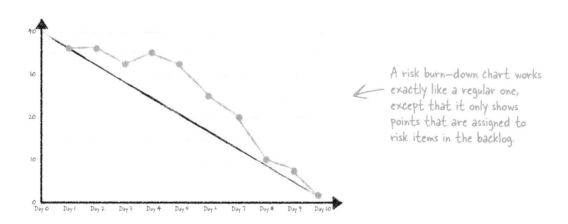

A risk burn-down chart works exactly like a regular one, except that it only shows points that are assigned to risk items in the backlog.

a few more tools and techniques that might be on the exam

A few last tools and techniques

There are just a few more tools and techniques that you might see on the PMI-ACP® exam. Luckily, they're very straightforward, and should fit in easily with what you've already learned.

> You probably won't need to know the details of how most of these games work, but you may see their names (most likely in an incorrect answer).

Collaboration games

Teams sometimes play collaboration games to help them work together to brainstorm, gain consensus, and make decisions as a group. There are many different collaboration games. A few that you might see on the exam include:

- **Planning poker** is an estimating game that you learned about in Chapter 4
- **Affinity estimating** (explained on page 326) is actually a kind of collaboration game, too
- **Mind maps** is a brainstorming game where four or five people write an item to focus on in a circle in the middle of a whiteboard, and draw a tree branching out from it that shows related ideas
- **Fist of five voting** is similar to "rock-paper-scissors" that teams use to gauge the group's opinion on a topic by showing the number of fingers that indicate how strongly they like or hate the idea
- **Dot voting** is a decision-making game where a set of options are evaluated by writing them on a large piece of paper, and distributing sheets of sticky dots so that everyone can stick their dots next to different options—the options with the most dots are the ones that are chosen by the group
- **100 point voting** is similar to dot voting, where team members get 100 points to distribute among the options

---

BYOQ – Bring Your Own Questions

Want to give yourself an edge on the exam? Try writing your own questions! You can use the "question clinic" templates from Chapters 2 through 6 as your guide. Try it out:

* Write a "which-is-BEST" question about holding a Daily Scrum meeting
* Write a "which-comes-next" question about value stream mapping
* Write a "red herring" question about adaptive planning
* Write a "which-is-NOT" question about refactoring
* Write a "least-worst-option" question about the Scrum value of openness

preparing for the pmi-acp

You might see any of the tools and techniques that you learned about in this book on the exam—but the questions won't necessarily refer to them by name. Instead, the question or answer might describe a tool or technique using words. In this exercise, we'll describe a tool or technique. Your job is to choose the right one from the bottom of the page and write it in the blank.

A game that team members play using cards to help them agree on estimates   ....

A minimal description of a feature that explains who needs it and why they need it   ....

Starting with an initial version of an artifact and updating it as more information is known   ....

A thought experiment where you imagine things failed and figure out what caused it   ....

Experimental work to determine the impact of a specific potential problem, issue, or threat   ....

Reviewing, re-estimating, and re-prioritizing the list of planned features and work items   ....

The list of planned features and work items that also includes problems, issues, and threats   ....

An estimate of work that assumes no interruptions, distractions, or problems   ....

A chart that shows the total daily estimated impact of problems, issues, and threats   ....

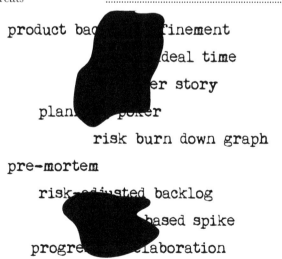

Oh no! Someone was careless and spilled a bottle of ink on the answers. Can you figure out the solution without seeing all of the words?

product backlog refinement
ideal time
user story
planning poker
risk burn down graph
pre-mortem
risk-adjusted backlog
risk-based spike
progressive elaboration

Answers on page 355

domain 5 exercise solutions

Exercise Solution

The PMI-ACP® exam focuses on specific scenarios, so we can use scenarios to explore adaptive leadership. Each of the following scenarios demonstrates one of the stages of team development. Write down which stage each scenario describes. Then write down the leadership style, and fill in either "High" or "Low" for the level of direction and support that leadership style provides.

1. Joe and Tom are both programmers on the Global Contracting project. They disagree on the overall architecture for the software they're building, and frequently get into shouting matches over it. Joe thinks Tom's design is too short-sighted and can't be reused. Tom thinks Joe's design is too complicated and probably won't work. They're at a point right now where they're barely talking to each other.

   Stage of development: _Storming_

   Leadership style: _Coaching_ Level of direction: _High_ Level of support: _High_

2. Joan and Bob are great at handling the constant scope changes on the Business Intelligence project. Whenever the stakeholders request changes, they shepherd them through the change control process and make sure the team doesn't get bothered with them unless it's absolutely necessary. That leaves Darrel and Roger to focus on building the main product. Everybody is focusing on their area and doing a great job. It seems like it's all just clicking for the group.

   Stage of development: _Performing_

   Leadership style: _Delegating_ Level of direction: _Low_ Level of support: _Low_

3. Derek just got to the team, and he's really reserved. Folks on the team aren't quite sure what to make of him. Everybody's polite, but it seems like some people are a little threatened by him.

   Stage of development: _Forming_

   Leadership style: _Directing_ Level of direction: _High_ Level of support: _Low_

4. Danny just realized that Janet is really good at developing web services. He's starting to think of ways to make sure that she gets all of the web service development work and Doug gets all of the client software work. Doug seems really happy about this too—he seems to really enjoy building Windows applications.

   Stage of development: _Norming_

   Leadership style: _Supporting_ Level of direction: _Low_ Level of support: _High_

354   Chapter 7

preparing for the pmi-acp

Tools Solution

You might see any of the tools and techniques that you learned about in this book on the exam—but the questions won't necessarily refer to them by name. Instead, the question or answer might describe a tool or technique using words. In this exercise, we'll describe a tool or technique. Your job is to choose the right one from the bottom of the page and write it in the blank.

Description	Answer
A game that team members play using cards to help them agree on estimates	*planning poker*
A minimal description of a feature that explains who needs it and why they need it	*user story*
Starting with an initial version of an artifact and updating it as more information is known	*progressive elaboration*
A thought experiment where you imagine things failed and figure out what caused it	*pre-mortem*
Experimental work to determine the impact of a specific potential problem, issue, or threat	*risk-based spike*
Reviewing, re-estimating, and re-prioritizing the list of planned features and work items	*product backlog refinement*
The list of planned features and work items that also includes problems, issues, and threats	*risk-adjusted backlog*
An estimate of work that assumes no interruptions, distractions, or problems	*ideal time*
A chart that shows the total daily estimated impact of problems, issues, and threats	*risk burn down graph*

Oh no! Someone was careless and spilled a bottle of ink on the answers. Can you figure out the solution without seeing all of the words?

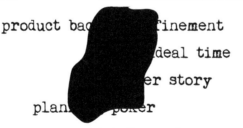

product backlog refinement
ideal time
user story
planning poker
risk burn down graph
pre-mortem
risk-adjusted backlog
risk-based spike
progressive elaboration

Exam Questions

1. An agile team just finished a sprint planning session for sprint 5 of an ongoing project. They used planning poker to come up with the following story point values for their stack-ranked backlog:

Product Backlog

Story 1: 8 points
Story 2: 5 points
Story 3: 3 points
Story 4: 5 points
Story 5: 13 points
Story 6: 5 points
Story 7: 2 points
Story 8: 3 points

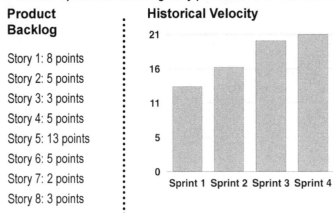

Based on the team's velocity histogram above, what is the last story they should expect to deliver in this sprint?

A. Story 3
B. Story 5
C. Story 6
D. Story 4

2. An agile team is four days into a two-week sprint when one of their stakeholders asks for the status of the current sprint. The team directs the stakeholder to the team's burn up chart. What can the stakeholder tell from the information in the chart?

A. The team is running behind
B. Scope has been added
C. The team is ahead of schedule
D. There isn't enough data to know status

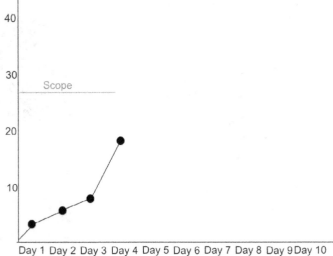

Exam Questions

3. You're an agile practitioner on a team that builds mobile apps. One of your teammates identified a problem during the last retrospective, pointing out that the team's velocity was getting lower with each two-week sprint. Afterward, you find that the user stories the team is working on are often too big to be completed in one sprint, often carrying over into the next three or four sprints before they are completed. What do you do next?

   A. Work with the product owner to identify a sprint goal that can be accomplished in the next sprint, and then decompose the stories in that sprint so that all of them can be delivered in two weeks

   B. Carry the stories over from sprint to sprint, but forecast that they will all be completed in the product's next major release

   C. Work with the team to get better at delivering big stories faster

   D. Stop committing to individual sprint goals, and commit to release goals instead

4. Sarah is a scrum master on a team that develops games. She's noticed that her team often commits to much more work in a sprint than they can accomplish. When she brought it up in the team's retrospective, she learned that her teammates are often pulled off of development tasks to deal with maintenance requests for production software, which interrupts the flow of work for them. What should she do next?

   A. Create backlog items for the maintenance requests, estimate them, and include them in sprint planning so that the team can include those requests in its sprint commitments

   B. Work with the team to create a buffer in the amount the team commits to, so that they don't over-commit when they know the maintenance requests will happen

   C. Tell the support group that the development team will no longer be responding to maintenance requests

   D. None of the above

5. Agile teams often create special tasks for teams to investigate approaches to solving high-risk design problems. What are these special tasks called?

   A. Exploratory work
   B. Buffers
   C. Slack
   D. Risk-based spikes

6. Teams will often prioritize risks, issues, and threats above other work because solving those items will mean success or failure for the whole project. What is the practice of prioritizing these items called?

   A. Risk-ranking
   B. Backlog refinement
   C. Risk-adjusted backlog
   D. None of the above

domain 5 exam answers

Exam Questions Answers

1. Answer: D

Since the stories are in priority order, the team should try to deliver stories 1-4. The total size of the effort for those four stories adds up to 21 story points, which is what the team delivered in the last sprint.

2. Answer: C

According to the chart, the team committed to deliver around 28 points in this sprint. On day 4 they were already at 20 points complete. It's very likely that they're ahead of schedule.

3. Answer: A

All of the stories in the sprint backlog should be sized so that they can be completed within a sprint, which according to the question lasts two weeks. The product owner should be working to help the team to deliver as much value as possible at the end of each increment. If the team stays focused on large releases, they won't get the benefits of early validation that come with releasing small increments more frequently.

4. Answer: A

The team can't plan their work if they don't put all of it in the backlog. Just because a work item is coming from a different stakeholder than usual (in this case, the support team), that doesn't mean that the team should ignore it. If the work is valuable to the organization, that work should be prioritized, estimated, and planned along with new feature work by the product owner as part of the backlog refinement and sprint planning process.

5. Answer: D

Risk-based spikes allow agile teams to time box their effort in researching high-risk functionality so that teams can "fail fast" if there is no solution to the problem.

6. Answer: C

A risk-adjusted backlog includes planned features and work items, but also includes problems, issues, and threats—in other words, a backlog that also includes risk items that can be prioritized by risk as well as value. This helps you and your team to focus on understanding your risks first, which can help prevent your project from running into trouble (and even failing!) later on.

preparing *for the* **pmi-acp**

AGILE TEAMS TRY TO DO THE MOST IMPORTANT WORK FIRST – AND THEY **FAIL FAST IF THEY'RE GOING TO FAIL.** WHEN A TEAM IDENTIFIES RISKS, THEY CAN SOMETIMES SINK A WHOLE PROJECT. BUT IT'S BETTER TO FAIL AFTER A SPRINT OR TWO THAN FIND OUT YOU'VE GONE DOWN THE WRONG ROAD AFTER THE TEAM'S PUT IN MONTHS OF EFFORT!

BRAIN POWER

How can you make sure that you work on the highest priority items first, even when you're fixing defects in your project?

domains 6 and 7 in your own words

Domain 6: Problem Detection and Resolution

IN YOUR OWN WORDS

In your own words, write down what you think Domain VI ("Problem Detection and Resolution") is about:

....

The tasks in Domain 6 are listed on page 11 of the content outline. Write down what you think each task is about:

....
Task 1

....
Task 2

....
Task 3

....
Task 4

....
Task 5

→ Answers on page 373

BYOQ – Bring Your Own Questions

Writing your own questions is a great way to get knowledge into your brain. And when you recognize different question strategies, it actually helps you keep calm during the exam.

* Write a "which-is-BEST" question about continuous integration
* Write a "which-comes-next" question about limiting work in progress
* Write a "red herring" question about holding a sprint planning meeting
* Write a "which-is-NOT" question about servant leadership
* Write a "least-worst-option" question about information radiators

Domain 7: Continuous Improvement

preparing for the **pmi-acp**

IN YOUR OWN WORDS

In your own words, write down what you think Domain VII ("Continuous Improvement") is about:

....

The tasks in Domain 7 are listed on page 12 of the content outline. Write down what you think each task is about:

....

Task 1

....

Task 2

....

Task 3

....

Task 4

....

Task 5

....

Task 6

→ Answers on page 374

BYOQ – Bring Your Own Questions

* Write a "which-is-BEST" question about osmotic communication
* Write a "which-comes-next" question about the Product Owner approving a sprint
* Write a "red herring" question about demonstrating working software at a sprint review
* Write a "which-is-NOT" question about risk burn-down graphs
* Write a "least-worst-option" question about the XP value of simplicity

you are here ▸ 361

Exam Questions

> We gave you questions for domains 6 and 7 back to back. See if you can do them all in a row, then look at the answers. This will help you get ready for the big final exam in Chapter 9.

1. Agile teams often maintain an ordered list of work items and put the highest-risk items at the top of the list so that they get worked on first. What is the BEST name of this practice?

   A. Rank ordering
   B. Mitigation plan
   C. Risk-adjusted backlog
   D. Weighted shortest job first

2. You are an agile practitioner on a software team. Your team is holding a meeting to plan the next sprint. During the team's estimation for one of the work items, two team members disagree about the technical approach to the problem. Both of their solutions seem plausible to the rest of the team. The team needs to come to a resolution quickly, because that specific work item is a core piece of functionality that later sprints will depend on.

   What is the BEST solution to the problem?

   A. Have the whole team work on solving this problem before starting any other work
   B. Create an architectural spike for each of the two solutions, building them both over the course of the sprint in order to learn which one works best
   C. Preserve your options by not starting the work until later in the project
   D. Write down the disagreement as a risk and decide a solution later in the project

3. You are an agile practitioner on a software team. Your team has been together for three sprints, but lately they've been having a hard time getting along. Some members of the team think that others don't have the technical experience to make changes to core sections of the code base that the team is working on. This results in many disagreements during Daily Scrum and sprint planning meetings. How would you BEST describe the team and the managing technique you should use to deal with it?

   A. Forming stage, Directing managing technique
   B. Storming stage, Coaching managing technique
   C. Norming stage, Supporting managing technique
   D. Performing stage, Delegating managing technique

4. When a Scrum team meets to inspect the plan that they are working on and make adjustments so that they can accomplish their goals, that practice is called _____.

   A. The plan inspection meeting
   B. The planning approval gate
   C. The Daily Scrum
   D. The status report

Exam Questions

5. Whenever a programmer commits code to the repository, tests are automatically run on it and the code is compiled. Which of the followng BEST describes this?

    A. Code review
    B. Breaking the build
    C. Using a build server
    D. Continuous deployment

6. You are an agile practitioner on a software team that has been together for the last two years now. You and your teammates have developed a rapport. Any time there's a problem, everyone generally knows who to call or what to do. Individual team members have different interests and focus on different kinds of problems, but they generally have no trouble coming together to accomplish their sprint goals. How would you BEST describe the team and the managing technique you should use to deal with it?

    A. Forming stage, Directing managing technique
    B. Storming stage, Coaching managing technique
    C. Norming stage, Supporting managing technique
    D. Performing stage, Delegating managing technique

7. You are an agile practitioner on a software team. This is the first sprint that your team has worked together, and they don't know much about each other. They all have their own skills and goals, and haven't really worked out a good way to communicate about the project just yet. How would you BEST describe the team and the managing technique you should use to deal with it?

    A. Forming stage, Directing managing technique
    B. Storming stage, Coaching managing technique
    C. Norming stage, Supporting managing technique
    D. Performing stage, Delegating managing technique

8. At the beginning of the project, a team identifies a list of possible threats to project success. They printed out all of the risks in a large font, and taped them to a whiteboard for everyone to see in a central location in the office. The team gathers periodically to discuss and review these identified risks. What practice is the team using?

    A. Weekly risk review
    B. Steering committee meeting
    C. Risk approval gate
    D. Information radiator

Exam Questions

1. Agile team members are always learning new skills to help their teams complete work. Instead of having a very narrow and deep focus, they are _____ who can do multiple things on a team.

    A. Generalizing specialists
    B. Shallow contributors
    C. Experienced development leads
    D. Highly specialized developers

2. The main artifact of a sprint retrospective is _____.

    A. A list of accomplishments
    B. A set of improvement actions
    C. A set of meeting minutes
    D. A list of challenges

3. You are an agile practitioner on a software team. At the end of each sprint, you and your team demonstrate to the product owner the changes the team has made, and listen to his or her feedback about those changes. What is the BEST way to describe this practice?

    A. Daily scrum
    B. Product demo
    C. Product owner approval
    D. Sprint review

4. You are an agile practitioner on a team that builds mobile games. A member of your team has a new technical design for one of the features that's in the product backlog. When she brings it up to the team, some teammates are skeptical that it will work. Everyone agrees that if it does work it will be a significant improvement to the game's performance, and will be much easier to maintain than the current design. What is the BEST thing for the team to do next?

    A. Suggest that the team member stick with the current design since the team already knows it will work
    B. Suggest that the team member write up a design document and run it by the stakeholders for approval
    C. Suggest that the team member create a spike and try out the solution to see if it will work
    D. Suggest that the team stop work on the existing design because it's slower than the one the team member described

Exam Questions

5. You are on a software development team. During a retrospective, one person says most of the work in the last three sprints has focused on part of the codebase that only half of the team is familiar with, and as a result those people have been doing most of the valuable work during those sprints, and that they've been forced to add lower-value items to the sprint backlog in order to keep the rest of the team busy. Another person agrees with this assessment, and adds that there were serious bugs caused by unfamiliarity with that code. What is the BEST way to improve this situation?

   A. Break the work down into smaller chunks so the team can feel like they're accomplishing more
   B. Start pair programming all of the work that the team does so that people get more familiar with the code and can help each other catch bugs earlier
   C. Plan the work so that everyone is busy
   D. Limit the amount of work in progress so that team members don't have so much to do

6. Which of the following is NOT a tool used in retrospectives to get team consensus on improvements?

   A. Dot voting
   B. MoSCoW
   C. Fist-of-five voting
   D. Ishikawa diagrams

7. An agile team has just completed a two-week sprint. One of the team's stakeholders asks for the status of the risks that were identified in planning. The team directs the stakeholder to the team's risk burn down chart. What can the stakeholder tell from the information in the chart?

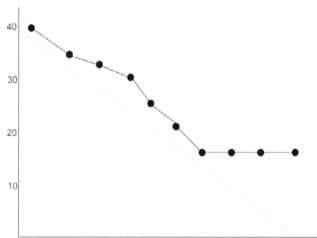

   A. The team did not close out all of the risks that they identified in planning
   B. Scope has been added
   C. The team is ahead of schedule
   D. There isn't enough data to understand the status of the risks

domain 6 exam answers

Exam ~~Questions~~ Answers

1. Answer: C

Agile teams maintain a risk-adjusted backlog to ensure that they work through the highest-risk items in the earliest sprints possible. Sometimes failing to solve a high-risk item can cause a whole project to fail and agile teams are always trying to fail as fast as possible, so they do those items first.

2. Answer: B

When the team has two equally viable solutions but can't decide on one, it often makes sense to pursue both of them at the same time. Creating an architectural spike to let the team explore both options will most likely give the team the information they need in order to make an informed decision in the next sprint planning session.

3. Answer: B

This team hasn't been together very long and they are arguing in Daily Scrums and planning meetings. It sounds like they're in the Storming phase of development. They will need the agile practitioner to take on a coaching role to help them get through their disagreements.

4. Answer: C

Scrum teams use the Daily Scrum to inspect their plan and make adjustments so that they can achieve their sprint goals. If some team members are having trouble with their work, others help them get rid of any impediments and finish what they set out to accomplish.

5. Answer: C

This is an example of a team using a build server. Whenever a team member makes a commit, all of the unit tests are run on the code and the entire repository is compiled. That way the whole team will know if there's a breaking change to the code as early as possible. If the tests fail or the code won't compile, the team knows that the problem is coming from the change that was just committed. Then the programmer who made that change can fix it before it causes other problems down the line.

> Were you expecting to see "Continuous integration" as one of the answers? Using a build server isn't the same thing as continuous integration, which is when team members continuously integrate code from their working folders back into the version control system to find problems early.

Exam Questions Answers

6. Answer: D

This team is performing. They all know what their teammates are capable of and have established a way of working that takes advantage of each other's strengths. It's typically most effective to delegate to them, and to let them keep doing what they're doing.

7. Answer: A

This team is forming. They will look to you to make most of the decisions until they become more comfortable with each other, so the directing style of management is most appropriate.

8. Answer: D

Displaying the risks up on a whiteboard for everyone in the project to see and review is an example of an information radiator. It's a good way to keep the risks that the team identified in everyone's mind so that everyone's aware of them, and can ask for help if they cause problems.

Exam ~~Questions~~ Answers

1. Answer: A

Agile teams value generalizing specialists who can apply a wide range of skills wherever they're needed, so that they can add value in many ways throughout a product's delivery timeline.

2. Answer: B

The goal of a retrospective should be specific actions that the team can take to make the process, system, or method that they're following more useful, streamlined, and collaborative for the team.

3. Answer: D

This question is describing the sprint review that occurs at the end of every Scrum sprint. In it, the product owner provides valuable feedback on the work product of the sprint. In some cases, that product owner will provide feedback, which will be added to the product backlog and incorporated into upcoming sprints in order of relative priority.

4. Answer: C

The team member should build a spike solution, in which she tries out her idea to see if it will work. Everyone on the team needs to be able to make mistakes. If the expectations on the team are so closely managed that team members have no freedom to decide as they go, they will not be able innovate or keep their options open.

5. Answer: B

When teams are highly specialized, pair programming can be a good way to break down barriers and help people to understand code they might not have worked with before. Pairing someone who's less familiar with the code with someone who knows it very well can spark great discussions between them; this is a very effective way to ramp people up on parts of the codebase that they are not familiar with yet. Pair programming also helps teams to catch bugs earlier and deliver a higher quality product, because pairs are constantly reviewing the code even as it's writtten.

Exam Questions ~~Questions~~ Answers

6. Answer: B

MoSCoW is a tool for prioritization. The rest of the answers offered are used in retrospectives.

7. Answer: A

The team identified more story points in risk reduction than they were able to complete within the sprint. In the last four days of the sprint the team did not make progress on any risk items. Instead, the sprint closed with roughly 15 points of risk mitigation effort still outstanding. Therefore, there are still planned risks that were not closed out.

ONE MORE THING...

THE PMI-ACP® EXAM IS **HARD**. IT'S FULL OF "LEAST-WORST-OPTION" QUESTIONS AND SEEMINGLY AMBIGUOUS SCENARIOS. **BUT WE KNOW YOU'RE UP TO THE TASK!** THE PRACTICE EXAM AT THE END OF THIS GUIDE IS **CAREFULLY CALIBRATED** TO MATCH THE DIFFICULTY OF THE REAL THING. JUST BE PATIENT... REMEMBER THAT WHEN YOU GET A PRACTICE QUESTION WRONG, THAT MEANS YOU'RE MORE LIKELY TO GET A SIMILAR QUESTION RIGHT ON EXAM DAY.

pmi-acp® crossword

Examcross

Get ready for one last test of your knowledge before you take the final practice exam! How much of this crossword can you do without looking at the answers?

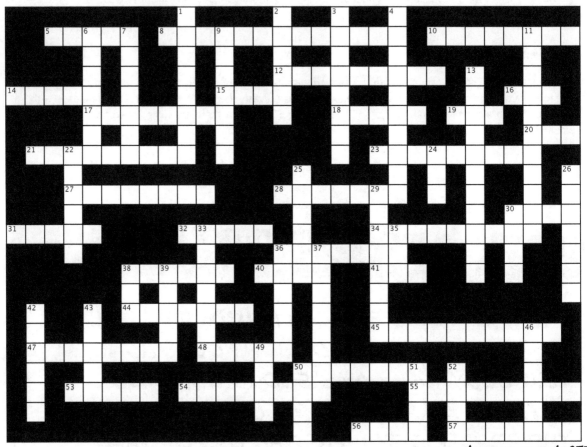

➡ Answers on page 375

Across

5. Scrum value that tells us it's most effective to work on one item at a time
8. Meeting at the end of the sprint where the team talks about how it went and what can improve
10. The team determines what work will be performed in the sprint during the sprint _____ meeting
12. Scrum artifact that includes all items completed during the sprint
14. Planning _____ is a collaboration game to help the team estimate
15. Agile teams value responding to change over following a _____

16. Amount of money the project will return to the company that is funding it
17. The kind of leadership where you adjust leadership style to match the stage of team formation
18. _____ time is an estimate that assumes no delays, interruptions, or problems
19. Money you expect to make from a project that you are building
20. Smallest possible piece of functionality that can be delivered
21. The style of leadership that matches the forming stage of team development
23. Low fidelity tool for sketching a user interface
27. The number of points' worth of work the team accomplishes on average per iteration

Across

28. Contains features and work items to be built, may contain risks
30. Kind of programming where two people work together at the same computer
31. Dot voting and fist-of-five voting are examples of collaboration _____
32. Kind of solution that is also referred to as "exploratory work"
34. Transparency, inspection, and adaptation are the three pillars of _____ process control
36. A description of a fictional user
38. Simple requirements prioritization technique where you decide if you must, could, should, or won't have each requirement
40. Product Owner, Scrum Master, and team member are examples of these
41. Actual value at a given time of the project minus all of the costs associated with it
44. The _____ backlog is the Scrum artifact that contains the single source of all changes and features
45. XP value that is helped by creating loosely coupled or decoupled code
47. In _____ estimating, team members create groups and take turns assigning items to them
48. Agile teams value individuals and interactions over processes and _____
50. Agile teams value _____ software over comprehensive documentation
53. The average time a work item spends in the process is its _____ time
54. An information _____ is posted in a visible area in the team space
55. Kind of testing that helps ensure you're making effective design choices
56. Scrum teams are _____ organizing
57. The kind of communication where you absorb information overheard in the team space

Down

1. The kind of cycle where you determine if you can improve efficiency and quality
2. The _____ backlog is the Scrum artifact that includes the items to be completed in the current iteration
3. The stage of development for which the coaching leadership style is best applied
4. The leadership style most appropriate for the performing stage of team formation
6. Scrum and XP value that helps team members stand up for the project
7. The kind of leadership the Scrum Master provides
9. XP and Scrum value that tells team members to treat each other the way they would want to be treated themselves
11. XP practice—the kind of design that helps the team embrace change
13. The kind of elaboration where an item is updated incrementally as more information is known
22. Meeting at the end of a sprint where work is demonstrated to the stakeholders
24. Actual cost in money or hours on value the product is delivering
25. Only the Product Owner has the authority to _____ a sprint, but it can cause the stakeholders to lose trust in the team
26. The stage of development for which the supporting leadership style is appropriate
29. Scrum value that tells us that team members should feel comfortable with everyone having visibility into their work
30. Working at a sustainable _____ means working 40 hours a week so the team doesn't burn out
33. The kind of feedback loop where you determine if what you're delivering meets expectations
35. A story _____ shows planned releases
36. Story _____ are a relative sizing technique
37. How you modify code structure without changing its behavior
38. Smallest possible product that the team can deliver which still satisfies the users' and stakeholders' needs
39. Brief description of functionality from a user's perspective
42. What many teams fear, but effective XP teams embrace
43. How often the team meets to inspect the work by answering questions about their progress, planned work, and roadblocks
46. Relative sizing technique where you assign XS, S, M, L, or XL to each item
49. The average time a stakeholder spends waiting for a work item to be completed is its _____ time
50. Limit _____ in progress in order to improve throughput of work items through your process
51. Product feedback methods reduce the _____ of evaluation, or the difference between what was asked for and what the team built
52. Type of analysis that shows you how features that once delighted users are now seen as basic requirements

domains 5 and 6 in your own words solution

Domain 5: Exam Questions — IN YOUR OWN WORDS SOLUTION

Here's how we interpreted the tasks in our own words. It's OK if you used different words!

Evolve your project plan as you learn more about your project, stakeholders, and roadblocks

The tasks in Domain 5 are listed on page 10 of the content outline. Write down what you think each task is about:

Iterate at every level of the project (daily meetings, sprints, quarterly cycles, etc.)
Task 1

Be completely transparent with your stakeholders about how you plan the project
Task 2

Start out with broad commitments, and make more specific ones as the project unfolds
Task 3

Use retrospectives and your understanding of deliverables to change how and how often you plan
Task 4

Use an inspection-adaptation cycle to stay on top of scope, priority, budget, and schedule changes
Task 5

Collaborate to understand the ideal size of work items before taking velocity into account
Task 6

Don't forget about maintenance and operations activities, which can affect your project plan
Task 7

Create initial estimates that take into account the fact there are still a lot of unknowns
Task 8

Keep refining your estimates as you learn more about how much effort the project will require
Task 9

Continue to update your plan as you get a clearer picture of the team's velocity and capacity
Task 10

Domain 6: Exam Questions

IN YOUR OWN WORDS SOLUTION

Here's how we interpreted the tasks in our own words. **It's OK** if you used different words!

In your own words, write down what you think Domain VI ("Problem Detection and Resolution") is about:

Watch for problems and fix them, then improve how you work to prevent them from reoccurring

The tasks in Domain 6 are listed on page 11 of the content outline. Write down what you think each task is about:

Give everyone on the team freedom to experiment and make mistakes
Task 1

Constantly watch for risks to the project, and make sure the whole team is aware of them
Task 2

When problems happen, make sure they're fixed—or, if they can't be fixed, set expectations
Task 3

Be completely transparent about the risks, problems, issues, and threats to the project
Task 4

Make sure risks and issues actually get fixed by including them in product and iteration backlogs
Task 5

Domain 7: Exam Questions

IN YOUR OWN WORDS SOLUTION

Here's how we interpreted the tasks in our own words. It's OK if you used different words!

The team works together to keep improving the way that they do project work

The tasks in Domain 7 are listed on page 12 of the content outline. Write down what you think each task is about:

Constantly look at the practices, values, and goals, and use that information to adapt the process
Task 1

Hold retrospectives frequently, and experiment with improvements to address issues you find
Task 2

Demonstrate working software at the end of every iteration, and genuinely listen to feedback
Task 3

Generalizing specialists are really valuable, so give everyone opportunities to improve their skills
Task 4

Use value stream analysis to discover waste, and make individual and team efforts to eliminate it
Task 5

When you gain knowledge from making improvements, share it with the rest of the organization
Task 6

preparing for the pmi-acp

Examcross Solution

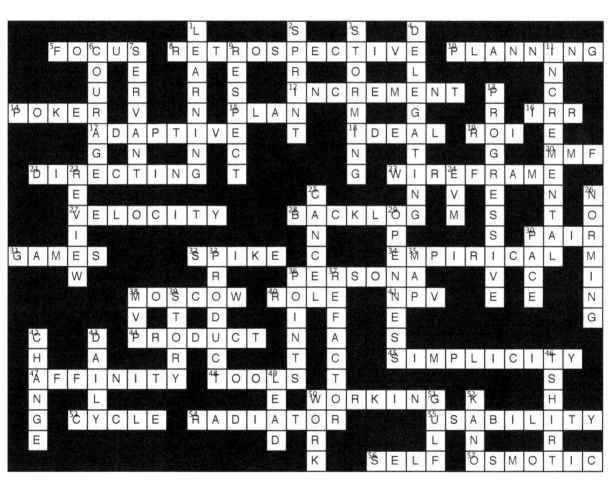

Are you ready for the final exam?

Congratulations! Take a minute and think about how much you've learned about agile. Now it's time to test your knowledge and see how well you're prepared to take the PMI-ACP® exam. The last chapter in this book is a full length, 120-question, simulated PMI-ACP® exam, carefully constructed using the same exam content outline as the real thing and using questions that have style and content similar to what you'll see on exam day. Here are a few tips to make it as effective as possible:

* **When you take the final exam in the next chapter, make it the only study activity that you do that day.**

* **Give yourself plenty of time to do it, and take the whole exam all at once.**

* **Make sure you drink lots of water while you're answering the questions.** ← *Your brain actually learns better when it's well-hydrated!*

* **As you're answering the questions, think about each answer and only mark down one response, even if you're not 100% sure.**

* **After every 10 questions, go back and read through each of them again, and see if you still agree with your answer.**

* **Don't look at any of the exam answers until you've gone through all of the questions.**

* **Make sure you get plenty of sleep the night after you take the practice exam. That helps the information stick in your brain!**

> COGNITIVE PSYCHOLOGISTS RECOGNIZE THAT SLEEP PLAYS A REALLY IMPORTANT ROLE IN HELPING YOUR BRAIN ORGANIZE AND CONSOLIDATE INFORMATION INTO YOUR LONG-TERM MEMORY.

Good luck!

8 Professional responsibility

Making good choices

KNOWING THE PMI CODE OF ETHICS AND PROFESSIONAL CONDUCT HAS MADE ME A **BETTER AGILE PRACTITIONER** ... AND A BETTER ROOMMATE!

OH, REALLY? WELL, DOES IT SAY ANYTHING ABOUT LEAVING HALF-EATEN BURRITOS IN THE SINK?

It's not enough to just know your stuff. You need to make good choices to be good at your job. Everyone who has the PMI-ACP credential agrees to follow the **Project Management Institute Code of Ethics and Professional Conduct**, too. The Code helps you with **ethical decisions** that aren't really covered in the body of knowledge—you may get a few questions about it on the PMI-ACP exam. Most of what you need to know is **really straightforward**, and with a little review, you'll do well.

meet the code of ethics and **professional conduct**

Doing the right thing

You'll get some questions on the exam that present situations that you might run into while managing your projects and then ask you what to do. Usually, there's a clear answer to these questions: ***it's the one where you stick to your principles.*** Questions will make the decisions tougher by offering rewards for doing the wrong thing (like money for taking a project shortcut), or they will make the infraction seem really small (like photocopying a copyrighted article out of a magazine). If you stick to the principles in the PMI Code of Professional Conduct regardless of the consequences, you'll **always** get the answers right.

The main ideas

In general, there are a few kinds of problems that the code of ethics prepares you to deal with.

1. **Follow all laws and company policies.**
2. **Treat everybody fairly and respectfully.**
3. **Have respect for the environment and the community you're working in.**
4. **Give back to the community by writing, speaking, and sharing your experience with other agile professionals.**
5. **Keep learning and getting better and better at your job.**
6. **Respect other people's cultures.**
7. **Respect copyright laws.**
8. **Always be honest with everyone on the project.**
9. **If you find that another person has done something to damage the PMI-ACP or any other PMI credential in any way, you must report it to PMI.**

> So if you find out that someone has stolen questions from the PMI-ACP exam, cheated on the PMI-ACP exam, falsely claimed to have a PMI-ACP certification, or lied about anything related to the PMI-ACP certification process, then you MUST report that to PMI.

The PMI-ACP exam content outline includes knowledge of ethics and professional conduct, because ethical behavior should be part of any agile practitioner's knowledge and skills. That means there may be a small number of questions about ethics and professional conduct scattered throughout the exam.

professional responsibility

COME ON. IS THIS REALLY ON THE TEST? I KNOW HOW TO DO MY JOB. DO I REALLY NEED A MORALITY LESSON?

Being PMI-ACP® certified means that you know how to do your job and that you will do it with integrity.

It might seem like it doesn't really matter how you will handle these situations, but think about it from an employer's perspective for a minute. Because of the PMI Code of Ethics and Professional Conduct, employers know that when they hire a PMI-ACP® certified agile practitioner, they are hiring someone who will follow company policies and do everything aboveboard and by the book. That means that you'll help to protect their company from litigation and deliver on what you promise, which is actually pretty important.

So you should definitely not be surprised to see at least a few questions about ethics and professional responsibility on the exam.

Keep your eye out for "red herring" questions that turn out to be about ethics and social responsibility. They might lay out a situation that sounds like a normal project management problem, but requires you to use one of the principles in the PMI Code of Ethics and Professional Conduct.

Can you think of some situations where you might need to make decisions using these principles in your own projects?

you are here ▶ 379

never take bribes

Keep the cash?

It is never, under any circumstances, OK to accept a bribe—even if your company and customer might benefit from it somehow. And bribes aren't always cash. They can be anything from free trips to tickets to a ball game. Any time you're offered anything to change your opinion or the way you work, you must decline the offer and disclose it to your company.

In some countries, even though you may be "expected" to pay a bribe, it's not OK to do it—even if it's customary or culturally acceptable.

KATE, YOU WERE SO GREAT TO WORK WITH. WE'D LIKE TO SEND YOU $1,000 AS A TOKEN OF OUR APPRECIATION.

AWESOME. I'VE BEEN WANTING TO GO SHOPPING FOR A WHILE. AND WHAT ABOUT THAT VACATION? ACAPULCO, HERE WE COME!

I WOULD NEVER ACCEPT A GIFT LIKE THAT. DOING A GOOD JOB IS ITS OWN REWARD!

The right way

The easy way

I'M SORRY, I CAN'T ACCEPT THE GIFT. I REALLY APPRECIATE THE GESTURE, THOUGH.

Fly business class?

Any time there's a policy in your company, you need to follow it. Even if it seems like no harm will be done if you don't follow the policy, and even if you will be able to get away with it, you should not do it. And that goes double for laws—under no circumstances are you ever allowed to break a law, no matter how much good it "seems" to do you or your project.

And if you ever see someone in your company breaking the law, you need to report it to the authorities.

> WE'VE GOT SOME EXTRA MONEY IN THE BUDGET AND YOU'RE REALLY DOING A GREAT JOB. I KNOW THE TRAVEL POLICY SAYS WE ALWAYS FLY COACH. BUT WE CAN AFFORD TO SPLURGE A BIT. WHY DON'T YOU BUY A BUSINESS TICKET THIS TIME?

> DID YOU KNOW THAT THOSE CHAIRS GO INTO TOTALLY FLAT BEDS? THIS IS SO COOL. I'VE WORKED SO HARD, I'VE TOTALLY EARNED IT!

> THERE'S NO EXCUSE FOR NOT FOLLOWING THE RULES. THE TRAVEL POLICY SAYS FLY COACH. NO EXCEPTIONS!

> WOW, BEN. THAT'S REALLY NICE OF YOU. BUT THE ECONOMY FARE WILL BE FINE.

respect *intellectual property*

New software

When it comes to copyright, it's never OK to use anything without permission. Books, articles, music, software…you always need to ask before using it. For example, if you want to use some copyrighted music in a company presentation, you should write to the copyright owner and ask for permission.

> HEY KATE, I JUST GOT A COPY OF THAT SCHEDULING SOFTWARE YOU WANTED. YOU CAN BORROW MY COPY AND INSTALL IT.

> ABSOLUTELY! THIS WILL TOTALLY MAKE MY JOB ABOUT 100 TIMES FASTER, AND IT'S FREE? IT'S MY LUCKY DAY.

> THAT SOFTWARE WAS CREATED BY A COMPANY THAT DESERVES TO BE PAID FOR THEIR WORK. IT'S JUST WRONG NOT TO BUY A LICENSED COPY.

> THANKS FOR LETTING ME KNOW IT'S AVAILABLE. I'LL GO BUY A COPY.

professional *responsibility*

Shortcuts

You might see a question or two asking if you really need to follow all of the processes. Or you might be asked by your boss to keep certain facts about your project hidden from stakeholders or sponsors. You have a responsibility to make sure your projects are run properly, and to never withhold information from people who need it.

> WE DON'T HAVE TIME FOR ALL OF THIS DOCUMENTATION. LET'S CUT OUT A COUPLE OF THESE PLANS TO KEEP OUR PROJECT ON SCHEDULE.

> ALL RIGHT, LESS WORK FOR ME! LET'S FACE IT, IT'S NOT LIKE I HAVE ALL THE TIME IN THE WORLD FOR WRITING PLANS!

> I WOULD NEVER DO A PROJECT WITHOUT FOLLOWING ALL OF THE 47 PROCESSES OUTLINED IN THE PMBOK GUIDE.

> I KNOW WE DON'T HAVE MUCH TIME, BUT TAKING THIS SHORTCUT COULD ACTUALLY COST US MORE TIME THAN IT SAVES IN THE END.

respect our environment

A good price or a clean river?

Being responsible to the community is even more important than running a successful project. But it's more than being environmentally aware—you should also respect the cultures of everyone else in your community, and the community where your project work will be done.

That means even though languages, customs, holidays, and vacation policies might be different from country to country, you need to treat people the way they are accustomed to being treated.

WE JUST FOUND OUT THAT ONE OF OUR SUPPLIERS DUMPS HARMFUL CHEMICALS IN THE RIVER. THEY'VE ALWAYS GIVEN US GREAT RATES, AND OUR BUDGET WILL GO THROUGH THE ROOF IF WE SWITCH SUPPLIERS NOW. THE WHOLE THING GIVES ME A HEADACHE. WHAT SHOULD WE DO?

WE CAN'T LET THE PROJECT FAIL FOR A BUNCH OF STUPID FISH.

THE EARTH IS OUR HOME AND IS SO MUCH MORE IMPORTANT THAN THIS PROJECT. WE HAVE TO DO WHAT'S RIGHT...

BEN, I KNOW IT COULD CAUSE US PROBLEMS, BUT WE'RE GONNA HAVE TO FIND ANOTHER SUPPLIER.

professional responsibility

We're not all angels

We know that the choices you make on your project are not always black and white. Remember that the questions on the exam are designed to test your knowledge of the PMI Code of Professional Conduct and how to apply it. A lot of situations you will run into in real life have a hundred circumstances around them that make these decisions a little tougher to make than the ones you see here. But if you know what the code would have you do, you're in a good position to evaluate those scenarios as well.

Seriously, it's a quick read—and it'll help you on the exam.

Now, go read PMI's Code of Ethics and Professional Conduct before you take these exam questions. To find it, search for "PMI Code of Ethics and Professional Conduct" using the search feature on the PMI website, or using your favorite search engine.

I MAY NOT BE THE LIFE OF THE PARTY, BUT THINK LIKE ME, AND YOU'LL NAIL THE ETHICS PART OF THE EXAM.

Exam Questions

1. You read a great article over the weekend, and you think your team could really benefit from it. What should you do?

    A. Photocopy the article and give it to the team members.

    B. Type up parts of the article and email it to the team.

    C. Tell everyone that you thought of the ideas in the article yourself.

    D. Buy a copy of the magazine for everyone.

2. You find out that a contractor that you're working with discriminates against women. The contractor is in another country, and it's normal in that country. What should you do?

    A. Respect the contractor's culture and allow the discrimination to continue.

    B. Refuse to work with the contractor, and find a new seller.

    C. Submit a written request that the contractor no longer discriminate.

    D. Meet with your boss and explain the situation.

3. You're working on a project when the client demands that you take him out to lunch every week if you want to keep his business. What's the BEST thing to do?

    A. Take the client out to lunch and charge it to your company.

    B. Refuse to take the client out to lunch because it's a bribe.

    C. Take the client out to lunch, but report him to his manager.

    D. Report the incident to PMI.

4. You are working on one of the first financial projects your company has attempted, and you have learned a lot about how to manage the project along the way. Your company is targeting financial companies for new projects next year. What's the BEST thing for you to do?

    A. Talk to your company about setting up some training sessions so that you can teach others what you have learned on your project.

    B. Keep the information you've learned to yourself so that you'll be more valuable to the company in the next year.

    C. Decide to specialize in financial contracts.

    D. Focus on your work with the project and don't worry about helping other people to learn from the experience.

5. You find out that you could save money by contracting with a seller in a country that has lax environmental protection rules. What should you do?

    A. Continue to pay higher rates for an environmentally safe solution.

    B. Take advantage of the cost savings.

    C. Ask your boss to make the decision for you.

    D. Demand that your current contractor match the price.

Exam Questions

6. You overhear someone on your team using a racial slur. This person is a critical team member and you are worried that if he leaves your company it will cause project problems. What should you do?

- A. Pretend you didn't hear it so that you don't cause problems.
- B. Report the team member to his boss.
- C. Bring it up at the next team meeting.
- D. Meet in private with the team member and explain that racial slurs are unacceptable.

7. You've given a presentation for your local PMI chapter meeting. This is an example of what?

- A. A PDU
- B. Contributing to the Project Management Body of Knowledge
- C. Donating to charity
- D. Volunteering

8. You are about to hold a bidder conference, and a potential seller offers you great tickets to a baseball game for your favorite team. What should you do?

- A. Go to the game with the seller but avoid talking about the contract.
- B. Go to the game with seller and discuss the contract.
- C. Go to the game, but make sure not to let him buy you anything because that would be a bribe.
- D. Politely refuse the tickets.

9. Your company has sent out a request for proposals for consulting work, and your brother wants to bid on it. What's the BEST thing for you to do?

- A. Give your brother inside information to make sure that he has the best chance at getting the project.
- B. Publicly disclose your relationship with him and excuse yourself from the selection process.
- C. Recommend your brother but don't inform anyone of your relationship.
- D. Don't tell anyone about your relationship, but be careful not to give your brother any advantage when evaluating all of the potential sellers.

exam answers

Exam ~~Questions~~ Answers

1. Answer: D

You should never copy anything that's copyrighted. Make sure you always respect other people's intellectual property!

2. Answer: B

It's never OK to discriminate against women, minorities, or others. You should avoid doing business with anyone who does.

3. Answer: B

The client is demanding a bribe, and paying bribes is unethical. You should not do it. If your project requires you to bribe someone, then you shouldn't do business with that person.

4. Answer: A

You should always try to help other people learn, especially about improving the way they run their projects.

5. Answer: A

You should never contract work to a seller who pollutes the environment. Even though it costs more to use machinery that doesn't damage the environment, it's the right thing to do.

6. Answer: D

You should make sure that your team always respects other people.

7. Answer: B

Any time you help share your knowledge with others, you are contributing to the Project Management Body of Knowledge, and that's something you should do any time you have any kind of PMI certification!

Exam Questions Answers

8. Answer: D

You have to refuse the tickets even if the game sounds like a lot of fun. The tickets amount to a bribe, and you shouldn't do anything that might influence your decision in awarding your contract.

9. Answer: B

You have to disclose the relationship. It's important to be up front and honest about any conflict of interest that could occur on your projects.

> EVEN IF I DON'T SEE MANY QUESTIONS ABOUT THIS ON THE PMI-ACP EXAM, IT'S STILL A GREAT TOPIC TO BE FAMILIAR WITH AS A PROFESSIONAL.

this page intentionally left blank

9 practice makes perfect

Practice PMI-ACP Exam

I KNOW WE'RE SUPPOSED TO BE STUDYING, BUT I CAN'T STOP THINKING ABOUT FUDGE.

Bet you never thought you'd make it this far! It's been a long journey, but here you are, ready to review your knowledge and get ready for exam day. You've put a lot of new information about agile into your brain, and now it's time to see just how much of it stuck. That's why we put together this full-length, 120-question PMI-ACP® exam simulation for you. We followed the same PMI-ACP® Examination Content Outline used to design the exam. And we <u>carefully calibrated the difficulty of the questions</u> to match the real PMI-ACP® exam, and we matched their style and substance as well, so they look *just like the ones you'll encounter* when you take the real thing. So take a deep breath, get ready, and let's get started.

The questions in this simulated exam are more <u>difficult</u> than the ones you've seen so far! Expect a lot of ambiguous, least-worst-option questions. You'll get similar ones on the real exam.

Just remember, if you get a question wrong in this simulation, you're <u>more</u> likely to get it right on exam day.

this is a new chapter

exam questions

1. A Scrum team's stakeholder discovers a new requirement and approaches a team member to build it. The team member builds a prototype for the stakeholder, who is able to start using it immediately. The product owner discovers this and demands that the team member include her in any future decisions, but the team member feels this is the most efficient way to work. The product owner and team member approach the scrum master to resolve the conflict. What should the scrum master do?

    A. Help the product owner understand how the team member is improving stakeholder communications
    B. Side with the product owner
    C. Help the team member follow the rules of Scrum by showing the correct use of user stories
    D. Help the product owner and team member compromise and find a middle ground

2. After the team gives a demonstration of the work they built during an iteration, a stakeholder complains that the software is difficult to use. What technique can the team employ to prevent this in the future?

    A. Develop user interface requirements that cover usability
    B. Define organizational usability standards
    C. Observe stakeholders interacting with a preliminary version of the user interface
    D. Create a wireframe and review it with the stakeholders

3. The main project stakeholder for a Scrum project emails the product owner to notify them that one of the primary deliverables must be delivered two months earlier than planned. The team members meet and agree on an approach that will achieve the goal, but it will cause a delay to several features of other deliverables. The product owner warns the team that this will be unacceptable to the stakeholder. How should the team proceed?

    A. Initiate the change control procedure
    B. Have the product owner meet with the stakeholder to discuss acceptable trade-offs
    C. Begin exploratory work on a spike solution
    D. Start working on the approach that the team collaboratively agreed on

4. You are reviewing your team's kanban board, and you discover that many work items tend to accumulate in a specific step in the process. What is the best way to handle this situation?

    A. Work with the team to remove work items from that step of the process
    B. Use Little's law to calculate the long-term average inventory
    C. Increase the arrival rate of work items into the process
    D. Work with project stakeholders to establish a WIP limit for that column

practice pmi-acp exam

5. A tester on a Scrum team has expressed interest in taking on some programming tasks. One of the developers is concerned that this may lead to quality problems. How should the team proceed?

   A. The scrum master should look for opportunities to provide development training for the tester
   B. The developer should serve as a full-time mentor to the tester
   C. The scrum master should call a meeting to get consensus on letting the tester do development work
   D. The tester should start to take on development tasks as they become available during the sprint

6. A new member of an agile team disagrees with the ground rules that team members currently adhere to. How should the team proceed?

   A. Explain the reason for the rule and encourage the new team member to try following it
   B. Throw out the rule and collaborate on a replacement
   C. Have the scrum master explain the rules of scrum to the new team member
   D. Show the new team member which principle in the Agile Manifesto the rule is based on

7. You are a leader on an agile team. Which of the following is not an action that you would take?

   A. Set an example by acting the way you would like other team members to act
   B. Make sure everyone on the team understands the goal of the project
   C. Make important decisions about how the team will design the software
   D. Prevent external problems from taking up too much of the team's time

8. You are in a sprint planning meeting. Two members of your team are arguing over one of the stories. They cannot agree on whether they will be able to reuse an existing aspect of the user interface, or if they will need to build out new user interface elements. What is the best way for the team to resolve this issue?

   A. The product owner resolves the issue by determining how the team will solve the problem
   B. The team adds buffers to the plan to account for the uncertainty
   C. The Scrum Master resolves the issue by determining how the team will solve the problem
   D. The team uses negotiation to reach an agreement on the specific acceptance criteria for the feature

exam questions

9. You are an agile practitioner working directly with several business stakeholders. One of the stakeholders has provided a requirement that is proving difficult to implement. Several team members have called a meeting to propose an alternative to the requirement that will be much less expensive to implement. How should you handle this situation?

   A. Use servant leadership to engage the team
   B. Invite the stakeholder to the next daily standup
   C. Explain the team's alternative to the stakeholder
   D. Explain the stakeholder's expectations and needs to the team and collaborate on a solution

10. An agile team is in their second iteration. Some team members' disagreements are starting to turn into arguments, and one team member recently accused another of shirking responsibility. What best describes this team?

    A. The team is in the storming phase, and needs directing
    B. The team is in the norming phase, and needs supporting
    C. The team is in the norming phase, and needs coaching
    D. The team is in the storming phase, and needs coaching

11. A Scrum team claims to be self-organizing. What does that mean?

    A. The team plans each sprint together and makes decisions about individual task assignments at the last responsible moment
    B. The team delivers working software at the end of each sprint, and adjusts the next sprint plan to maximize value delivered to stakeholders
    C. The team does not need a manager, and instead relies on the scrum master to provide servant leadership
    D. The team only needs to plan on a sprint-by-sprint basis, and does not have to commit to any deadline beyond the length of the sprint

12. An agile practitioner is working with a vendor to implement an important product feature. The practitioner is concerned that the vendor is working on low-priority features in early iterations, while neglecting higher priority features. What is the best way to handle this situation?

    A. The practitioner raises the issue at the next daily standup
    B. The practitioner raises the issue at the next iteration planning meeting
    C. Value of the deliverables are optimized through collaboration between the practitioner and the vendor
    D. The practitioner moves high priority items into the backlog

13. You are an agile practitioner on a team at a vendor of software services. One of your clients is having trouble planning their iterations. Your team ran into a similar problem, and used a specific technique to resolve it. What action should you take?

   A. Explain the practice to your contacts at the client
   B. Create a document that describes the improvement
   C. Offer to attend the client's daily standup meetings
   D. Do nothing in order to respect the organizational boundaries

14. A scrum master on another team asks you for advice about how to handle a user who keeps changing his mind about what the team should build. You should:

   A. Show the scrum master the company's standards for creating a project plan and implementing a change control process
   B. Show how your own users have changed their minds in the past, and that you worked with the team to make adjustments during sprints and in sprint planning
   C. Offer to run the other team's daily standup and retrospective meetings
   D. Explain that agile teams value responding to change, and that the scrum master should help the team understand this principle

15. Which of the following is not valuable for fostering an effective team environment?

   A. Pay attention during retrospectives and contribute wherever possible
   B. Make it clear that it's OK to make mistakes
   C. When team members disagree on an approach, have a constructive argument
   D. Be very careful that you follow all of the company's ground rules for working on projects

16. You are an agile practitioner on a team using Kanban for process improvement. What metrics would you use to measure the effectiveness of your improvement effort, and how would you visualize the data?

   A. Use a resource histogram to visualize resource allocation over the course of the project
   B. Use a burndown chart to visualize velocity and points completed per day
   C. Use a value stream map to visualize time worked versus time spent waiting
   D. Use a cumulative flow diagram to visualize arrival rate lead time, and work in progress

exam questions

17. Your team can choose between two feasible solutions to a technical problem. One solution uses encryption but runs more slowly, the other solution does not use encryption and runs more quickly. What is the best way to choose between the two solutions?

   A. Choose the faster solution
   B. Use a spike solution to determine which approach will work
   C. Choose the more secure solution
   D. Elicit relevant non-functional requirements from stakeholders

18. You are an agile practitioner in the scrum master role. Your team is holding a retrospective. What is your responsibility?

   A. Observe the team identify improvements and create a plan to implement them, and help team members understand their roles in the meeting
   B. Make yourself available to the project team if they have questions about the rules of Scrum
   C. Participate in identifying improvements and creating a plan to implement them, and help team members understand their roles in the meeting
   D. Help the team understand the needs of the stakeholders and represent their viewpoint

19. You are a scrum master. A member of your team is concerned that there are too many team meetings, and would like to skip the daily standup meeting once a week. What should you do?

   A. Explain that the rules of Scrum require that everyone attend the meeting
   B. Help the team member understand how the daily meeting helps everyone on the team find problems early and fix them
   C. Partner with the team member's manager because attending the daily standup is a job requirement
   D. Work with the team to set ground rules that everyone attend the daily standup

20. Your agile team is working with a vendor to build some of the product components, including a component that will be used in the next sprint. After meeting with the representative from the vendor to discuss the project's goals and objectives, the vendor representative emails you their scope and objectives document, explaining that they use a waterfall process and this is how the high-level vision and supporting objectives are communicated in their organization. What is the best way to handle this situation?

   A. Request that the vendor team creates user stories to express requirements
   B. Advocate for agile principles and explain that your team values working software over comprehensive documentation
   C. Carefully read the scope and objectives document and follow up with the vendor representative about any discrepancies with your team's understanding of the project
   D. Invite the vendor representative to the sprint review meetings

practice pmi-acp exam

21. You are an agile practitioner on a Scrum team developing financial analytics software. You and your teammates are very interested in trying a new technology. The product owner expresses concern that the extra time required to ramp up on it will cause delays. How should the team proceed?

A. Have the scrum master negotiate an agreement between the team and the product owner to use the new technology
B. Have the product owner explain the new technology to the primary stakeholders
C. Reject the new technology and stick with technology familiar to the team to avoid the extra time required
D. Have the team members collaborate with the product owner to find ways to align their technology goals with the project objectives

22. You are an agile practitioner on an XP team. A teammate discovered a serious problem with the architecture of the software that the team has been working on which will require a major redesign of several large components. What should the team do next?

A. Refactor the code and practice continuous integration
B. Use pair programming to help everyone understand the scope of the problem
C. Use incremental design and delay design decisions until the last responsible moment
D. Work with the stakeholders to help them understand the impact on the project

23. The timebox for doing the work for an iteration has expired. What is the next action the team takes?

A. Conduct a demonstration of all fully and partially completed features to the stakeholders
B. Conduct a retrospective to enhance the effectiveness of the team, project, and organization
C. Begin planning the next iteration
D. Conduct a demonstration of all features that were fully completed to the stakeholders

24. You are the product owner on a Scrum team building software that will be used by a team of financial services analysts. At the last two sprint reviews, the manager of the financial services analyst team was angry that your team did not build all of the features that she was expecting. What is the appropriate response?

A. Meet with the manager throughout the next sprint to discuss each story's acceptance criteria, and update the sprint backlog based on that discussion
B. Send a daily email to each stakeholder with the latest version of the sprint backlog
C. Invite the manager to the next daily standup meeting
D. Invite the manager to the next sprint planning meeting

exam questions

25. You overhear two senior managers discussing a company-wide problem with software teams that deliver software late, and that the software often fails to deliver much value. What is the best way to handle this situation?

   A. Take the opportunity to evangelize about Scrum and insist that more teams be required to use it
   B. Offer to speak with other teams about your own team's past success with agile
   C. Engage your product owner to determine how best to take advantage of this situation
   D. Explain that agile teams always follow the values and principles of agile

26. What is the most effective way to communicate progress in a team space?

   A. Visualize project progress and team performance information
   B. Position desks so everyone is face to face
   C. Hold a retrospective
   D. Communicate progress at the daily standup

27. You are an agile practitioner working on exploratory work that your team included in the current iteration plan. The goal of this work is to find problems, issues, and threats. The output of this exploratory work is that certain results should be surfaced to the team. Which of the following is not a valid reason to surface a specific issue?

   A. It will slow down progress
   B. It might prevent the team from delivering value
   C. It isn't the result that you expected
   D. It is a problem or impediment

28. A software team at a company with a strict waterfall process is having engineering problems which are causing them to build features that do not adequately meet users' needs. How can this team address the situation?

   A. Assign team members to the product owner and scrum master role and mange the work using sprints
   B. Use quarterly and weekly cycles, refactoring, test-driven development, pair programming, and incremental design
   C. Use Kaizen and practice continuous improvement
   D. Establish a team space that uses caves and commons, osmotic communication, and information radiators.

29. A stakeholder calls a meeting halfway through the sprint and explains that due to a change in business priorities one of the backlog items is no longer needed. What is the best way for the team to handle this?

   A. The product owner and stakeholder present the change to the team at the sprint review
   B. The product owner works with the team to remove the item from the sprint backlog, and the team delivers any other working software they have built when the sprint ends
   C. The product owner removes the item from the sprint backlog, and extends the end date of the sprint to accommodate the change in plan
   D. The product owner cancels the sprint and the team starts planning a new sprint

30. You are an agile practitioner on a team that uses a Scrum/XP hybrid. Two team members disagree on how much effort it will take to implement a story in the current sprint. Which of the following is not an effective action to take?

   A. Use wideband Delphi to generate an estimate for the story
   B. Have the product owner decide if the longer estimate is acceptable to the stakeholders
   C. Have an informal group discussion about the factors that cause the estimates to differ
   D. Call a team meeting to play a round of planning poker

31. Your company implements a requirement that teams create highly detailed documentation as part of the company-wide software development lifecycle. What is the correct response?

   A. Use negotiation techniques to help the organization become more agile
   B. Agile teams do not value comprehensive documentation, so the team should not produce it
   C. Select a process for the team that delivers the highly detailed documentation without sacrificing delivery of customer value
   D. Ensure that the team is delivering working software, while still producing the minimal documentation needed to build the software

32. You are an agile practitioner. Several members of your team have expressed concern that the project is not progressing as well as they would like. What is the best course of action?

   A. Post a burn-down chart in a highly visible part of the team space
   B. Consult the communications plan and distribute project performance information
   C. Discuss the status of the project at the next daily standup meeting
   D. Distribute status reports that include burn-down charts

exam questions

33. During a retrospective, a Scrum team finds that their velocity was reduced significantly at the same time that two teammates were taking vacations that had been planned for a long time. How is this most likely to affect the release plan?

    A. The team must reduce the size or number of deliverables that they committed to in the release plan
    B. The release plan will not be affected
    C. The team can increase the size or number of deliverables that they committed to in the release plan
    D. The team must change the frequency of releases in the release plan

34. The team is planning the next iteration. They just finished reviewing the overall list of features that will eventually be delivered. What is the next thing that they should do?

    A. Have each team member answer questions about work completed, future work, and known impediments
    B. Define a release plan that includes the correct level of detail
    C. Extract individual requirements to focus on for the next increment
    D. Establish communication with the appropriate stakeholders

35. A team member is working on an important deliverable. At the retrospective, she says it is less complex than expected. Which of the following is not true?

    A. The release plan should be adjusted to reflect changes to expectations about the deliverable
    B. The team should expect more progress to be made on the deliverable in the next iteration
    C. The velocity should increase in the next iteration
    D. The effort required to create the deliverable should be less than the team originally expected

36. You are on a team using Kanban. What of the following is best used as the main indicator of project progress for specific increments?

    A. Value stream map
    B. Task board
    C. Kanban board
    D. Cumulative flow diagram

37. A junior team member suggests a new way of estimating user stories. How should you respond?

    A. Try the new technique at the next opportunity
    B. Help coach the junior team member by explaining how the team currently estimates user stories
    C. Use Kaizen to improve the process
    D. Encourage the team member to respond to change rather than following a plan based on estimates

practice pmi-acp exam

38. Your Scrum team just completed inspecting the project plan. One team member raised a potential issue that is likely to require a change to the planned work. What is the next step?

   A. The team will hold a sprint retrospective and discuss the impact to the product and sprint backlogs
   B. The product owner will alert the stakeholder that the team has discovered an important issue
   C. Knowledgeable team members will meet to determine what changes need to be made to the sprint backlog
   D. The scrum master raises a change request to modify the plan while the team proceeds with planned work so they meet their commitments

39. Your team has completed a brainstorming session to identify risks, issues, and other potential problems and threats to the project. Which of the following is not a useful next step?

   A. Assign a relative priority to each of the issues, risks, and problems
   B. Assign owners to each of the problems and risks and keep track of the status
   C. Use Kano analysis to prioritize the requirements for the project
   D. Encourage action on specific issues that were raised

40. Your project is changing frequently, and you are concerned that you are not delivering business value as effectively as possible. How do you make sure that your team is delivering value and increasing that value throughout the project?

   A. Use information radiators
   B. Meet with executives after each increment
   C. Meet with executives every day
   D. Brainstorm improvement ideas with the team

41. A member of your project team suddenly leaves the company. She was the only one on the team with the expertise to solve a major technical problem, and without her the team has no way to meet commitments promised to the stakeholders. How should you proceed?

   A. The team should continue working on the project as usual and only alert the stakeholders at the last responsible moment
   B. The team members should collaborate together to get past the obstacle
   C. The product owner should work with stakeholders to reset their expectations because the issue cannot be resolved
   D. The scrum master should educate the stakeholders on the rules of Scrum

42. Your team discovers that the velocity decreased by 20% three iterations ago, and that it has stayed steady at that lower level since then. How is this most likely to affect the release plan?

    A. The team must reduce the size or number of deliverables that they committed to in the release plan
    B. The release plan will not be affected
    C. The team can increase the size or number of deliverables that they committed to in the release plan
    D. The team must change the frequency of releases in the release plan

43. Two stakeholders disagree on important product requirements. How should the product owner handle this situation?

    A. Schedule a meeting with the stakeholders and attempt to establish a working agreement
    B. Practice servant leadership to encourage collaboration
    C. Have two team members perform separate spike solutions for each requirement
    D. Choose the stakeholder requirement that delivers the most value early

44. A product owner reports that an important stakeholder is concerned that the team's implementation of a business requirement may not take certain external factors into account. Several team members acknowledged that this is a potential issue, but agree that it is extremely unlikely. How should the team members handle this situation?

    A. Reassure the product owner that the risk is very low, so no action needs to be taken
    B. Add the issue to an information radiator that tracks the status and ownership of threats and issues
    C. The team members should work together to resolve the problem on their own so that the product owner has plausible deniability
    D. Calculate the net present value of the issue and use it to reprioritize the risk register

45. You are an agile practitioner on a team that uses 30-day iterations. A stakeholder requests a forecast of the next six months. What do you do?

    A. Use a story map to build a release plan for the next six months from the current product backlog
    B. Explain that the team uses 30-day timeboxed iterations, and cannot forecast so far in advance
    C. Create a Gantt chart that has a high level of detail for the next sprint and milestones for the following ones
    D. Hold a team meeting and use face-to-face communication to collaborate on a strategy

46. You are an agile practitioner who maintains a prioritized list of requirements for the team. You receive a company-wide memo that your division has been restructured, and that there are now new senior managers. One of those managers will be directly impacted by one of the project's requirements. What do you do?

   A. Engage with the senior manager
   B. Raise the issue at the next daily standup meeting
   C. Update the product backlog to reflect the new priorities
   D. Add the senior manager to the stakeholder register

47. During a sprint planning meeting the team is working for stories written on index cards. They discuss the acceptance criteria for a story and write it on the back of the card. This is repeated for every story in the increment. Which of the following describes this activity?

   A. Collaborating to deliver maximum value
   B. Refining requirements based on relative value
   C. Gaining consensus on a definition of done
   D. Refining the backlog

48. A senior manager is forming a committee to decide on a company-wide methodology. You are asked to speak to this committee. What should you do?

   A. Put up an information radiator on the floor where the committee meets
   B. Show how previous Scrum projects made the organization more effective and efficient
   C. Explain that waterfall methodologies are bad, and agile methodologies are good
   D. Insist that the organization follow Scrum because it's an industry best practice

49. An agile practitioner is attending a daily scrum meeting. She is expected to report on the status of tasks that were assigned to her by the scrum master. What best describes this situation?

   A. This team is not self-organizing
   B. This team is self-organizing
   C. The agile practitioner is emerging as a team leader
   D. The scrum master is showing servant leadership

exam questions

50. During a review of the deliverables, several team members identified a problem with software quality that could increase overall project cost. What is the best way to figure out the next steps for the team to take?

    A.   Refactor the code

    B.   Hold a retrospective

    C.   Use an Ishikawa diagram

    D.   Limit work in progress

51. A Scrum team is planning their fourth sprint. Team members who disagreed in the previous sprints are starting to see eye to eye, and previous clashes have given way to an emerging spirit of cooperation. What best describes this team?

    A.   The team is in the norming phase, and needs coaching

    B.   The team is in the storming phase, and needs directing

    C.   The team is in the storming phase, and needs coaching

    D.   The team is in the norming phase, and needs supporting

52. An office manager is working with an agile team to optimize their workspace. Which is the most efficient approach?

    A.   Adopt an open office plan that eliminates individual desks and promotes a shared environment

    B.   Give individuals or pairs private offices next to a shared meeting room

    C.   Adopt an open office plan that eliminates partitions and positions team members face to face

    D.   Give individuals or pairs semi-private cubicles that open into a shared meeting space

53. What is the most effective way for an agile team to prioritize the work to do during the increment?

    A.   The team decides which stories are most valuable and gives them a higher priority in the product backlog

    B.   The product owner determines the relative priority at the sprint review

    C.   The team selects the features with the highest NPV

    D.   The business representative collaborates with stakeholders to optimize value

54. You are an agile practitioner on a team that just completed work for an iteration. The team just finished demonstrating the working software to the stakeholders. One of the stakeholders is upset that one of the features he expected the team to deliver was pushed to the next sprint. How can this be avoided in the future?

    A.   Send daily status reports to all stakeholders

    B.   Review the lines of communication and update the communication management plan

    C.   Keep the stakeholders up to date on any changes to the deliverables and trade-offs the team has made

    D.   Require the stakeholders to attend all daily standups

55. A senior manager has announced that she is going to attend a sprint planning session. The Scrum team often has unrestrained discussion about which items should go into the sprint that often include disagreements and sometimes even arguments. The product owner is concerned that even though disagreements typically end with a positive result, being exposed to this will cause that manager to lose trust in the team's ability to meet their commitments. How should this situation be handled?

   A. The senior manager should be discouraged from attending the sprint planning session
   B. Team members should be encouraged to openly disagree, even if it involves arguing with each other
   C. The team should plan to hold a second sprint planning session without the senior manager
   D. The product owner should encourage the team to be on their best behavior for the senior manager

56. An agile practitioner needs to get feedback to determine whether the work currently in progress and the work planned for future iterations needs correction. What is the best way to accomplish this?

   A. Prioritize the backlog based on work item value and risk
   B. Hold daily standup meetings to get feedback from the team
   C. Checkpoint with stakeholders at the end of each iteration
   D. Use a Kanban board to visualize the workflow

57. One of your project's stakeholders indicates a potentially severe risk to the current sprint. Which approach is most appropriate?

   A. Increase the length of the timebox
   B. Re-estimate the backlog for the project
   C. Include fewer stories in the sprint
   D. Include the stakeholders in the daily standup

58. You are a scrum master on a 14-person Scrum team. You notice that the team is having difficulty concentrating during the daily scrum. What should the team do?

   A. Establish a group norm that requires everyone to concentrate during the daily scrum
   B. Hold two separate 7-person daily scrum meetings
   C. Replace the daily scrum with a virtual meeting that uses comments in a social media platform
   D. Divide team members into two smaller teams

exam questions

59. You are informed that a new stakeholder for your project is in a region with a time zone difference of eight hours from the rest of your team. What is the best way for your Scrum team to interact with this person?

   A. Use digital videoconferencing tools and accommodate to the stakeholder's time zone
   B. Primarily communicate with email so that people can work in their comfortable time zones
   C. Request a conference call at a time that's convenient for the whole team to attend
   D. Fly the stakeholder to the team space to co-locate with the team for several weeks

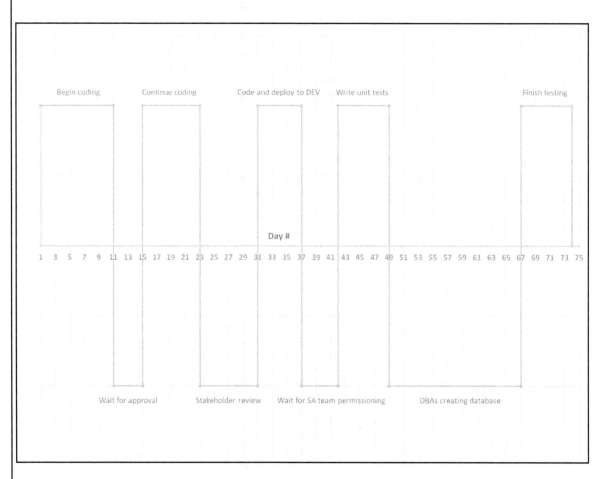

60. How should an agile practitioner on a Scrum team interpret this chart?

   A. The team finished coding after only 50% of the project calendar time had elapsed, which is an opportunity to eliminate waste
   B. The team spent 38 days working and 35 days waiting, so there are many opportunities to eliminate waste
   C. The project took 74 days to complete, without many opportunities to eliminate waste
   D. The project is behind schedule

61. A team member approaches the scrum master for guidance on the best way to forecast how much work the team can accomplish in the next sprint. What should the scrum master advise?

   A. Use planning poker to estimate the actual time in hours for each story in the sprint
   B. Assign a relative numeric size to each story in past sprints and use it to estimate the average velocity
   C. Hold a wideband Delphi estimation session to generate data for a detailed Gantt chart
   D. Use a story map to build a release plan for the rest of the project

62. You are the scrum master on an agile team. Two team members are having a disagreement about an important project issue. Assuming all of these actions resolve the disagreement equally well, what is the best way for you to proceed?

   A. Collaborate with the product owner to find a solution to the problem and present it to the team members
   B. Allow the team members to come to their own resolution, even if it involves an argument with strong opinions
   C. Step in and help the team members find common ground
   D. Create ground rules that prevent disagreements from turning into arguments

63. You are a member of a Scrum team. You discover an important problem that directly affects one of the stakeholders, and the team needs feedback from that stakeholder as quickly as possible in order to avoid a delay. Your team's product owner meets with this stakeholder once a week, but due to a schedule conflict this week's meeting is postponed. What do you do?

   A. Schedule a meeting with the product owner
   B. Have a face-to-face meeting with the stakeholder as quickly as possible
   C. Send an email to the stakeholder with details about the problem
   D. Invite the stakeholder to the next daily standup meeting

64. The team determines that one of the work items is low priority, but could lead to serious problems in the final product if the implementation does not work. How should they handle this situation?

   A. Add a highly visual indicator in the team space so they don't lose track of the issue
   B. Use the internal rate of return to evaluate the work item
   C. Refactor the software to remove the problem
   D. Increase the priority of the work item in the product backlog

exam questions

65. Which of the following is not a benefit of encouraging team members to become generalizing specialists?

   A. Reducing team size
   B. Creating a high performing, cross-functional team
   C. Improve the team's ability to plan
   D. Reducing bottlenecks in the project work

66. You are an agile practitioner on a project that has been running for several iterations. The team's understanding of the effort required to complete several major deliverables has changed over the last three iterations, and it continues to change. How do you handle this situation?

   A. Add story points to each work item in the backlog to reflect the increased complexity
   B. Add buffers to account for the uncertainty of the project
   C. Hold planning activities at the start of each iteration to refine the estimated scope and schedule
   D. Increase the frequency of retrospectives to gather more information

67. Several bugs were discovered by users, and the product owner has determined that they are critical and must be fixed as soon as possible. How should the team respond?

   A. Create a change request and assign the bug fixes to a maintenance team
   B. Stop other project work immediately and fix the bugs
   C. Add items for fixing the bugs to the backlog and include them in the next iteration
   D. Add buffers to the next iteration to account for maintenance work

68. A senior manager asks the team for a project schedule that shows how team members will spend their time. The team meets and creates a highly detailed schedule that shows how each team member will spend every hour of the next six months. What should the scrum master do?

   A. Use servant leadership and recognize that the team members actually get work done
   B. Ask the team to build a less detailed schedule
   C. Post the schedule in a highly visible place in the team space
   D. Send the schedule directly to the senior manager

practice pmi-acp exam

69. A team member complains to the scrum master that their manager calls several meetings every week to give updates that are not relevant to their project. The team member is frustrated because the interruptions are slowing down work. What should the scrum master do?

   A. Give the team member permission to skip the meetings and focus on the work
   B. Raise the issue with the product owner
   C. Prepare a report on the impact that the meetings are having on the work
   D. Approach the manager to discuss alternatives to interrupting the team

70. A member of a Scrum team routinely arrives late to the daily standup meeting. How should the team handle this situation?

   A. Hold a face-to-face meeting between the scrum master and the team member to discuss the issue
   B. Have the product owner bring up the issue with the stakeholders
   C. Hold a meeting to collaborate on establishing team norms that includes a penalty for being late
   D. Raise the issue at the next retrospective and create a plan for improvement

71. An agile practitioner is starting a new project. The team has the first meeting for planning work. What should the practitioner expect this meeting to produce?

   A. A detailed project plan that shows how the team will create working software
   B. A shared understanding of deliverables defined by units of work that the team can produce incrementally
   C. Information radiators that show the progress of the project in a highly visible part of the team space
   D. An informal project plan that describes agreements and face-to-face meetings

72. Which of the following best describes the level of commitment made by agile teams?

   A. Agile teams make commitments to deliver all deliverables at the beginning of the project
   B. Agile teams commit to deliverables for the current iteration, but are not required to make long-term commitments
   C. Agile teams commit only to a minimum viable product at the start of the project
   D. Agile teams commit to broad deliverables early in the project, and make more specific commitments as it unfolds

exam questions

73. A team member reports at a daily standup that she ran into a serious technical problem that will delay the story that she is working on. The product owner has decided to remove the story from the sprint. This story was specifically requested by a senior manager who is the main project stakeholder. What action should be taken next?

    A. Update the information radiators
    B. Re-estimate the items in the backlog
    C. Share the information with the primary stakeholders
    D. Bring up the problem at the retrospective

74. You are a scrum master on a team in the financial services industry. Your PMO director sends an organization-wide email about a regulatory compliance change that will require you to adjust the way requirements are managed. The PMO has provided several possible alternative methods for managing requirements that are in compliance with this regulation change, but none of them match the way that your team currently manages requirements. What should you do?

    A. Do not make any changes that would violate the rules of Scrum
    B. Support organizational change by educating the PMO about Scrum
    C. Review the new techniques at the next sprint planning meeting
    D. Use Kaizen and practice continuous improvement

75. In a sprint review, one of the team members raises a serious issue. He's known about this issue for some time, but this is the first the rest of the team has heard about it. What is the next thing that you should do?

    A. Use an Ishikawa diagram
    B. Speak with the team member about raising issues like this as soon as they are known
    C. Schedule the sprint retrospective
    D. Arrange the team space to encourage osmotic communication

76. A Scrum team is planning their next sprint. How can they best establish a shared vision of what they plan to accomplish during the sprint?

    A. Post information radiators and keep them updated
    B. Set ground rules for the team
    C. Re-estimate the items in backlog
    D. Agree on a sprint goal

77. Your team has delivered fewer items than expected for the third iteration in a row. You suspect that there is a significant amount of time wasted waiting for development, operations, and maintenance work to be completed by other teams. What is the best way to detect where this waste is occurring?

    A. Perform a value stream analysis
    B. Create a more detailed iteration plan
    C. Use an Ishikawa diagram
    D. Impose limits on work in progress

78. What is the most effective strategy for prioritizing stories in the sprint backlog?

    A. Plan an early product release that has just enough features
    B. Prioritize high-risk items first
    C. Collaborate with stakeholders to maximize early delivery of value
    D. Identify high-value features and develop them in early iterations

79. The product owner of a Scrum team discovers that stakeholder priorities have changed, and a deliverable they have not yet started is now more important than the one they are currently working on. How can the team best handle this situation?

    A. Complete the current sprint and adapt the plan during the next sprint planning session
    B. Reduce the number of bottlenecks by limiting work in progress
    C. Cancel the current sprint immediately and create a new plan to reflect the new priorities
    D. Begin refactoring the code to reflect the updated priorities

80. During a daily scrum meeting a team member raises a serious risk as a potential problem. Which of the following is not a useful action for the team to take?

    A. The team should refactor the source code and perform continuous integration
    B. The team should consider doing exploratory work during the next sprint to mitigate the risk
    C. The stakeholders should be kept informed of any potential threat to the team's commitments
    D. The product owner should incorporate activates in the product backlog to manage the risk

81. The team members and product owner are having an argument about whether or not a feature can be accepted. They are unable to agree on an answer, and the project is now in danger of being late. How can this problem be prevented in the future?

    A. Agree on a strict chain of command
    B. Agree on a process for conflict resolution
    C. Agree on a definition of "done" for each work item
    D. Agree on a timebox length for discussions about feature acceptance

82. An XP team is notified that an important server upgrade will be delayed by six months due to budget constraints. The upgrade included several important features that their plan depended on, and the delay will require two team members to spend three entire weekly cycles on a workaround. How should the team account for this?

    A. Track the work done by the two team members separately from the rest of the project work
    B. Update the release plan to reflect the change in the team's capacity to work on main deliverables
    C. Use a risk-based spike to reduce the uncertainty
    D. Expect the velocity to be reduced and update the release plan accordingly

83. Halfway through the sprint, the team discovers a serious problem while testing the code. It's critical that they fix this problem as soon as possible, but it will take more time than they have left in the sprint. What should they do?

    A. The product owner adds items to the sprint and product backlogs
    B. The product owner extends the sprint deadline to accommodate the fix
    C. The product owner should call a team meeting and discuss potential solutions
    D. The product owner adds a high-priority item to the product backlog to fix the problem

84. A stakeholder asks the product owner of a scrum team for a list of features, stories, and other items to be delivered during the sprint. What team activities are used to create this information?

    A. Hold daily scrum meetings
    B. Hold sprint planning meetings
    C. Perform product backlog refinement
    D. Perform sprint retrospectives

practice *pmi-acp* **exam**

85. At a retrospective, several members of a Scrum team raised serious potential risks to the project. How can this best be managed by the team?

   A. Add stories to the next sprint backlog to handle every risk that was raised
   B. Keep an up-to-date information radiator that shows the priority and status of each risk
   C. Handle each risk at the last responsible moment by delaying any action until it becomes a real problem
   D. Create a risk register and add it to the project management information system

86. A team is estimating the size of the items in the product backlog using ideal time. What does this mean?

   A. The team determines the actual calendar date that each item will be delivered
   B. The team estimates the actual time required to build each item without taking velocity or interruptions into account
   C. The team assigns a relative size to each item using units specific to the team
   D. The team applies a formula to determine the size of each item based on its complexity

87. Two members of your XP team are arguing about which engineering approach will lead to a better solution. They are unable to reach a conclusion, and the conflict is starting to create a negative environment. How should you handle this situation?

   A. Use fist-of-five voting to determine the correct approach
   B. Encourage them to begin pair programming on a minimal first step that will support both approaches
   C. Refactor the code and practice continuous integration
   D. Set team ground rules that prohibit arguments between team members

88. You discover that you have made a serious mistake when refactoring code, and it's going to cause your team to miss an important deadline. Which of the following is not an acceptable response?

   A. Keep working on the highest priority tasks and bring up the issue in the retrospective
   B. Tell your teammates and make every attempt to correct the problem
   C. Bring up the problem in the next Daily Scrum
   D. Send an email to the rest of your team letting them know that there are going to be consequences for the timebox

89. You are a team lead on an XP team holding a retrospective meeting. One of your team members says that the team could have done a better job planning the work if they had tried a different planning technique, and that the project would benefit from using it next time. What is an appropriate response?

   A. Determine whether the technique is compliant with the practices and principles of XP
   B. Use Kaizen to improve the process
   C. Determine the impact of using the new technique
   D. Suggest that the team member take the lead in working with the team to try out the new approach

90. Which of the following is not an effective way to encourage an effective environment for your team?

   A. Let team members experiment and make mistakes without negative consequences
   B. Help team members trust each other when they talk about their own mistakes
   C. Use mistakes as opportunities for improvement
   D. Allow team members' mistakes to go uncorrected

91. You are the project manager for a team at a vendor of software services. One of your clients sent you a value stream map that indicates that the team working on a feature spent significant non-working time waiting for the legal departments of both companies to reach agreements on scope changes. How does this affect the project?

   A. The non-working time is extra work in progress that might be able to be limited
   B. The clients and vendor have a contract negotiation relationship
   C. The non-working time is project waste that might be able to be eliminated
   D. No meaningful conclusion can be drawn

92. member says that he did not make much progress because he was interrupted by five phone calls throughout the day from stakeholders. This is the third time that he has made this complaint. What is the best way for the team to handle this situation?

   A. Use a "caves and commons" office layout to limit interruptions
   B. Implement a policy barring the stakeholders from reaching out to team members directly
   C. Establish a daily "no-call" window during which team members working on project tasks can turn the ringers on their phones down and ignore calls
   D. Adjust the sprint backlog to account for the decrease in productivity

practice pmi-acp exam

93. During a sprint review, the project sponsor feels that a feature was built incorrectly and gets angry at the team member who coded it, but gives no constructive feedback. How should the scrum master respond?

  A. Work with the product owner to update the product backlog
  B. Speak with the sponsor about encouraging a safe environment
  C. Make sure the sponsor does not know which team member coded each feature in the future
  D. Getting angry at the team member is a mistake, and the sponsor should be free to make mistakes

94. Your team is in the early stages of planning, and work has not yet begun. Several key stakeholders have been identified and engaged, but there is still uncertainty about particular kinds of users, what they need from the project, and how to best meet those needs. What is the best way to handle this situation?

  A. Create a kanban board to visualize the workflow
  B. Use agile modeling to envision a high-level architecture
  C. Hold a brainstorming session to create personas
  D. Create user stories to document and manage requirements

95. The team has identified a severe database problem that can only be addressed by the infrastructure administrators performing a database server upgrade. How should the team handle this situation?

  A. A developer with database experience is made responsible for upgrading the server
  B. The team should identify the issue in the daily standup meeting
  C. The product owner refines the backlog and adds a high-priority work item for the upgrade
  D. The team must perform the database server upgrade themselves

96. Two senior managers sponsoring your project have expressed disappointment with the current release plan. Which of the following is not an effective strategy for engaging them?

  A. Have the product owner engage the senior managers to better understand their needs
  B. Invite both senior managers to periodic working software demonstrations
  C. Invite both senior managers to a project planning meeting and require their sign-off on a project plan before proceeding
  D. Call a team meeting to discuss the senior managers' interests and expectations

exam questions

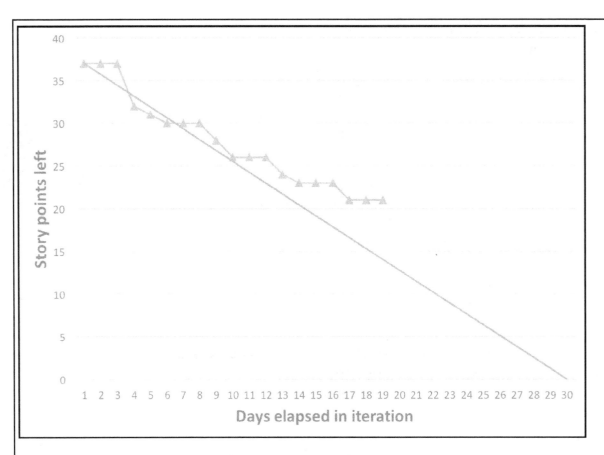

97. How should an agile practitioner on a Scrum team interpret this chart?

    A. The velocity is constant
    B. The sprint goal is in jeopardy
    C. The team did a poor job planning
    D. The velocity is increasing

98. Members of your team are arguing about whether they are required to make a change that the product owner asked for. As an agile practitioner, what is your response?

    A. Review the change control procedure
    B. Work with each team member so that everyone understands the way that your team responds to change
    C. Allow the team to come up with ground rules to manage this situation
    D. Re-estimate the items in the backlog and work with team to self-organize and meet the new goals

99. Partway through a sprint, the product owner gets an email from the DevOps group responsible for deploying the software built by the team with a reminder about a new policy that requires a modification to the installation scripts that must be included in all future deployments. Deployment is required for the sprint review. Modifying the script will delay other work past the end of the sprint. What should the product owner do next?

   A. Add the script modification to the sprint backlog and move the lowest priority item to the product backlog
   B. Add the script modification to the product backlog
   C. Extend the sprint deadline to include the script modification
   D. Schedule a face-to-face meeting with the manager of the DevOps group

100. Your team needs to determine what stories to work on in the next iteration. Which is not an effective way to proceed?

   A. The team starts the iteration by working on riskiest or most valuable stories
   B. The scrum master helps the team understand the methodology they use to decompose stories and identify tasks
   C. The product owner helps everyone understand the relative priority of each story
   D. The scrum master helps lead the team through planning by deciding the order that the team works on the stories

101. Which primary XP practice facilitates osmotic communication?

   A. Whole team
   B. Continuous integration
   C. Sit together
   D. Pair programming

102. An agile team is defining a release plan. What is the best way to organize the requirements so that value is delivered early?

   A. Define minimally marketable features
   B. Re-estimate the items in backlog
   C. Post a visible burn-down chart in the team space
   D. Use an Ishikawa diagram

exam *questions*

103. Team members are concerned that a technical problem will cause a serious issue later in the project. One team member points out that if the problem occurs, then they will need to find a different technical approach. What should the team do next?

    A. Have the product owner add an item to the list of long-term features and deliverables
    B. Update the release plan to reflect a delay
    C. Perform exploratory work in an early sprint in order to determine whether their solution will work
    D. Alert the stakeholders to the impact the technical problem will have on the team's commitments

104. You are a product owner on a Scrum team. One of your stakeholders is a senior manager who just joined the company. She did not come to the last two sprint reviews. What should you do?

    A. Collaborate with the scrum master to help educate the stakeholder on the rules of Scrum
    B. Meet with the stakeholder's manager and explain the rules of Scrum require her to attend the sprint review
    C. Set up a meeting with the stakeholder to bring her up to date and get her feedback on the project
    D. Post an information radiator about the project outside the stakeholder's office

105. You are meeting with several stakeholders partway through in iteration late in the project. One of them mentions that a senior manager you have not met would disagree with one of the features that the team has built into the software. What is the next thing you should do?

    A. Reprioritize the backlog to reflect the potential risk of requirements change
    B. Identify the potential issue at the next daily standup
    C. Schedule a meeting with the senior manager
    D. Add the issue to the risk register

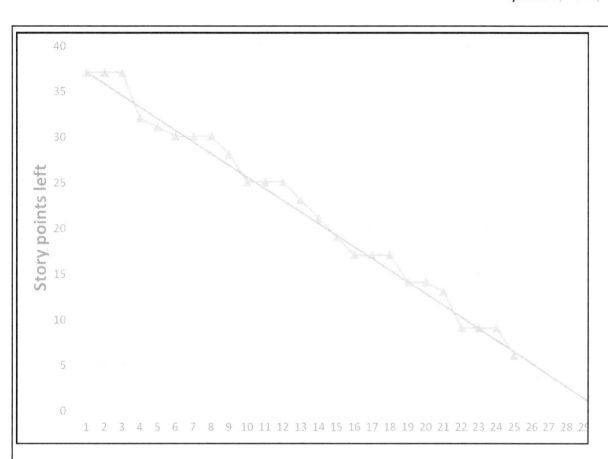

106. How should an agile practitioner on a Scrum team interpret this chart?

- A. The velocity is increasing
- B. The velocity is constant
- C. The team did a poor job planning
- D. The sprint goal is in jeopardy

107. After a retrospective, one of the other team members tells you in confidence that he is concerned the team is making poor design and architecture decisions. How should you respond?

- A. Offer to raise the issue yourself so he doesn't have to feel like he's causing problems
- B. Tell the product owner and scrum master in confidence
- C. Encourage the team member to bring this up with the whole team
- D. Promise that you will not break his confidence so that you don't threaten team cohesion

exam questions

108. The team has just completed planning activities for an iteration. What should they do next?

   A. Review the project management plan
   B. Hold a daily standup meeting
   C. Determine the length of the timebox
   D. Update stakeholders about the expected deliverables

109. An agile practitioner on a hybrid Scrum/XP team discovers that several team members spend many hours each week resolving commit conflicts, and feels that improving the way they perform continuous integration will fix the issue. Which of the following is not a useful next step?

   A. Create and distribute a detailed process document that covers continuous integration best practices
   B. Engage the team throughout the project to help them learn better continuous integration techniques
   C. Educate the rest of the team on how out-of-date working folders can lead to commit conflicts
   D. Help the team improve their overall continuous integration process

110. You are a member of an XP team planning the next weekly cycle. There is a database design task that needs to be done. One of your team members is an expert in database design, and says that he is the only team member who should be allowed to do that. What is the best way to proceed?

   A. Encourage the application of individual expertise in order to increase productivity
   B. Encourage the team to use pair programming
   C. Encourage the expert to serve as a mentor to a junior member of the team
   D. Raise the issue at the next daily standup meeting

111. An agile practitioner encounters a stakeholder who insists that the team create a complete, highly detailed plan before any work begins. The agile practitioner should:

   A. Correct the stakeholder, because agile teams only use working software and not comprehensive documentation
   B. Show that the team has had success in the past with periodic product demonstrations and changing course mid-stream
   C. Review the product backlog with the stakeholder and identify the stories that are likely to go into each release
   D. Create the complete, highly detailed plan to satisfy the stakeholder

112. An agile team is starting a new project. What is the best way to provide a starting point for managing the project?

   A. Create a release plan that includes buffers to account for maintenance
   B. Create a release plan that reflects a high-level understanding of the effort
   C. Create a Gantt chart that has a high level of detail
   D. Create a story map based on highly detailed effort estimates

113. An agile team is conducting a periodic review of their practices and team culture. What is the purpose of this review?

    A. To adhere to a methodology that requires periodic retrospectives
    B. To review and update the list of features, stories, and tasks that comprise the team's long-term work
    C. To improve their project process in order to increase the effectiveness of the team
    D. To identify the root cause of a specific problem

114. An agile practitioner discovers that an important project stakeholder feels that she cannot trust the team to meet their commitments. What is the best way to improve this situation?

    A. Work with the team to improve how they communicate success criteria and collaborate with stakeholders on product trade-offs
    B. Meet with the stakeholder and make a strong commitment to delivering specific features
    C. Work with the team to set up a binding service-level agreement with the stakeholder on acceptance criteria for each increment
    D. Meet with the stakeholder to explain that the rules of Scrum require her to trust the team

115. You are an agile practitioner and you just finished meeting with a stakeholder who identified several important priority changes. You have assigned a relative value to each item in the list of planned features, but the team is not yet able to prioritize them. What is the next action that the team should take?

    A. Re-estimate the items in the backlog
    B. Perform an architectural spike
    C. Update the information radiators
    D. Initiate the change control procedure

116. Which of the following is not a benefit of co-location?

    A. Osmotic communication
    B. Ability to create an informative workspace
    C. Increased access to teammates
    D. Reduced distractions

17. What is the best way to ensure that the work products being delivered have the maximum value?

    A. The scrum master collaborates with the stakeholders
    B. The team collaborates with the product owner
    C. The product owner collaborates with stakeholders
    D. The project manager collaborates with senior managers

exam *questions*

118. The team has identified a problem that caused a delay in the previous iteration. They now want to understand exactly what went wrong, and all of the factors that led up to the problem, so that they can improve their overall method for running their projects. What is an appropriate tool for this?

    A. Ishikawa diagram
    B. Spike solution
    C. Information radiator
    D. Burn-down chart

119. You are an agile practitioner on a team in a company that builds medical devices. Quality, specifically with regards to patient safety, is the most important factor in your project's success. What is the most effective way to ensure product quality?

    A. Use root cause analysis to identify the source of problems
    B. Include quality items in the iteration backlog
    C. Maximize value by periodically meeting with stakeholders
    D. Inspect, review, and test work products frequently and incorporate identified improvements

120. Company-wide budget cuts require your team to reduce the schedule by three months. The product owner indicates that stakeholders will be angry about the lack of delivery. What should you do next?

    A. Follow your methodology's rules to inspect the plan and adapt it to reflect the change in budget and schedule
    B. Present an alternative to senior management to make a case for increasing the budget
    C. Alert the product owner to scope and schedule changes just before they happen, so you can make decisions at the last responsible moment
    D. Find a way to keep the project going without alerting the product owner to serious problems

practice *pmi-acp* exam

Before you look at the answers...

Before you find out how you did on the exam, here are a few ideas to help make the material stick to your brain. Remember, once you look through the answers, you can use these tips to help you review anything you missed.

> This is especially useful for conflict resolution questions—the ones where you're presented with a disagreement between team members, and asked how you'd handle it.

❶ Don't get caught up in the question.

If you find yourself a little confused about a question, the first thing you should do is try to figure out exactly what it is the question is asking. It's easy to get bogged down in the details, especially if the question is really wordy. Sometimes you need to read a question more than once. The first time you read it, ask yourself, "What's this question *really* about?"

❷ Try this stuff out on your job.

Everything you're learning about for the PMI-ACP® exam is really practical, and **based on real-world agile ideas**. If you're actively working on projects, then there's a really good chance that some of the ideas you're learning about can be applied to your job. Take a few minutes and think about how you'd use these things to make your projects go more smoothly.

> When you write your own question, you do a few things:
> - You reinforce the idea and make it stick to your brain.
> - You think about how questions are structured.
> - By thinking of a real-world scenario where the concept is used, you put the idea in context and learn how to apply it.
>
> And all that helps you recall it better!

❸ Write your own questions.

Is there a concept that you're just not getting? One of the best ways that you can make it stick to your brain is to write your own question about it! We included Question Clinic exercises in *Head First Agile* to help you learn how to write questions like the ones you'll find on the exam.

❹ Get some help!

If you're not a member of PMI yet, join today! There are **local PMI chapters** all around the world. They're a great way to connect with the PMI community. Most chapters feature speakers and study groups that help you learn.

Looking for a great way to meet the PMI-ACP® exam's training requirements? Check out Safari Live Online Training from O'Reilly! Courses on agile pop up there all the time. It's included with your Safari membership:

http://www.safaribooksonline.com/live-training/

exam *answers*

1. Answer: B

This is a case where the product owner is right, and the team member is doing something that is potentially dangerous. The reason that Scrum teams have a product owner role is so that someone can stay on top of all stakeholder communications. There's nothing wrong with team members working directly with stakeholders, but they should never cut the product owner out of the discussion.

> Did it bother you that "scrum master" was not capitalized in the exam question? Get used to it! Questions on the actual exam might not have capitaliztion that matches your expectations.

2. Answer: C

Usability testing is an important way that teams can test their software to make sure it is easy to use, and agile teams conduct frequent reviews by testing the software and incorporating the improvements back into the deliverables. A very common way to perform usability testing is to observe users while they interact with early versions of the software.

> Capturing user interface requirements and using wireframes to plan the user interface are both valuable ways to improve the usability of the software, and agile teams use both of them. But agile teams also value working software over comprehensive documentation, so they typically opt for usability testing over UI requirements and wireframes.

3. Answer: B

When problems occur, agile teams work closely with their stakeholders to understand acceptable trade-offs. On a Scrum team, the product owner is responsible for interacting with the stakeholders to help them understand how the project is going. So when a problem happens on a Scrum project that will impact what the team delivers, the product owner needs to meet with the stakeholder and discuss exactly how the team will proceed. Agile teams work with their stakeholders to maintain a shared understanding of important trade-offs that affect delivery, which helps build a mutual trust between them.

> A spike solution doesn't make sense here, because the question didn't mention anything about exploring a potential technical solution.

> The stakeholders need to be involved because the team will need to change their behavior when the WIP limit for the step is reached—and that often affects the stakeholders. This helps everyone get to the root cause of the flow problem more quickly.

4. Answer: D

When teams use a kanban board to visualize their workflow, they use columns to represent workflow steps, and typically use sticky notes or index cards to show individual work items flowing through the process. If items tend to accumulate in one column, it tells the team that step is a potential root cause for the process flow slowing down. The way to fix it is to work with the stakeholders to impose a work in progress (WIP) limit, usually by writing the maximum allowable number of work items for that step.

Did you notice that the practice exam had a LOT of which-is-BEST questions and least-worst-option questions? One of the most difficult aspects of the PMI-ACP® exam is choosing the best answer when several might be correct, or when none seem to be.

5. Answer: D

> This is especially true of Scrum teams. Because they're self-organizing, they can make decisions about who can do the work at the last responsible moment.

Generalizing specialists can come in really handy, and agile teams do everything that they can to help encourage people to broaden their skills. When everyone on the team has a broader skill set, it lets the team do more work with fewer people, and helps them to avoid bottlenecks. Agile teams try to provide as many opportunities as possible for their team members to develop generalized skills. So when there's an opportunity for a team member to expand their skills—like a tester taking on development work—agile teams take advantage of it.

6. Answer: A

The main reason for teams to set ground rules is so that they can foster coherence, and continue to increase their collective commitment to the project's goals and to delivering value to the stakeholders. Teams should always have good, sensible, sound reasons for setting ground rules. So the best way to help the new team member fit in with the new team is to explain those reasons, and encourage him or her to try following the new rule.

> If there isn't a good, sensible reason for the rule, then the new team member might be right and the rule might not be a good idea. But that person should still try it out first, because keeping an open mind about the team's culture is the best way to encourage team coherence.

7. Answer: C

Leaders on agile teams practice servant leadership. This means making sure that the individual team members get credit for their work, feel appreciated, and get work done. Servant leaders spend a lot of time working behind the scenes to remove roadblocks that will cause problems down the road. Servant leaders do not typically assign work or decide how the team should build their products.

8. Answer: D

One of the most important aspects of how agile teams manage their requirements is that they gain consensus among the whole team on the definition of "done" for each item in an iteration. Everyone on the team needs to agree on clear and specific acceptance criteria for every feature that they are going to deliver at the end of the iteration. An effective way to reach consensus on acceptance criteria is to use negotiation.

> A common way for teams to negotiate this is to have "give and take" where the current iteration's definition of "done" includes some of the work, but agrees to include the rest of the work in a future iteration.

9. Answer: D

An agile practitioner working directly with multiple stakeholders is in the product owner role. A product owner meets periodically with stakeholders to identify expectations and requirements, and works with the team to help them understand those requirements. In this case, a stakeholder has a requirement, so the product owner's job is to make sure that the team is knowledgeable about that stakeholder's needs and expectations.

10. Answer: D

Once teams have been together long enough to get into the work they sometimes enter a phase—referred to as "storming"—in which team members often develop strong negative opinions about each other's character. Adaptive leadership, where leaders modify their style based on the stage of group development, tells us that teams in the "storming" phase need supportive leadership, which involves high levels of direction and high levels of support.

> *This question is based on Tuckman's model of group development and Hershey's situational leadership model, theories developed in the 1960s and 1970s about how teams form and how leaders should adapt to them. But it's more important to understand the ideas of what happens to teams when they form and how effective leaders should adapt to them than to remember the names Tuckman or Hershey.*

11. Answer: A

Self-organizing teams plan the work together and make decisions about who does specific tasks at the last responsible moment. During sprint planning meetings Scrum teams typically break down the stories, features, or requirements from the sprint backlog into individual tasks and work items. But because they are self-organizing, instead of assigning those tasks to team members at the beginning of the sprint, most Scrum teams rely on the individuals to assign the tasks to themselves during the Daily Scrum.

> *Assigning work at the last responsible moment doesn't necessarily mean that tasks are only self-assigned during the Daily Scrum. If there's a really important reason to assign a task to a team member during sprint planning, it wouldn't be responsible to delay that assignment until the first Daily Scrum.*

12. Answer: C

The primary focus of an agile team is to deliver value early, and the way the team does that is by collaborating with the stakeholder and prioritizing the highest value work. However, in this question the agile practitioner is the stakeholder, not the team member—the agile team doing the work is at the vendor, and the agile practitioner will work with that team's product owner. So in this case, the product owner on the vendor's team must collaborate with the agile practitioner.

> *When you see a question asking about working with a vendor, part of your job is to figure out if the practitioner is the stakeholder and the vendor's team members are filling the product owner, scrum master, and team roles.*

13. Answer: A

One reason that agile teams are able to improve over time is that they pay attention not just to their individual projects, but to the entire system that they're working within. One way that they do that is by disseminating knowledge and practices—not just across their own organization, but across organizational boundaries.

practice *pmi-acp* exam

14. Answer: B

> You should always favor showing examples of agile principles from successful projects over simply explaining them.

Agile practitioners must always advocate for agile principles, and one of the core principles of agile is that agile teams value responding to change over following a plan. Explaining the value to the scrum master is a good idea, but the best way to advocate for agile principles is by modeling those principles.

15. Answer: D

Every individual person on an agile team is encouraged to show leadership. To do this, agile teams foster an environment where it's safe to make mistakes, and where everyone is treated with respect. However, the company does not typically set ground rules for the team.

> However, following the company's rules for project management isn't usually a particularly effective way to bring a team together.

> **THIS IS AN *ESPECIALLY TOUGH* QUESTION. NONE OF THE ANSWERS SEEM LIKE THEY'RE PARTICULARLY GOOD EXAMPLES OF SOMETHING NOT TO DO IF YOU WANT TO FOSTER AN EFFECTIVE TEAM ENVIRONMENT.**

> Yes, you really will get a lot of tough, ambiguous, least-worst-option questions like this on the real PMI-ACP® exam!

> **THE *LEAST WORST OPTION* IS BEING CAREFUL ABOUT FOLLOWING THE COMPANY'S GROUND RULES FOR PROJECT MANAGEMENT. THAT CAN (BUT DOESN'T ALWAYS) HELP YOUR PROJECT RUN MORE SMOOTHLY, BUT IT'S *NOT REALLY A GREAT WAY* TO HELP MAKE YOUR TEAM MORE COHESIVE OR EFFECTIVE.**

16. Answer: D

Kanban teams typically use cumulative flow diagrams to visualize the flow of work through the process. This allows them to get a visual sense of the average arrival rate (how frequently work items are added), lead time (the amount of time between when a work item is requested and when it's delivered), and work in progress (the number of work items in the process at any time).

17. Answer: D

Security and performance requirements (like the use of encryption or how quickly the software runs) are good examples of non-functional requirements. Agile teams elicit non-functional requirements that are relevant to their project by considering the environment that the code will run in, and they work with stakeholders to understand and prioritize those requirements.

When you're presented with several potential technical approaches, a spike solution is a good way to determine which one will work. However, in this case the team already knows that both solutions are feasible and what the results of each approach will be, so they wouldn't actually learn anything from a spike solution.

18. Answer: C

Agile teams conduct frequent retrospectives so that they can improve the way they do their work. On a Scrum team, everyone—including the scrum master, who participates as a peer, just like the other team members—participates in the retrospective by identifying improvements and working on a plan to implement those improvements. The scrum master also has an additional job, to help teach the rest of the team the rules of Scrum, including how to fill their roles in the meeting and maintain the meeting timebox.

19. Answer: B

The scrum master is a servant leader whose job it is to help ensure that everyone has a common knowledge of the practices used by the team. When a servant leader is approached by a team member with a question or misunderstanding about a practice, he or she helps that person understand how the practice works and why it helps the team achieve the project's goals.

20. Answer: C

When an agile team works with a vendor, that vendor will typically use a methodology that is different from the one used by the agile team. In this case, the vendor is using a waterfall methodology—and that's OK. What's important here is that agile teams establish a shared vision of each project increment. In this question, the roles are flipped around, so you are the stakeholder, but aligning your expectations with the team doing the work and building trust with that team is still critically important to the project's success. So if the vendor uses scope and objectives documents to do that, then your job is to make sure that your team's view of the high-level vision and supporting objectives matches the view of the vendor team, and take action to fix any disagreements.

Agile teams might value working software over comprehensive documentation, but they still value documentation. Don't assume that an answer is wrong just because it involves working with documentation.

21. Answer: D

People and teams always have their own professional and personal goals on every project. One reason that agile teams are so effective is because they take this into account by making sure that the team goals and the project goals are aligned. Scrum teams, for example, write down a simply stated, straightforward goal for every sprint. When the team has their own specific goal, they should collaborate to find common ground so that they accomplish the sprint goal while still making progress towards their team goal.

There are often disagreements between team members—in this case, between the product owner and the rest of the team. Collaboration will almost always work better than negotiation in a situation like this.

22. Answer: D

When teams run into serious problems, one of the first things that they should do is make sure that everyone—especially the stakeholders—understands the impact of the problem. And when that problem is going to cause serious delays, they need to reset everyone's expectations in order to make sure they still deliver as much value as possible.

23. Answer: D

When an agile team—and especially a Scrum team—completes a timeboxed iteration, the next step is to get feedback on the work that they completed by holding a demonstration for the stakeholders. However, agile teams only demonstrate work that is fully completed. If the work has not been completed, the team will usually include it as the first thing to be done when planning the next iteration.

24. Answer: A

When your team's stakeholders' expectations are in line with the working software that your team delivers, it builds trust. That trust grows over time as each stakeholder sees that the working software increasingly incorporates his or her requirements, and that the team is able to adjust as those requirements change. The product owner plays a very important part in this on a Scrum team by making sure that each stakeholders' expectations about what the team will deliver is always in line with the work they are doing.

> When the stakeholder and team agree on the definition of "done" for the increment, it prevents nasty surprises at the sprint review. A really effective way to do that is for the product owner and stakeholder to review the acceptance criteria for each story.

25. Answer: B

Part of your job as an agile practitioner is to always keep an eye out for ways to support change at the organization level. One of your goals is educating and influencing people in the broader organization, and the best way to do this is to speak about your own team's success.

> When you're trying to influence others, it's much more effective to talk about your own team's success, rather than simply explaining how agile works or acting like a pushy agile zealot.

This is an especially tough question. Did you choose the incorrect answer about engaging the product owner? Understanding why that answer is wrong requires you to be really familiar with what a product owner does—and doesn't do—on a Scrum team. The product owner role is entirely focused on the project and the project's specific stakeholders. This question asked about the company as a whole, not about the specific project. So this is really a question about the agile practitioner's responsibility to support change at the organizational level by educating others in his or her company.

---

26. Answer: A

Information radiators are an effective tool that agile teams use to create an informative workspace. An information radiator is a highly visual display (like a chart posted in a central location in the team space) that shows real progress and team performance.

27. Answer: C

Agile teams are encouraged to do experimentation in order to surface problems and impediments to the team, and exploratory work (like spike solutions) is a really good way to do that. The results of that work should be surfaced to the team when they are problems or impediments that might slow the team down or impact the team's ability to deliver value to the stakeholders.

> **This is an especially tough question**, because it requires you to understand a very specific task in one of the domains in the examination content outline, specifically task #1 in domain VI (Problem Detection and Resolution): "Create an open and safe environment by encouraging conversation and experimentation, in order to surface problems and impediments that are slowing the team down or preventing its ability to deliver value." This question is worded in a way that references specific parts of that task (slow down progress, prevent its ability to deliver value). This problem domain accounts for 10% of the scored questions on the test, and there are only five tasks in that domain, so you may potentially see two scored questions based on this task.

28. Answer: B

Agile teams select and tailor their process based not only on agile practices and values, but also on the characteristics of the organization. This team is having engineering problems, which is one hint that XP is the right solution for them. They might want to switch to Scrum, but assigning a team member to the product owner role is not an effective way to do that because the product owner will not have the authority to accept items on behalf of the team.

Kaizen and continuous improvement are generally a good approach for improving a team, but that answer is not very specific. It is better to go with the answer that offers specific improvements that the team can make.

This is an especially tough question. Did you choose the incorrect answer about assigning team members to the product owner and scrum master role? This sounds like a good idea! The problem is that it's almost never a good idea to simply choose an existing team member to be the product owner, because product owners must have enough authority to adequately make decisions and accept features as done on behalf of the company, and it's extremely unlikely that someone like that is already on the team. Instead of simply assigning a team member to the product owner role, teams must work with their users, stakeholders, and senior managers to find a product owner with that level of authority. Since that answer is incorrect, the next best answer is to adopt XP's delivery-focused practices—especially quarterly and weekly cycles—because that is an effective way to solve the team's problem.

29. Answer: B

When a stakeholder needs a change to an item the team is currently working on, the product owner has the authority to immediately make that change. The most important thing is that the team is working on maximizing the value, so the item should be removed from the sprint backlog and the sprint should continue as usual: work on the other items continues, and the team holds a sprint review with the stakeholder when the timebox expires.

↖ *Technically, the product owner has the authority to cancel a sprint, but it should be done in very rare occasions because it can seriously damage the trust that the team has built up with the stakeholders.*

30. Answer: B

Teams that have been using agile methodologies effectively for a long time tend to be really good at estimating, and there are many different ways to estimate. The important thing is that deciding on estimates, like any other decision made by an agile team, is most effective when it's done collaboratively. Planning poker and wideband Delphi are methods for collaborative estimation that allow several team members to work together to come up with an estimate. Having an informal discussion is also a good way to collaborate. But simply leaving it in the hands of the product owner isn't collaborative at all. And simply taking the maximum estimate generated by a team member is a great way to pad your schedule, but it is definitely not open or transparent, which goes against the Scrum value of openness.

31. Answer: C

When your company has a requirement for all teams, you need to comply with it. That's why agile teams tailor their process based on how the wider organization functions. But they still make sure that they are focused first and foremost on delivering value to the customer.

↑ *Also, agile teams do value comprehensive documentation. They just value working software more.*

32. Answer: A

An agile practitioner should practice visualization of important project information by maintaining information radiators that are highly visible. It's important that they show the team's real progress, and a burn-down chart is a great way to do that.

33. Answer: B

Teams often experience temporary drops in velocity, especially when multiple team members are on vacation. If those vacations have been planned for a long time, then that information should have been taken into account already in the release plan, so it should not change.

34. Answer: C

This question is describing a Scrum team's sprint planning meeting. During that meeting, the team first reviews the product backlog, which involves reviewing the overall list of features that will be delivered. The next thing the team should do is create the sprint backlog, which involves extracting items from the product backlog to deliver in the increment for the sprint.

exam *answers*

35. Answer: C

The complexity of deliverables plays a major role in how much work it will require to build. When the team member discovers that a deliverable was less complex than anticipated, the team should use that information to adapt the way they plan their project. Since the deliverable will require less work than expected, it means they'll make more progress during each iteration toward completing the deliverable, and they can plan to release the deliverable earlier. But their velocity shouldn't increase in the next iteration, because the team should take the reduced complexity into account when calculating the velocity for that iteration.

> The velocity of the iteration that the team just completed probably increased temporarily because the team member got more work done than she anticipated due to the unexpectedly low complexity of the deliverable. But now that they know it's less complex, they'll adjust their plan, and the velocity should return to normal.

36. Answer: B

Kanban is a method for process improvement, not project management. So while kanban boards, cumulative flow diagrams, and value stream maps are valuable tools for visualizing and understanding the workflow for your process, they aren't tools for tracking project progress. A task board, on the other hand, is a great tool for tracking project progress.

37. Answer: A

Agile teams enhance their creativity by experimenting with new techniques whenever they can. This helps them discover ways of working that can improve efficiency and effectiveness. The only way to determine whether or not this new technique is an improvement is to try it out.

38. Answer: C

> It's important for the product owner to keep the stakeholders up to date. However, the team hasn't even determined whether this is a real issue, so it's premature to alert stakeholders.

In this question, a Scrum team just finished inspecting the project plan, and Scrum teams always do that during their daily scrum meeting. When issues are raised at the daily scrum, team members with knowledge of the issue schedule a follow-up meeting so that they can figure out how to adapt to the change, which almost always involves modifying the sprint backlog.

```
File  Edit  Window  Help  Ace the Test
    This is a tough question. It requires you to have a good understanding not
just of how Scrum teams hold their daily scrum meetings, but also why they do
   it. The rules of Scrum don't explicitly include an artifact called "project
   plan," but teams still do planning, so you need to understand how that works.
   Scrum teams meet every day as part of the process of transparency, inspection,
   and adaptation. The purpose of the daily scrum is to inspect the current plan
   and the work being done. If there are any potential issues, team members with
      knowledge about the issue have a follow-up meeting to figure out whether or
      not they need to adapt the plan. By doing this every day, Scrum teams are
   able to constantly adjust their plan to keep it up to date with changes to the
         schedule, budget, and stakeholder requirements and priorities.
```

practice pmi-acp exam

39. Answer: C

When teams have identified threats and issues, they should maintain a prioritized list that they keep visible and constantly monitor. The reason for this is to encourage the team to take action on the issues (rather than ignore them), and to make sure that each issue has an owner and that the team keeps track of the status of each issue.

40. Answer: B

The first sentence is a red herring.
That's true of every project!

Getting frequent feedback from users and customers is an effective way to confirm that you're delivering business value and enhancing that value. You get that feedback at the sprint review, which is the meeting where you review the increment.

41. Answer: C

Sometimes teams run into issues that simply cannot be resolved. When this happens, the most important thing is to make sure everyone—especially the stakeholders—understands as soon as possible exactly how this will impact the commitments.

42. Answer: A

When velocity drops, it's often temporary. For example, the amount of work the team produces in an iteration might decrease temporarily if a team member is on vacation or if a specific work item turns out to be more difficult or complex than anticipated. But if the velocity drops significantly and stays at that lower level for several iterations, the team needs to adjust their release plan to reflect the fact that they won't get deliverables done as quickly. That way they can maintain commitments to their stakeholders that are realistic, and not overly optimistic because they're based on outdated information.

Agile teams typically schedule releases that align with the end of their iterations, releasing work that's been completed during the iteration. Often, a lower velocity won't require the team to change the frequency of those releases. They'll just deploy fewer deliverables at each release. That way the steady flow of completed deliverables will continue (even if the project takes longer).

43. Answer: A

An important part of stakeholder engagement on an agile team is to help the stakeholders to establish their own relationships so that they can more effectively collaborate. Meeting with them to set up a working agreement for the sake of the project is an effective way to accomplish this.

Servant leadership typically refers to the way someone in a leadership position—often the scrum master—relates to the rest of the team, recognizing that they're the ones actually getting the work done.

44. Answer: B

When teams encounter risks, issues, and threats to the project, an important priority should always be to communicate the status of those issues. An information radiator is a very good tool for doing that.

you are here ▸ 433

exam answers

45. Answer: A

A story map gives your team a way to collaborate with each other and create a visual release plan by organizing stories into releases. This helps your team provide forecasts for future releases to your stakeholders. And it does it at a level of detail that gives them enough information to plan effectively, without including specific details that the team can't possibly know or honestly commit to this early on.

46. Answer: A

Agile teams—and especially Scrum teams—work so well because they maintain a very high level of stakeholder involvement. One way that product owners do that is by constantly looking for changes in the project and the organization, and immediately acting on those changes to see if that change affects the project's stakeholders. In this case, an organizational change created a new project stakeholder, so the product owner needs to engage with that person as soon as possible.

> *This question starts off by describing the product owner role: "an agile practitioner who maintains a prioritized list of requirements for the team"—in other words, the person who maintains the product backlog.*

47. Answer: C

Teams refine the requirements for the software that they build by gaining consensus on the acceptance criteria for each feature or work item, and these acceptance criteria combine to form the definition of "done" for the product increment.

> *A lot of people will have endless arguments disagreeing on how the terms "definition of 'done'" and "acceptance criteria" differ slightly in meaning. Some people believe that definition of "done" applies only to the increment, while acceptance criteria apply only to individual stories or features. But for the exam, you may see the terms used interchangeably, and you will probably not be asked a question that requires you to differentiate between them.*

48. Answer: B

It's part of an agile practitioner's job to support change at the organization level, to educate people in the organization, and to influence behaviors and people in order to make the organization more effective and efficient.

49. Answer: A

When one person assigns work to the team and expects them to report status, that's the opposite of self-organizing, and the Scrum implementation is broken. On a self-organizing team, individual team members are empowered to make decisions together about what tasks they work on next. The daily scrum is where the whole team reviews these decisions

> *In an effective daily scrum, the agile practitioner would tell the rest of the team what task she plans to work on next. If this doesn't seem like an effective approach, another team member will raise that as an issue, and they'll meet together after the daily scrum to work out the details.*

50. Answer: C

Determining the root cause of a quality problem is an important first step to fixing a problem, and an Ishikawa (or fishbone) diagram is an effective tool for doing root cause analysis.

51. Answer: D

After teams have been together for a while, they often enter a phase—referred to as "norming"—in which they start to resolve their differences and personality clashes, and a cooperative quality starts to emerge among the team members. According to adaptive leadership, a management and leadership approach that involves changing the way that leaders work with teams as they move through their stages of formation, the "norming" stage requires supporting, or leadership that features a lot of support but allows the team more freedom to determine their own direction.

> This question is about adaptive leadership, which is based on Tuckman's theory of group development and Hershey's situational leadership model, which were developed in the 1960s and 1970s. It's more important to understand the ideas of what happens to teams when they form and how effective leaders should adapt to them than to remember the names of these management theories.

52. Answer: D

The "caves and commons" office layout, in which developers or pairs have semi-private spaces adjacent to a shared meeting space, is effective because it limits interruption while still allowing for osmotic communication (where team members learn important project information from overheard conversations). Open plans—especially ones where team members sit facing each other—can be very distracting, which makes it difficult to focus. And while closed-door offices do a great job of limiting interruptions (and team members definitely prefer them because they provide both privacy and status), they don't allow for osmotic communication.

53. Answer: D

The product owner is responsible for maximizing the value of the deliverables. The main way that he or she does this is by prioritizing the units of work in the product backlog so that the team delivers the most valuable ones first, and he or she determines that value by collaborating with stakeholders. The team does not determine the value of the work items by themselves—this is only done by the product owner in collaboration with the stakeholders.

> You might see terms like "business representative" or "proxy customer"—they're referring to the Product Owner.

54. Answer: C

No stakeholder likes to be told that a feature he or she is expecting to be done at the end of the current iteration will be delayed until the next one or later. That's why agile teams work especially hard to establish a clear picture of exactly what they will deliver at the end of the iteration—and they work really hard to maintain that shared understanding between the team and the stakeholders. So when the definition of "done" for the increment changes (in other words, when the team discovers a change in what they're planning to deliver at the end of the iteration), they need to let the stakeholders know immediately.

55. Answer: B

Constructive disagreement—and even the occasional argument—is normal and even valuable for teams. That's why agile teams always strive to create an open and safe environment by encouraging conversation, disagreement, and even constructive arguments. The presence of a senior manager should not change this.

exam answers

56. Answer: C

Feedback and corrections to planned work and work in progress is done using periodic checkpoints with stakeholders. Most agile teams accomplish this by holding a review at the end of each iteration.

57. Answer: C

Making the increment size smaller is an effective way to identify risks and respond to them as early as possible in the project. Including fewer stories in each iteration is a good way to limit the increment size.

58. Answer: D

There is an upper limit on the number of people who can be on a Scrum team—it typically can support a maximum of nine people (but some teams make it work with up to twelve). Fourteen is definitely too large for a Scrum team, and an early sign that the team is too large is that people have trouble concentrating during the daily scrum. The best thing for this team to do is to split into two smaller teams.

59. Answer: A

Agile teams always prefer face-to-face communications whenever possible, and digital videoconferencing tools are a great way to facilitate face-to-face communications. The team should always accommodate stakeholders whenever possible, but should not expect stakeholders to necessarily accommodate them (so requiring a stakeholder to fly out and co-locate with the team for several weeks is an unreasonable thing for a team to ask).

60. Answer: B

The value stream map displayed in the chart shows time that the team spent working on the top, and time that the team wasted waiting on the bottom. If you add up the days, the team spent a total of 38 days actively working on the project, and 35 days waiting for approvals, stakeholders, and SA and DBA activities. That is a very large portion of the project spent waiting, which means there are plenty of opportunities to eliminate waste.

61. Answer: B

Velocity is a very effective way to use the team's actual performance from past sprints to understand their actual capacity for doing work, and using that information to forecast how much work they can accomplish in future iterations. Teams do this by assigning a relative size—typically using made-up units like story points—to each story, feature, requirement, or other item being worked on, and using the number of points per iteration to calculate the team's capacity.

62. Answer: B

It's normal and healthy for team members to have constructive disagreements. It happens all the time on effective teams, especially when the team members feel personally committed to the project. While leaders sometimes need to step in and prevent arguments from getting out of hand, letting the team members resolve their own disagreements is always better for the team, because it creates cohesion and lets them reach common ground together.

63. Answer: A

When you're working on a Scrum team, it's the product owner's job to meet with the stakeholders, help them understand problems, and communicate the solutions to the team. Team members should never go directly to stakeholders with problems; they need to make sure the product owner is always involved.

> *The part of the question about the product owner's schedule conflict is a red herring. There's only one answer to this question that doesn't have the team member exclude the product owner.*

64. Answer: D

Agile teams are concerned not just with delivering high-value features, but with maximizing the total value that's delivered to the stakeholders. That's why they balance delivery of high-value work items with reducing risk. An important way agile teams do that is to increase the priority of high-risk work items in the backlog. This particular work item presents a high risk because it's a low-priority work item, but if there's a problem it will have a large impact.

65. Answer: C

A generalizing specialist, or someone who has expertise in a specific area but is also improving in several other areas of expertise, is very valuable to an agile team. Generalizing specialists can help reduce team size by filling several different roles. Bottlenecks are less likely to occur, because one source of project bottlenecks comes from having only one team member able to do a certain task but not being available to perform it. Generalizing specialists help to create high-performing, cross-functional teams. However, they don't necessarily have better planning skills than any other team member.

66. Answer: C

Agile teams recognize that they learn a lot about the work that they will do along the way, so they expect their plans to improve as the project progresses. They do this by adapting their plan at the start of each iteration, and meeting every day to find and address any issues with that plan. This is how they refine their estimates of the scope and the schedule so that their plans always reflect a current understanding of what's going on in the real world.

67. Answer: C

Agile teams handle maintenance and operations work exactly the same way that they handle any other work. If bug fixes are critical, the team will work on them at the next opportunity. And the next opportunity, in most cases, is the start of the next iteration.

> *Stopping work immediately to change directions introduces chaos, and is not an effective way to change priorities. Agile teams use iterations so that they can respond to change quickly without letting their projects spin out of control.*

exam answers

68. Answer: B

> When you give stakeholders a schedule that has an unrealistically high level of detail, you're basically lying to them. That's definitely not something agile teams do!

One reason that agile teams are easy to work with is that they provide their stakeholders with forecasts and schedules that are at a level of detail that gives the stakeholders the information that they need without an unrealistically high level of detail. The scrum master should understand this, and recognize that there's absolutely no way that the team could possibly know how each person will spend each hour for the next six months.

69. Answer: D

Scrum teams value focus because even a small number of interruptions every week can cause significant delays, and the frustration from interruptions can seriously demotivate the team. As a servant leader, the scrum master needs to pay attention to anything that demotivates the team in order to keep morale high and the team productive. So while a servant leader typically doesn't have the authority to grant permission to skip meetings called by the manager, it's absolutely within the scrum master's role to approach that manager and find ways to keep the interruptions to a minimum.

70. Answer: C

Dealing with a non-cooperative team member is always difficult. On an agile team it's especially hard because agile, more than most other ways of working, relies on a shared mindset among the team. That's why it's so important for team members to cooperate with each other. One way that they do that is to come up with ground rules that help improve the team's coherence and strengthen each other's shared commitment to the project's goals and to the team.

> One way that a lot of Scrum teams handle a situation like this is to create a rule where anyone who arrives late to the daily scrum twice in a row has to wear a silly hat for the rest of the day or put a small amount of money into a "tip jar" that pays for a pizza or a round of drinks when it gets full.

71. Answer: B

The first step in planning an agile project is defining deliverables. In other words, the team needs to know what they're building. Agile teams typically use incremental methodologies, so the deliverables are defined by identifying specific units that the team will build incrementally.

72. Answer: D

Managing the expectations of stakeholders is an important part of how agile teams work. One way that they do it is to make broad commitments at the beginning of the project, typically by coming up with general goals for the project deliverables. As the project unfolds and project uncertainty reduces, they can make more and more specific commitments. This helps give their stakeholders a good idea of exactly what will be delivered, without the team overcommitting or agreeing to deliver something that turns out to be impossible or unrealistic within the project's time and cost constraints.

73. Answer: C

Agile teams always provide as much transparency as they can to their primary stakeholders, especially when it comes to problems that could impact the project. Keeping the primary stakeholder informed is more important than updating information radiators, refining the backlog, or holding a retrospective.

> Any time a stakeholder is impacted, he or she needs to be kept informed. This is especially true on Scrum teams, where openness is highly valued.

> THIS IS A *TOUGH QUESTION*. ALL OF THE ANSWERS TO THIS QUESTION SEEM LIKE PRETTY GOOD OPTIONS, SO WHICH ONE DO YOU CHOOSE? THE KEY TO REASONING YOUR WAY THROUGH A QUESTION LIKE THIS IS UNDERSTANDING THE PRINCIPLES THAT DRIVE AN AGILE MINDSET... ESPECIALLY CUSTOMER COLLABORATION.

74. Answer: C

When you experiment with new techniques and process ideas, it helps you and your team discover more efficient and effective ways to get your project done, and this is an important way that agile teams enhance creativity. So when you are presented with a set of alternative techniques to use, you should consider them. On a Scrum team, the appropriate time for doing this is during the sprint planning meeting.

> The rules of Scrum are important and give you a highly effective way to manage projects and build software, but if they specifically conflict with company-wide rules, you'll need to find a way to work within your company's guidelines.

75. Answer: B

It's really important to encourage all of the team members to share knowledge. Agile teams collaborate and work together, because sharing knowledge is an important way that agile teams avoid risks and improve productivity.

> It's true that the sprint retrospective typically comes after the sprint review. However, there's a more pressing issue that you have to handle first.

76. Answer: D

When a Scrum team plans the next sprint, one thing that they do is craft a sprint goal. This is their objective for the sprint that they'll meet by completing the work in the sprint backlog and delivering the increment. The sprint goal is how they stablish a shared, high-level vision of what they will accomplish for their stakeholders by delivering the increment.

> Information radiators are a good way to communicate information about how the project is going, but they don't really do a lot to establish a shared vision for the sprint.

77. Answer: A

Value stream analysis is a very valuable tool for detecting waste, especially waste that is caused by waiting for other teams.

> An Ishikawa (or fishbone) diagram can help you describe the root cause of project problems, but it isn't tailored to finding specific causes of waste due to waiting time.

> If you see a question where several answers look like they could be correct, choose the answer that's most specific to the question being asked.

78. Answer: A

All of these answers are good ideas. But the question specifically asked about the most effective strategy for prioritizing stories in the sprint backlog. Agile teams need to deliver stakeholder value early, which is why they plan their releases around minimally marketable features or minimally viable products. An early product release that has just enough features is the definition of a minimally viable product. The other answers are good strategies to get there.

79. Answer: A

Scrum teams plan their work by dividing the project into increments, and delivering a "done" increment at the end of each sprint. Scrum teams typically don't make major adjustments to their long-term plans mid-sprint. Instead, they make sure they are working on the most valuable deliverables they can during any individual sprint, so that even if priorities changes, they can meet the commitments they made for the current sprint and still deliver value. They'll adapt their plans to the new priorities as soon as the current sprint is done.

> Completing the current sprint isn't the same thing as stubbornly sticking to an outdated plan. But if the alternative is cancelling the sprint, it's much better to complete the current sprint and deliver the backlog items that the team promised the stakeholders at the last sprint review.

80. Answer: A

When teams discover risks or other issues that could threaten the project, they need to communicate the status of those issues to the stakeholders, and if possible, incorporate activities into the backlog to deal with the risk. One useful activity is exploratory work, where team members take time during a sprint to build a risk-basked spike solution to help mitigate the risk. But while refactoring the source code and performing continuous integration might be useful for lowering risk due to technical debt, it is unlikely to help with this situation.

> When you see a "which-is-NOT" question, be really careful to read all of the answers, and make sure you pick the WORST answer, not the BEST.

81. Answer: C

When the team doesn't have a consensus on what it means for a work item to be done, it can lead to problems, arguments, and delays late in the iteration. This is why the team needs to determine a definition of "done" that can be used as acceptance criteria. This is usually done on a "just-in-time" basis by leaving the decision for the last responsible moment—but for the team in this question, they waited too long to make that decision.

82. Answer: B

The team was notified of an operations problem, and they need to modify their plan to take it into account. They have an estimate for the impact: two team members will need to spend three iterations on the workaround. So they'll treat this change the way they treat any other change, by adding stories to their weekly cycles, and adjusting their release plan to reflect the change. Since this workaround is just more project work, it won't reduce the velocity, because work on the stories for the workaround will count towards the velocity just like any other work.

> There's no need to run a risk-based spike, because there is no uncertainty. The team knows that the server upgrade will be delayed, and that they'll have to spend time and effort on the workaround.

83. Answer: A

When a serious risk happens early on in the project, that's when iteration is most important. In this case, the team discovered a problem that needs to be fixed as soon as possible, so work needs to start right away—that means the product owner should add an item to the sprint backlog to start that work immediately. But the work will continue into the next sprint, so he or she also adds another item to the product backlog to make sure the fix is completed.

84. Answer: B

Agile teams plan their projects at multiple levels. For example, Scrum teams use the product backlog to do long-term strategic planning, hold sprint planning meetings at the beginning of each sprint to build the sprint backlog, and review their plan every day at the daily scrum. In this case, the stakeholder wants to know about the sprint backlog, which is created at the sprint planning meetings.

> This question doesn't use the term "sprint backlog" but instead describes it ("a list of features, stories, and other items to be delivered during the sprint").

85. Answer: B

Agile teams should always think about risks and potential issues that could threaten the project. When they encounter them, the team should maintain them in a way that ensures that the status and priority of each risk is visible and monitored.

> It's a great idea to add items to the backlog in order to deal with risks. However, the team should not necessarily do it for every single risk that was raised in the retrospective. Sometimes risks can be accepted, and sometimes it's enough just to be aware of them.

86. Answer: B

Teams often size the items that they will work on using ideal time. This means working together to figure out how much time it would take for a team member to work on each item in an "ideal" situation: he or she has everything needed to complete the work, there are no interruptions, and no other external factors or issues that could get in the way of completing the work. Unlike relative size techniques (like assigning story points to each item), ideal time is the team's best estimate of the absolute time required.

> *Fist-of-five voting is a way for teams of people to express their opinions. But in this case the team is arguing over which technical approach is superior, and swaying opinions is not necessarily the best way to reach the best technical solution.*

87. Answer: B

People on teams have conflicts all the time. The difference on an agile team is that they genuinely try to collaborate with each other. In this case, the XP team practice incremental design by finding a minimal first step that leaves the design open to either person's approach. Having the two team members use pair programming to build that approach together is a highly collaborative way to handle the situation. (Also, setting team ground rules to prohibit arguments is a terrible idea. Some arguments are healthy, and can lead to a better product and a more cohesive team.)

88. Answer: B

A really important part of an agile team is that everyone is allowed to experiment and make mistakes. When you make a mistake, you need to be open and public about it with your team. It's tempting to try to cover up the problem, but when problems happen you can't shield the rest of the team from the consequences. You need to be open about what happened, and work through the problem together.

> *When you're open about your own mistakes, it helps build a safe and trustful team environment.*

89. Answer: D

When new leaders emerge on an agile team, your job is to encourage that leadership. It's often difficult to try out new techniques, so your job as an agile practitioner is to establish a safe and respectful environment for that.

90. Answer: D

Agile teams are highly innovative because they create a safe environment where they're allowed to make mistakes so they can improve. An important part of the mindset of allowing mistakes is to think of them as problems that need to be corrected, rather than learning experiences. And it's important to be open about mistakes that you've made, and encourage others to do the same.

> *If you "allow" a mistake to go "uncorrected" you're still viewing it as a mistake that you were generous enough to let slip by. Part of developing an effective agile mindset is learning to see mistakes as genuine opportunities for improvement.*

91. Answer: C

A value stream map is the result of value stream analysis. Typically, a value stream map shows the flow of an actual work item (such as a product feature) through a process, with each step categorized as either working or waiting (non-working) time. One goal of value stream analysis is identification of waste in the form of non-working time that can be eliminated.

92. Answer: C

Having the product owner approach the stakeholders makes sense, but if the stakeholders need to talk to team members, it's unreasonable to ask them to go through an intermediary. Agile teams value face-to-face (or phone) conversations, and those conversations can be very important to the project.

Interruptions can be extraordinarily damaging to the team's productivity. Even a brief interruption can take a team member—especially a developer writing code—out of his or her state of "flow," and it can take up to 45 minutes to get back into it. So four or five phone calls a day might not sound too bad, but that level of interruption can cause someone to sit at their desk all day and get literally no work done. It's unrealistic to change the office layout (and that won't fix the phone call problem, anyway). And while it makes sense to adjust the sprint backlog, that doesn't fix the problem. So the best option is to establish a daily "no-call" window to limit interruptions.

> THIS IS AN **ESPECIALLY TOUGH QUESTION.** ALL OF THOSE ANSWERS HAVE POTENTIAL DOWNSIDES, SO YOU NEED TO FIGURE OUT WHICH IS THE "LEAST WORST" OPTION. IN THIS CASE, THE "NO-CALL" WINDOW WILL LIMIT THE INTERRUPTIONS WITHOUT PLACING UNREASONABLE DEMANDS ON THE TEAM OR THE STAKEHOLDERS.

93. Answer: B

Teams work best when they have a safe and trustful environment where people are allowed to experiment and make mistakes. As a servant leader, the scrum master must do everything that he or she can to establish that environment, even when it means having uncomfortable conversations with senior managers.

This is going to be a difficult discussion for the scrum master, and it's a good example of how it's not always easy for Scrum teams to value courage.

exam answers

94. Answer: C

A persona is a profile of a made-up user that includes personal facts and often a photo. It's a tool that a lot of Scrum teams use to help them understand who their users and stakeholders are and what they need. Agile teams need to identify all of their stakeholders—including future ones they don't necessarily know about today. Personas are a great tool for doing that.

95. Answer: C

Agile teams don't work in a vacuum—they constantly look at all of the infrastructure, operational, and environmental factors that could affect their project, even when those factors happen outside of the team. When they run across the problem, it's handled like any other problem: the product owner prioritizes it in the backlog based on its value. In this case, this is a severe problem, so the work item that the product owner adds to the backlog must be given high priority so that the team resolves it quickly.

96. Answer: C

Agile team members work hard to identify their project's business stakeholders and make sure that everyone on the team has a good understanding about what they need and expect from the project. But requiring stakeholders to attend planning meetings and requiring a sign-off on the plan does the opposite—it will make them feel less engaged, and create bureaucratic hurdles that prevent the team from responding to change.

Read every question carefully, and especially watch out for "which-is-not" questions.

97. Answer: B

This is a burn down chart for a team whose current sprint is running into trouble. They're two-thirds of the way through the current 30-day iteration, and the velocity has slowed down significantly. If they don't remove stories from their sprint backlog, it's unlikely that they will meet their sprint goal.

You can't determine that the team did a poor job planning just because the velocity is slower than expected. There are plenty of problems that teams can't anticipate—for example, a team member could have gotten sick. This is why Scrum teams constantly inspect and adapt, and why agile teams value responding to change over following a plan.

98. Answer: B

Your job as an agile practitioner includes helping to ensure that everyone on your team shares a common understanding of the agile practices that you are using. Common knowledge of agile practices is a basic part of working together effectively. So in this situation, you need to sit down with each team member and make sure they understand the practices that you use to respond to change.

99. Answer: A

Product owners must prioritize any relevant non-functional requirements exactly like they do with all other requirements, and this includes operational requirements that might come from a DevOps group. In this case, the script needs to be modified in order to hold the sprint review, so the change has to be included in the current sprint—and since that will cause some work to be delayed past the end of the sprint, that work must be moved back to the sprint backlog.

> Any time work will extend past the end of a sprint, it needs to be moved back to the sprint backlog and planned for a future sprint. It's never an option to break the timebox and extend the length of the sprint to include additional work.

100. Answer: D

Agile teams are self-organizing and empowered to make decisions about how to meet their iteration goals. This means that they work together to determine what tasks they need to perform in order to meet the sprint goals, and they'll often prioritize the stories with the most risk early in the iteration. The scrum master can help them self-organize and understand the methodology that they use, but he or she does not decide the order of the work, because that's not part of servant leadership.

101. Answer: C

Osmotic communication happens when team members absorb important project information from the discussions that take place around them. The XP primary practice of sitting together in a shared team space is an effective way to encourage osmotic communication.

102. Answer: A

Agile teams organize their requirements into minimally marketable features that they can deliver incrementally. By planning releases that deliver the most valuable features first, they can deliver value to the stakeholders as early as possible.

> You might also see the exam mention "minimally viable products," which are very closely related to minimally marketable features.

103. Answer: C

This team is concerned about a potential problem, but currently there has not been any actual impact on their project—and there won't be an impact if the problem turns out not to exist. This is a good opportunity to perform exploratory work (which some people refer to as spike solutions). That's a useful way for teams to determine if a technical problem can be resolved, or if they need to find a different approach.

104. Answer: C

One of the most important jobs that a product owner has on a Scrum team is making sure that new stakeholders are appropriately engaged in the project. Ideally, all stakeholders will attend every sprint review. However, there's no rule that says that every stakeholder must attend all sprint review meetings. Some stakeholders don't have time to attend them, or are in another time zone that makes it difficult for them to attend, or simply don't want to. It's the product owner's job to do whatever it takes to make sure those stakeholders are involved, using whatever manner works best for them.

exam *answers*

105. Answer: C

Agile teams—and especially the product owners on those teams—must identify all of the stakeholders and engage them throughout the whole project. In this case, you are meeting with stakeholders partway through an iteration, which means you are in the product owner role, so when you hear that there is a stakeholder who might impact the requirements for the project, you have identified a new stakeholder, so the next thing you should do is engage with that person.

106. Answer: B

This burn down chart shows a 30-day sprint that's going exactly as the team expects it to. They've probably been working together for a long time, because the velocity is constant. You can tell that because the burn down line is always very close to the guideline. There might be a few days where it's just above or below the line, but when you're looking at a burn down chart you care more about the trend than about individual days.

107. Answer: C

People work most effectively when they're in an open and safe environment where they're encouraged to talk about anything related to the project—especially issues that could potentially cause problems.

108. Answer: D

Once the team has finished planning an iteration, it's important that they make the results public to all of the project stakeholders. That's a really effective way to build trust between the team and the business because it shows that the team has committed to specific goals for the iteration. It also helps reduce uncertainty by making it clear exactly what the team intends to accomplish.

109. Answer: A

When people on an agile team discover issues that could affect the project, they make sure that their team members know—and more importantly, work with them to find ways to fix the problem. In fact, they'll do two things: they'll fix the problem today, and they'll make sure the process or methodology they follow addresses the issue so it doesn't happen again in the future.

110. Answer: B

Agile teams members should always be encouraged to collaborate with each other and share their knowledge. Pair programming is a highly effective practice for both collaboration and knowledge sharing.

111. Answer: B

Agile teams value working software over comprehensive documentation, and the best way to help stakeholders understand this is to show that past projects have gone well when they followed this value. However, it's always better to show success than to simply insist on a certain way of working.

Agile teams value working software over comprehensive documentation. But that doesn't mean they never use comprehensive documentation! They just value working software more.

446   Chapter 9

112. Answer: B

A story map is a great way to build a release plan for a team that uses stories. However, that plan should not be based on highly detailed effort estimates, especially at the beginning of the project.

Agile teams working on a new project need a starting point that they can use going forward. A good first step is to create a release plan, or a high-level plan of when specific deliverables will be released. Creating this plan involves creating very broad estimates of the scope of the items being delivered and the work required to build them, and using that information to come up with a very rough schedule. This schedule will not have a lot of detail, because that reflects the team's current high level understanding of the project.

113. Answer: C

Agile teams are always working to improve their effectiveness by continuously tailoring and adapting their project process. One way that they do this is by periodically reviewing the practices that they use, the culture of the team and the organization, and their goals.

114. Answer: A

One of the most effective ways for a team to build trust with a stakeholder is to establish a shared understanding of exactly what will be delivered during each sprint, and genuinely collaborate with him or her when trade-offs need to be made for technical or schedule reasons.

Agile teams value collaborating with their stakeholders over setting up contract-like agreements with them.

115. Answer: A

The question mentioned "the list of planned features"—this is the definition of the product backlog.

You are an agile practitioner meeting with stakeholders, which means that you are in the product owner role—and the product owner's job is to collaborate with stakeholders to understand the value of each deliverable, and use that information to prioritize the items in the backlog. Product owners need to take two things into account when they prioritize the backlog: the relative value of each feature, and the amount of work required to build it. Since you are able to assign relative value to each item in the backlog but you don't yet know how to prioritize them, the missing information is the amount of work required. The way to get that information is to re-estimate the items in the backlog.

> **This is an especially tough question.** A lot of questions on the exam ask about a specific tool, technique, or practice—this one is about the product backlog. But a lot of questions won't specifically mention it by name. Instead of calling it a product backlog, this question describes it ("list of planned features"). The key to questions like this is to break them down using terms that you know. "You have just assigned a relative value to each item in the list of planned features"—that means you just finished assigning a relative business value to the items in the product backlog. It also means that you must be the product owner, because that's the only person on the team who meets with stakeholders and assigns a business value to backlog items. So if the product owner has assigned a relative business value to each item in the product backlog, what's the next step for the team to take in order for the plan to work? Scrum teams plan their work based on business value and effort, so the next step for the team is to re-estimate the backlog.

exam *answers*

116. Answer: D

Co-location—or team members working in close proximity to each other in a shared team space—is a great way to encourage osmotic communication (or absorbing of important project information from overheard conversations). It makes it easier to create an informative workspace (for example, by posting information radiators), and team members benefit because they have access to their teammates. However, one downside of co-located teams is that there's a lot more potential for distractions.

> *There's no "perfect" way to organize your team space, and there are trade-offs that come with every strategy. However, the benefits of co-located teams in a shared team space far outweigh the costs.*

117. Answer: C

Agile teams maximize and optimize the value of the deliverables that they build by collaborating with stakeholders. On a Scrum team, it's the role of the product owner to collaborate with the stakeholder, understand the value, and help the team to deliver that value.

118. Answer: A

This team is attempting to do root cause analysis on a problem that they ran into so that they can fix the underlying problem and prevent it from happening in the future. An Ishikawa (or fishbone) diagram is an effective tool for performing root cause analysis.

119. Answer: D

Agile teams use frequent verification and validation to ensure product quality. This means doing product testing and conducting frequent reviews and inspections. These verification steps will help the team identify improvements, which must then be incorporated in the product.

120. Answer: A

> *Sometimes it may seem like working around a product owner is a good idea. It's not—you always need the product owner in the loop on every change so that the stakeholders can be kept in the loop. That's how agile teams make sure they deliver the most valuable products.*

Agile teams value responding to change, even when those changes are bad news—like a budget cut that will require the team to scale back the scope of what they'll deliver. And they value collaborating with their stakeholders, even when it means delivering that bad news. That's why every agile methodology includes some sort of mechanism or rule that lets them inspect the plan that the team is currently following (like holding daily standup meetings and retrospectives), changing the plan any time it becomes unrealistic, and alerting their stakeholders to the change.

So how'd you do?

The PMI-ACP® handbook (which you can download from the PMI.org website) explains how they use subject matter experts from all around the world to determine the passing score. That makes a lot of sense—it's a sound technique that lets PMI establish the examination point of difficulty with a lot of precision. That does make it a little hard to predict exactly how many questions you'll need to get right to achieve a passing score, but if you're scoring in the 80% to 90% range on this exam, then you're in really good shape

Index

A

adaptation (Scrum pillar) 93, 95, 98–99, 151

adaptive leadership 349

adaptive planning
    practice exam exercises 348–359, 372
    Scrum framework on 93, 95, 97–99
    self-organizing teams and 97

adjourning stage (team formation) 349

affinity estimating 326, 352

Agile Manifesto. *See also* specific values and principles
    about 25–26, 33
    chapter exercises 34, 66
    four values 25–33
    principles behind 41–54
    Scrum and 96–98

agile mindset
    about 10
    chapter exercises 15, 17, 20–21, 36–37, 49, 57, 67–69
    chapter questions and answers 14, 35, 50, 54
    Lean mindset and 262
    practice exam exercises 58–65, 314–321
    principles and practices 1–21
    values and principles 23–69

agile practitioner 310–311

Agile Retrospectives (Derby and Larsen) 154

amplify learning (Lean principle) 249, 253

Anderson, David 280

API (application programming interface) 205

application programming interface (API) 205

architectural spike 215

B

artifacts (Scrum) 76–77, 79, 86. *See also* specific artifacts

automated builds 208, 210

backbone (product) 149

backlog
    about 47, 54
    Increment. *See* Increment
    PBR meeting and 148, 151
    Product Backlog. *See* Product Backlog
    refinement. *See* PBR (product backlog refinement)
    risk-adjusted 351
    Sprint Backlog. *See* Sprint Backlog
    story maps visualizing 149
    types of items as 139
    XP and 182

backlog grooming 323. *See also* PBR (product backlog refinement)

Beck, Kent 181

bribes, accepting 380

Brooks, Fred 25

build automation 208, 210

build in integrity (Lean principle) 250

burndown charts
    about 11, 13, 19, 123, 143, 151
    release 151
    risk-based 351
    story points and 143
    velocity and 143–146

burn-up charts 147

C

card (user story) 125, 131–133, 138
cause and effect diagrams 155, 157
caves and commons design 193
ceremony (Scrum) 95
CFDs (cumulative flow diagrams) 292
change management
    about 15, 20, 32–33
    Agile Manifesto on 41, 43, 54–55, 68, 98
    over following a plan 25, 32–33, 36
    Scrum framework and 93–95, 99
    XP and 15, 20, 204–207, 229
check-ins (retrospective tool) 156
chunking 313
coaching style (situational leadership) 349
code bloat 226
code maintenance 223
code monkey trap 195, 198
Cohn, Mike 123, 151
collaboration, customer. *See* customer collaboration
collaboration games 352
collaborative improvement (Kanban) 280, 282, 295
collective commitment 84
color code dots (data gathering tool) 156
commitment-driven planning 151
commitment (Scrum value) 84–85, 188
commit (version control system) 206–207, 211
communication
    agile on 4–9, 11, 13, 31, 50, 52–53
    GASP on 124, 138, 154, 160, 172
    Scrum on 74, 86, 95–99, 113
    XP on 182, 188–190, 192–195, 198–199
complex code 223–227, 230
conceptual integrity (Lean tool) 256
confirmation (user story) 125, 131, 138
conflict (version control system) 206–207, 211
content outline (PMI-ACP exam) 309–313
continuous attention to excellence and design 41
continuous delivery of software
    about 54, 99
    Agile Manifesto on 41–42, 46, 96, 325
continuous improvement
    practice exam exercises 361–369
continuous integration 19, 209, 211, 219, 229
contract negotiation 25, 31, 33
conversation (user story)
    about 131
    agile principle on 15, 20, 41, 53–54, 127
    in XP teams 192–193
copyrighted material 382
cost of delay (Lean tool) 256
courage
    as Scrum value 83, 88, 188
    as XP value 184, 188
cumulative flow diagrams (CFDs) 292
customer collaboration
    about 31
    Agile Manifesto on 41, 52, 54
    over contract negotiation 25, 31, 33, 36
customer-valued prioritization tools 325
cycle time (Lean measurement) 257

D

Daily Scrum
    about 12, 14, 86
    as a ceremony 95
    in Scrum Sprint 78
    task board example 145–146
    timeboxed events and 74, 99
daily standup 4–11, 14, 16
decide as late as possible
    Lean principle 249, 253, 271
    Scrum framework 94, 99

defects
    as Lean waste category 261
    as manufacturing waste 268
delegating style (situational leadership) 349
deliver as fast as possible (Lean principle) 249
delivery cadence (Lean) 256–257, 262
Derby, Esther 154
Development Team role (Scrum) 12, 75, 78, 148
directing style (situational leadership) 349
Done column (task boards) 136–137
done, definition of 92, 99
dot voting 352

E

early delivery of software
    about 54, 99
    Agile Manifesto on 41–42, 96, 325
Earned Value Management (EVM) 326
elapsed time 135
eliminate waste (Lean principle)
    about 249
    categories of waste 260–262
embracing change. See change management
empirical process control theory 95, 99, 185
empiricism 95
empower the team (Lean principle) 250
energized work (XP) 196, 198–199
estimatable (INVEST acronym) 131
estimation considerations
    concepts used 135
    in INVEST acronym 131
    PBR meeting 148, 151
    planning poker 123, 132–134, 138, 151, 352
    Product Backlog 76, 79, 108, 138
    Sprint Backlog 138, 142, 151
    story maps 149
    story points 132–135, 138, 151
    trust and respect in 82

T-shirt sizes 126, 138, 151, 326
user stories 126–127, 131–134, 138
velocity in 135, 138, 151
ESVP technique 156
ethical decisions. See professional responsibility
events (Scrum) 74, 86
EVM (Earned Value Management) 326
exam question help
    getting brain to think about exam xxiii–xxv
exploratory testing 214
explorers (ESVP technique) 156
extra features (waste category) 260
extra processes
    as Lean waste category 260
    as manufacturing waste 268
extreme programming. See XP

F

face-to-face conversations. See conversation (user story)
fail fast concept 215, 359
feedback and feedback loops
    agile teams and 33, 42, 62, 65, 67
    Kanban method 280, 282, 294
    Lean mindset tool 253
    Scrum teams and 131, 157
    XP teams and 205, 208–209, 214–215
Fibonacci sequence 326
firmware 29–30
fishbone diagrams 155, 157
fist of five voting 352
five why's technique 269
flow efficiency (Lean measurement) 257, 264
flow management (Kanban)
    cumulative flow diagrams and 292
    process overview 274–275, 280–286, 288
    queuing theory and 255
    Toyota Production System 268–269
flow state (XP) 195–196, 199

the index

focus
    as Scrum value 83, 88, 186
    as XP value 185–186, 189
forming stage (team formation) 349

G

Gantt charts 262
GASPs (Generally Accepted Scrum Practices)
    about 122–123
    burndown charts. *See* burndown charts
    burn-up charts 147
    chapter exercises 128–130, 141, 159–162
    chapter questions and answers 138, 151
    personas 15, 20, 150
    practice exam exercises 164–173
    product backlog refinement 148, 151, 323
    red herring questions 140–141
    story maps 149, 151
    story points. *See* story points
    task boards. *See* task boards
    user stories. *See* user stories
Generally Accepted Scrum Practices. *See* GASPs
grooming 323. *See also* PBR (product backlog refinement)

H

hack (code) 177, 223, 230, 272

I

ideal time 135
improve collaboratively (Kanban) 280, 282, 295
incremental design 15, 20, 228–229
Increment (Scrum) 76–77, 79
independent (INVEST acronym) 131
individuals and interactions. *See* team performance
information radiators
    about 194, 199, 239, 429
    as communication tool 433, 440
    example of 367
    importance of 431

informative workspace (XP) 194, 199, 219, 239, 429, 448
In Progress column (task boards) 136–137
inspection (Scrum pillar) 93, 95, 98–99
integrity built in (Lean principle) 250
internal rate of return (IRR) 326
inventory (manufacturing waste) 268
INVEST acronym 131
IRR (internal rate of return) 326
Ishikawa diagrams 155, 157
iteration
    about 46–47, 54
    Lean mindset tool 253
    timeboxed 46, 54, 78, 188
    XP on 182–183, 205
iterationless development 262

J

Jeffries, Ron 181
just the facts ma'am question 19

K

Kaizen 269
Kanban boards 281, 285–286
Kanban method
    about 12–14, 245, 274, 286
    applying practices 15, 20, 280
    chapter exercises 289–290, 294, 296–299
    chapter questions and answers 288
    Lean mindset and 288
    practice exam exercises 300–306
    process improvement 280–286, 291–295
kanban (signal card) 288
Kano analysis 325
Kano, Noriaki 325
kludge (code) 223, 230

L

Larsen, Diana 154

last responsible moment
    Lean mindset and 249, 253, 271
    Scrum framework on 94, 99

leadership styles
    adaptive leadership 349
    servant leaders 75
    situational leadership theory 349

lead time (Lean measurement) 257

Lean mindset. *See also* specific principles and tools
    about 12–14, 245, 248, 262
    chapter exercises 251–253, 263, 265–266, 279, 296–299
    chapter questions and answers 262, 276, 288
    Kanban method and 288
    practice exam exercises 300–306
    principles and thinking tools 249–250, 253–257
    Scrum framework and 248
    XP and 248

Lean Software Development (Poppendieck) 268

learning amplification (Lean principle) 249, 253

least-worst-option question 278

low fidelity wireframes 214

M

maintaining code 223

managing flow (Kanban). *See* flow management (Kanban)

manufacturing waste 268

mind maps 352

mindset
    Agile Manifesto and 25–26
    daily standups and 7–9, 14
    Lean mindset. *See* Lean mindset
    methodology versus 10, 50
    PMI-ACP exam exercises 314–321
    XP teams 184

Minimally Marketable Feature (MMF) 255

Minimally Viable Product (MVP) 255

mistakes
    agile principles on 13, 33, 35, 52, 54
    Scrum framework on 82, 138
    XP on 190, 196, 199, 216

MoSCoW method 325

motion
    as Lean waste category 261
    as manufacturing waste 268

motivation 52, 54

muda (TPS waste type) 268

mura (TPS waste type) 268

muri (TPS waste type) 268

N

negotiable (INVEST acronym) 131

net present value (NPV) 326

norming stage (team formation) 349

NPV (net present value) 326

O

Ohno, Taiichi 268–269

100 point voting 352

openness (Scrum value) 82, 88, 188

options thinking (Lean tool) 256, 262, 270–271, 276

osmotic communication 194, 198–199

overproduction (manufacturing waste) 268

P

pair programming 208, 216, 219–220, 229

partially done work (waste category) 260

PBR (product backlog refinement) 148, 151, 323

perceived integrity (Lean tool) 256

performing stage (team formation) 349

personas 15, 20, 150

planning game (XP) 189

the index

planning poker 123, 132–134, 138, 151, 352

PMI-ACP exam
    about 18–19, 307–309, 376
    adaptive planning exercises 348–359, 372
    agile practitioner 310–311
    agile principles and mindset exercises 314–321
    chapter questions and answers 312
    content outline 309–313
    continuous improvement exercises 361–369
    examcross 370–371, 375
    practice exam answers 424–448
    practice exam questions xviii–xxx, 391–422
    problem detention and resolution exercises 360, 362–369, 373
    professional responsibility 386–389
    stakeholder engagement exercises 336, 338, 340–347
    team performance exercises 337, 339–347
    value-driven delivery exercises 322–335

PMI Code of Professional Conduct 378

Poppendieck, Mary and Tom 268–269

practices. *See also* GASPs
    agile methodology 1–21
    Kanban method 15, 20, 280
    principles versus 48, 50
    real-world challenges 26
    XP supported 181–182, 188–199, 202–230

pre-mortems 351

principles
    Agile Manifesto 41–54
    agile methodology and 1–21
    Lean mindset 249–250
    PMI-ACP exam exercises 314–321
    practices versus 48, 50

prioritize with dots technique 157

prisoners (ESVP technique) 156

problem detention and resolution exercises 360, 362–369, 373

process policies (Kanban) 280, 282, 292

product backlog refinement (PBR) 148, 151, 323

Product Backlog (Scrum)
    about 12, 79
    estimation considerations 76, 79, 108, 138
    evaluating from task boards 137
    PBR meeting 148, 151
    prioritizing features 120, 142
    Product Owner and 75–76, 96, 148
    timeboxed events and 74

Product Owner role (Scrum)
    about 12, 31, 42, 47, 50, 75
    authority and 84, 86, 96
    identifying user stories 131
    PBR meeting and 148, 151
    prioritizing features 120
    Product Backlog and 75–76, 96, 148
    Sprint Retrospective and 78, 101
    Sprint Review and 79, 86, 96

professional responsibility
    about 377, 385
    accepting bribes 380
    community responsibility and 384
    copyrighted material 382
    doing the right thing 378–379
    following company policy 381
    practice exam exercises 386–389
    process shortcuts and 383

progressive elaboration 32, 67, 355

project plans 134

pull systems 249, 255, 274–275, 280, 286

Q

quarterly cycle (XP) 182, 193, 198

questions
    just the facts ma'am 19
    least-worst-option 278
    red herring 140
    which-comes-next 90
    which-is-best 36
    which-is-not 200

queuing theory (Lean) 255, 276

R

red/green/refactor 209
red herring question 140–141
refactoring code 11, 19, 209, 224, 226–227
reject update (version control system) 206–207
relative prioritization/ranking 325
relative sizing techniques 326
release burndown chart 151
release plan, visualizing 149, 151
remove local optimizations (Lean) 272
repository (version control system) 207, 211
respect
 as Scrum value 82, 88, 188
 as XP value 184, 188
responding to change. *See* change management
retrospectives. *See* Sprint Retrospective
return on investment (ROI) 326
reusable code 225, 229
reworking code 39, 44, 67, 203–204
risk-adjusted backlog 351
risk-based spike 215
risk burndown charts 351
ROI (return on investment) 326
roles. *See also* specific roles
 in Scrum framework 12, 31, 47, 75–76
 in XP teams 190–193, 198
root cause analysis 269

S

Safari Books Online xxix
satisfied users 42
Schwaber, Ken 75
scientific method 15, 20
Scrum framework
 about 12–14, 47, 71–73, 81, 86–89, 102
 Agile Manifesto and 96–98
 artifacts in 76–77, 79, 86
 chapter exercises 80, 91, 100, 103, 112–115, 119–120
 chapter questions and answers 79, 86, 93, 101
 events in 74, 86
 GASPs. *See* GASPs
 important guideline 174–175
 last responsible moment in 94, 99
 Lean mindset and 248
 practice exam exercises 90, 104–111
 roles in 12, 31, 47, 75–76. *See also* specific roles
 Sprint overview 78
 tasks in 77–79, 83–84, 86, 92–93
 three pillars of 93–95, 99
 values in 82–86, 88, 188
 XP differences 185–188
Scrum Guide 75, 97
Scrum Master role
 about 12, 75, 84, 99
 moderating planning poker 132–133
 Sprint Retrospective and 78
Scrum Team role 75, 82
seeing waste (Lean tool) 254, 264
see the whole (Lean principle) 250, 272, 276
self-organizing teams 41, 97–101
servant leadership 75
set-based development (Lean tool) 257
seven wastes of software development 260–262
shoppers (ESVP technique) 156
shortcuts, process 383
short subjects (decision-making tool) 157
shotgun surgery 206, 212
signal card (kanban) 288
simplicity
 Agile Manifesto on 41, 64, 224–225
 XP on 224–230
situational leadership theory 349
slack (XP) 182–183, 188, 193, 226
small (INVEST acronym) 131

Snowbird ski resort 24, 40
sources of waste 268
spaghetti code 204
spike solutions 214–215, 229
Sprint Backlog (Scrum)
    about 12, 76–79
    Development Team and 148
    estimation considerations 138, 142, 151
    Sprint Planning session and 77, 98
    timeboxed events and 74
Sprint Goal (Scrum) 77–78, 83, 151
Sprint Planning session (Scrum)
    about 12, 14, 78, 86
    "Done" tasks and 92
    last responsible moment and 94
    Sprint Backlog and 77, 98
    timeboxed events and 74
Sprint Retrospective (Scrum)
    about 11–12, 19, 78, 86, 101
    outline for 154–155
    Product Owner and 78, 101
    timeboxed events and 74
    tools supporting 156–157
Sprint Review (Scrum)
    about 12, 14, 78
    Product Owner and 79, 86, 96
Sprint (Scrum)
    about 12, 14, 78, 86
    iteration practice and 47
    timeboxed events and 74
    tracking progress during 136–137, 142–147, 151
    walking skeleton and 149
stages of team formation 349
stakeholder engagement
    adjusting to agile practices 139
    practice exam exercises 336, 338
stories. *See* user stories
storming stage (team formation) 349
story maps 149, 151

story points
    about 122, 126, 135
    burndown charts and 143
    estimation considerations 132–135, 138, 151
    T-shirt sizes 126, 138, 151, 326
supporting style (situational leadership) 349
sustainable development
    Agile Manifesto on 41–42, 196
    XP on 190, 196, 198–199, 202–230
Sutherland, Jeff 75
systems thinking (Lean) 250, 272, 276

T

task boards
    about 11, 13, 122, 151
    Daily Scrum example 145–146
    Kanban boards and 281, 286
    tracking progress on 136–137
tasks (Scrum) 77–79, 83–84, 86, 92–93
task switching (waste category) 260
TDD (test-driven development) 209, 212–214, 219, 229–230
team performance
    Agile Manifesto on 25, 27, 41, 50, 52–53, 55, 68, 97
    daily standup meetings 4–11, 14, 16
    Lean mindset on 250, 262
    mindset and 7–9
    motivation and 52, 54
    over processes and tools 25, 27, 33
    practice exam exercises 337, 339–347
    Scrum framework on 75, 82–84, 96–98
    Scrum/XP hybrid 186–189, 193
    self-organizing teams 41, 97–101
    stages of team formation 349
    working agreements 293
    XP and 184, 190–199
team space 192–193
technical debt 226–227
10-minute build 208–210, 229

testable (INVEST acronym) 131

test-driven development (TDD) 209, 212–214, 219, 229–230

themes (XP) 182, 188, 193

thinking tools (Lean) 249–250, 254–258, 262

throughput rate 292

tightly coupled code 223, 225

timeboxed iterations 46, 54, 78, 188

timeline (data gathering tool) 156

To Do column (task boards) 136–137

Toyoda, Kiichiro 268

Toyota Production System (TPS) 268–269

TPS (Toyota Production System) 268–269

transparency (Scrum pillar)
    about 93, 99
    Daily Scrum and 95
    stakeholder engagement and 139
    task boards and 136

transportation (manufacturing waste) 268

travel policies 381

trust
    as Scrum value 82
    as XP value 190

T-shirt sizes 126, 138, 151, 326

Tuckman, Bruce 349

U

unit tests 205, 209, 212–213

usability testing 214

user interface 214

user stories
    about 11, 122, 124–125, 138, 151
    estimation considerations 126–127, 131–134, 138
    INVEST acronym 131
    purpose of 131
    task boards tracking status 136–137
    walking skeleton 149
    XP and 131, 182, 188, 193

V

vacationers (ESVP technique) 156

valuable (INVEST acronym) 131

value-driven delivery
    agile principles and 42–47
    practice exam exercises 322–335

value(s)
    about 10, 14
    Agile Manifesto on 25–34
    calculating for projects 326
    Scrum framework on 82–86, 96
    XP and 181, 184–186, 189, 193, 204–205

value stream maps (Lean) 254–255, 264, 267, 273

velocity
    about 126, 135, 144, 151
    burndown charts and 143–146
    in estimation considerations 135, 138, 151

velocity-driven planning 151

Venn diagrams 187–188, 251–252

version control systems 206–207, 211

visualize workflow (Kanban) 280–285

W

waiting
    as Lean waste category 261
    as manufacturing waste 268

Wake, Bill 131

walking skeleton (product) 149

waste elimination (Lean principle)
    about 249
    categories of waste 260–262
    types of manufacturing waste 268
    types of TPS waste 268

waterfall process 24, 35, 53

weekly cycle (XP) 182–183, 193, 198, 227

which-comes-next question 90

which-is-best question 36

the index

which-is-not question 200

whole team (XP) 190–191, 193, 198

WIP (work in progress) limits 274–275, 280–283, 286, 288, 291

wireframes 214

Wooden, John 181

workflow (Kanban). *See* flow management (Kanban)

working agreements 293

working software
  about 15, 20
  Agile Manifesto on 25, 28–30, 41, 44, 46, 55, 68
  continuous delivery of 41, 46, 54, 96, 99, 325
  early delivery of 41–42, 54, 96, 99, 325
  over comprehensive documentation 25, 28–30, 33

work in progress (WIP) limits 274–275, 280–283, 286, 288, 291

work items (Kanban boards) 281, 285

work-life balance (XP) 183, 198–199, 230

XP (extreme programming)
  about 12, 14, 177, 181, 193
  burndown charts. *See* burndown charts
  change management and 15, 20, 204–207, 229
  chapter exercises 187–188, 197, 200–201, 215, 217–218, 231, 233, 242–244
  chapter questions and answers 189, 198, 219, 230
  feedback and 205, 208–209, 214–215
  incremental design 15, 20
  iterative development 182–183
  Lean mindset and 248
  planning in 182–183, 193, 198
  practice exam exercises 234–241
  practices supported 181–182, 188–199, 202–230
  Scrum differences 185–188
  sustainable development 202–230
  team mindset and 184, 190–199
  user stories and 131, 182, 188
  values and 181, 184–186, 188–189, 193, 204–205
  work-life balance and 183, 198–199, 230

Learn from experts.
Find the answers you need.

Sign up for a **10-day free trial** to get **unlimited access** to all of the content on Safari, including Learning Paths, interactive tutorials, and curated playlists that draw from thousands of ebooks and training videos on a wide range of topics, including data, design, DevOps, management, business—and much more.

Start your free trial at:
oreilly.com/safari

(No credit card required.)

CPSIA information can be obtained
at www.ICGtesting.com
Printed in the USA
BVHW081006020619
549634BV00031B/63/P